Inorganic Chemistry

The INSTANT NOTES series

Series editor
B.D. Hames
School of Biochemistry and Molecular Biology, University of Leeds, Leeds, UK

Animal Biology
Molecular Biology
Ecology
Genetics
Microbiology
Chemistry for Biologists
Immunology
Biochemistry 2nd edition

Forthcoming titles
Molecular Biology 2nd edition
Neuroscience
Developmental Biology
Plant Biology
Psychology

The INSTANT NOTES Chemistry Series
Consulting editor: Howard Stanbury

Organic Chemistry
Inorganic Chemistry

Forthcoming titles
Physical Chemistry
Analytical Chemistry

Instant Notes

Inorganic Chemistry

P. A. Cox
Inorganic Chemistry Laboratory,
New College, Oxford, UK

First published 2000

A CIP catalogue record for this book is available from the British Library.

ISBN 1 85996 163 0

BIOS Scientific Publishers Ltd
9 Newtec Place, Magdalen Road, Oxford OX4 1RE, UK
Tel. +44 (0)1865 726286. Fax +44 (0)1865 246823
World Wide Web home page: http://www.bio.co.uk/

Published in the United States of America, its dependent territories and Canada by Springer-Verlag New York Inc., 175 Fifth Avenue, New York, NY 10010-7858, in association with BIOS Scientific Publishers Ltd

Published in Hong Kong, Taiwan, Cambodia, Korea, The Philippines, Brunei, Laos, and Macau only by Springer-Verlag Hong Kong Ltd, Unit 1702, Tower 1, Enterprise Square, 9 Sheung Yuet Road, Kowloon Bay, Kowloon, Hong Kong, in association with BIOS Scientific Publishers Ltd.

Consulting Editor: Howard Stanbury

Production Editor: Andrea Bosher
Typeset by J&L Composition Ltd, Filey, UK
Printed by Biddles Ltd, Guildford, UK

CONTENTS

ABBREVIATIONS

3c2e	three-center two-electron
3c4e	three-center four-electron
3D	three dimensional
ADP	adenosine diphosphate
An	actinide
AO	atomic orbital
ATP	adenosine triphosphate
bcc	body-centered cubic
BO	bond order
BP	boiling point
CB	conduction band
ccp	cubic close packing
CN	coordination number
Cp	cyclopentadienyl (C_5H_5)
E	unspecified (non-metallic) element
EA	electron affinity
EAN	effective atomic number
EDTA	ethylenediamine tetraacetate
Et	ethyl (C_2H_5)
fcc	face-centered cubic
hcp	hexagonal close packing
HOMO	highest occupied molecular orbital
HSAB	hard and soft acid-base
IE	(first) ionization energy
I_n	nth ionization energy ($n = 1, 2, \ldots$)
IUPAC	International Union of Pure and Applied Chemistry
L	unspecified ligand
LCAO	linear combination of atomic orbitals
LFSE	ligand field stabilization energy
LMCT	ligand-to-metal charge transfer
LUMO	lowest unoccupied molecular orbital
Ln	lanthanide
M	unspecified (metallic) element
Me	methyl (CH_3)
MLCT	metal-to-ligand charge transfer
MO	molecular orbital
MP	melting point
Ph	phenyl (C_6H_5)
R	organic group (alkyl or aryl)
RAM	relative atomic mass
SN	steric number
UV	ultraviolet
VB	valence band
VE	valence electron
VSEPR	valence shell electron pair repulsion
X	unspecified element (often a halogen)
Z	atomic number

PREFACE

Inorganic chemistry is concerned with the chemical elements (of which there are about 100) and the extremely varied compounds they form. The essentially descriptive subject matter is unified by some general concepts of structure, bonding and reactivity, and most especially by the periodic table and its underlying basis in atomic structure. As with other books in the *Instant Notes* series, the present account is intended to provide a concise summary of the core material that might be covered in the first and second years of a degree-level course. The division into short independent topics should make it easy for students and teachers to select the material they require for their particular course.

Sections A–F discuss the general concepts of atomic structure, periodicity, structure and bonding, and solution chemistry. The following Sections F–I cover different areas of the periodic table in a more descriptive way, although in Section H some concepts that are peculiar to the study of transition metals are also discussed. The final section describes some aspects of inorganic chemistry in the world outside the laboratory.

I have assumed a basic understanding of chemical ideas and vocabulary, coming, for example, from an A-level chemistry course in the UK or a freshman chemistry course in the USA. Mathematics has been kept at a strict minimum in the discussion of atomic structure and bonding. A list of further reading is given for those interested in pursuing these or other aspects of the subject.

Many people have contributed directly or indirectly to the production of this book. I would particularly like to thank the following: Howard Stanbury for introducing me to the project; Lisa Mansell and other staff at BIOS for their friendliness and efficiency; the anonymous readers and my colleagues Bob Denning and Jenny Green for their helpful comments on the first draft; my students past and present for their enthusiasm, which has made teaching inorganic chemistry an enjoyable task; and Sue for her love and understanding.

A1 THE NUCLEAR ATOM

Key Notes

Electrons and nuclei

An atom consists of a very small positively charged nucleus, surrounded by negative electrons held by electrostatic attraction. The motion of electrons changes when chemical bonds are formed, nuclei being unaltered.

Nuclear structure

Nuclei contain positive protons and uncharged neutrons. The number of protons is the atomic number (Z) of an element. The attractive strong interaction between protons and neutrons is opposed by electrostatic repulsion between protons. Repulsion dominates as Z increases and there is only a limited number of stable elements.

Isotopes

Isotopes are atoms with the same atomic number but different numbers of neutrons. Many elements consist naturally of mixtures of isotopes, with very similar chemical properties.

Radioactivity

Unstable nuclei decompose by emitting high-energy particles. All elements with $Z > 83$ are radioactive. The Earth contains some long-lived radioactive elements and smaller amount of short-lived ones.

Related topics

Actinium and the actinides (I2)

Origin and abundance of the elements (J1)

Electrons and nuclei

The familiar planetary model of the atom was proposed by Rutherford in 1912 following experiments by Geiger and Marsden showing that nearly all the mass of an atom was concentrated in a positively charged **nucleus**. Negatively charged **electrons** are attracted to the nucleus by the **electrostatic force** and were considered by Rutherford to 'orbit' it in a similar way to the planets round the Sun. It was soon realized that a proper description of atoms required the quantum theory; although the planetary model remains a useful analogy from the macroscopic world, many of the physical ideas that work for familiar objects must be abandoned or modified at the microscopic atomic level.

The lightest atomic nucleus (that of hydrogen) is 1830 times more massive than an electron. The size of a nucleus is around 10^{-15} m (1 fm), a factor of 10^5 smaller than the apparent size of an atom, as measured by the distances between atoms in molecules and solids. Atomic sizes are determined by the rad[illegible] of the electronic orbits, the electron itself having apparently no size at all. [illegible] between atoms alters the motion of electrons, the nuclei rem[illegible] Nuclei retain the 'chemical identity' of an element, and the o[illegible] cal elements depends on the existence of stable nuclei.

Nuclear structure

Nuclei contain positively charged **protons** and uncharged **r**[illegible] particles with about the same mass are known as **nucleo**[illegible] protons is the **atomic number** of an element (Z), and is matched in a neutral atom by the same number of electrons. The total number of nucleons is the **mass number** and is sometimes specified by a superscript on the symbol of the element. Thus ^{1}H has a nucleus with one proton and no neutrons, ^{16}O has eight protons and eight neutrons, ^{208}Pb has 82 protons and 126 neutrons.

Protons and neutrons are held together by an attractive force of extremely short range, called the **strong interaction**. Opposing this is the longer-range electrostatic repulsion between protons. The balance of the two forces controls some important features of nuclear stability.

- Whereas lighter nuclei are generally stable with approximately equal numbers of protons and neutrons, heavier ones have a progressively higher proportion of neutrons (e.g. compare ^{16}O with ^{208}Pb).
- As Z increases the electrostatic repulsion comes to dominate, and there is a limit to the number of stable nuclei, all elements beyond Bi ($Z = 83$) being radioactive (see below).

As with electrons in atoms, it is necessary to use the quantum theory to account for the details of nuclear structure and stability. It is favorable to 'pair' nucleons so that nuclei with even numbers of either protons or neutrons (or both) are generally more stable than ones with odd numbers. The **shell model** of nuclei, analogous to the orbital picture of atoms (see Topics A2 and A3) also predicts certain **magic numbers** of protons or neutrons, which give extra stability. These are

2 8 20 28 50 82 126

^{16}O and ^{208}Pb are examples of nuclei with magic numbers of both protons and neutrons.

Trends in the stability of nuclei are important not only in determining the number of elements and their isotopes (see below) but also in controlling the proportions in which are they are made by nuclear reactions in stars. These determine the abundance of elements in the Universe as a whole (see Topic J1).

Isotopes

Atoms with the same atomic number and different numbers of neutrons are known as **isotopes**. The chemical properties of an element are determined largely by the charge on the nucleus, and different isotopes of an element have very similar chemical properties. They are not quite identical, however, and slight differences in chemistry and in physical properties allow isotopes to be separated if desired.

Some elements have only one stable isotope (e.g. ^{19}F, ^{27}Al, ^{31}P), others may have several (e.g. ^{1}H and ^{2}H, the latter also being called **deuterium**, ^{12}C and ^{13}C); the record is held by tin (Sn), which has no fewer than 10. Natural samples of many elements therefore consist of mixtures of isotopes in nearly fixed proportions reflecting the ways in which these were made by nuclear synthesis. The **molar mass** (also known as **relative atomic mass**, RAM) of elements is determined by these proportions. For many chemical purposes the existence of such isotopic mixtures can be ignored, although it is occasionally significant.

- Slight differences in chemical and physical properties can lead to small variations in the isotopic composition of natural samples. They can be exploited to give geological information (dating and origin of rocks, etc.) and lead to small variations in the molar mass of elements.

- Some spectroscopic techniques (especially **nuclear magnetic resonance, NMR**) exploit specific properties of particular nuclei. Two important NMR nuclei are ^{1}H and ^{13}C. The former makes up over 99.9% of natural hydrogen, but ^{13}C is present as only 1.1% of natural carbon. These different abundances are important both for the sensitivity of the technique and the appearance of the spectra.
- Isotopes can be separated and used for specific purposes. Thus the slight differences in chemical behavior between normal hydrogen (^{1}H) and deuterium (^{2}H) can be used to investigate the detailed mechanisms of chemical reactions involving hydrogen atoms.

In addition to stable isotopes, all elements have unstable **radioactive** ones (see below). Some of these occur naturally, others can be made artificially in particle accelerators or nuclear reactors. Many radioactive isotopes are used in chemical and biochemical research and for medical diagnostics.

Radioactivity

Radioactive decay is a process whereby unstable nuclei change into more stable ones by emitting particles of different kinds. **Alpha, beta** and **gamma** (α, β and γ) radiation was originally classified according to its different penetrating power. The processes involved are as follows.

- An α particle is a ^{4}He nucleus, and is emitted by some heavy nuclei, giving a nucleus with Z two units less and mass number four units less. For example, ^{238}U ($Z = 92$) undergoes α decay to give (radioactive) ^{234}Th ($Z = 90$).
- A β particle is an electron. Its emission by a nucleus increases Z by one unit, but does not change the mass number. Thus ^{14}C ($Z = 6$) decays to (stable) ^{14}N ($Z = 7$).
- γ radiation consists of high-energy electromagnetic radiation. It often accompanies α and β decay.

Some other decay processes are known. Very heavy elements can decay by **spontaneous fission**, when the nucleus splits into two fragments of similar mass. A transformation opposite to that in normal β decay takes place either by **electron capture** by the nucleus, or by emission of a positron (β^{+}) the positively charged antiparticle of an electron. Thus the natural radioactive isotope ^{40}K ($Z = 19$) can undergo normal β decay to ^{40}Ca ($Z = 20$), or electron capture to give ^{40}Ar ($Z = 18$).

Radioactive decay is a statistical process, there being nothing in any nucleus that allows us to predict when it will decay. The probability of decay in a given time interval is the only thing that can be determined, and this appears to be entirely constant in time and (except in the case of electron capture) unaffected by temperature, pressure or the chemical state of an atom. The probability is normally expressed as a **half-life**, the time taken for half of a sample to decay. Half-lives can vary from a fraction of a second to billions of years. Some naturally occurring radioactive elements on Earth have very long half-lives and are effectively left over from the synthesis of the elements before the formation of the Earth. The most important of these, with their half-lives in years, are ^{40}K (1.3×10^{9}), ^{232}Th (1.4×10^{10}) and ^{238}U (4.5×10^{9}).

The occurrence of these long-lived radioactive elements ha
quences. Radioactive decay gives a heat source within the Eart
fuels many geological processes including volcanic activity ar
ation and movement of the crust. Other elements result fron
including helium and argon and several short-lived radioacti
from the decay of thorium and uranium (see Topic I2).

All elements beyond bismuth ($Z = 83$) are radioactive, and none beyond uranium ($Z = 92$) occur naturally on Earth. With increasing numbers of protons heavier elements have progressively less stable nuclei with shorter half-lives. Elements with Z up to 110 have been made artificially but the half-lives beyond Lr ($Z = 103$) are too short for chemical investigations to be feasible. Two lighter elements, technetium (Tc, $Z = 43$) and promethium (Pm, $Z = 61$), also have no stable isotopes.

Radioactive elements are made artificially by bombarding other nuclei, either in particle accelerators or with neutrons in nuclear reactors (see Topic I2). Some short-lived radioactive isotopes (e.g. ^{14}C) are produced naturally in small amounts on Earth by cosmic-ray bombardment in the upper atmosphere.

A2 ATOMIC ORBITALS

Key Notes

Wavefunctions — The quantum theory is necessary to describe electrons. It predicts discrete allowed energy levels and wavefunctions, which give probability distributions for electrons. Wavefunctions for electrons in atoms are called atomic orbitals.

Quantum numbers and nomenclature — Atomic orbitals are labeled by three quantum numbers n, l and m. Orbitals are called *s*, *p*, *d* or *f* according to the value of l; there are respectively one, three, five and seven different possible m values for these orbitals.

Angular functions: 'shapes' — *s* orbitals are spherical. *p* orbitals have two directional lobes, which can point in three possible directions. *d* and *f* orbitals have correspondingly greater numbers of directional lobes.

Radial distributions — The radial distribution function shows how far from the nucleus an electron is likely to be found. The major features depend on n but there is some dependence on l.

Energies in hydrogen — The allowed energies in hydrogen depend on n only. They can be compared with experimental line spectra and the ionization energy.

Hydrogenic ions — Increasing nuclear charge in a one-electron ion leads to contraction of the orbital and an increase in binding energy of the electron.

Related topics — Many-electron atoms (A3); Molecular orbitals: homonuclear diatomics (C3)

Wavefunctions

To understand the behavior of electrons in atoms and molecules requires the use of **quantum mechanics**. This theory predicts the allowed **quantized** energy levels of a system and has other features that are very different from 'classical' physics. Electrons are described by a **wavefunction**, which contains all the information we can know about their behavior. The classical notion of a definite trajectory (e.g. the motion of a planet around the Sun) is not valid at a microscopic level. The quantum theory predicts only **probability distributions**, which are given by the square of the wavefunction and which show where electrons are more or less likely to be found.

Solutions of **Schrödinger's wave equation** give the allowed energy levels and the corresponding wavefunctions. By analogy with the orbits of electrons in the classical planetary model (see Topic A1), wavefunctions for atoms are known as **atomic orbitals**. Exact solutions of Schrödinger's equation can be obtained only for one-electron atoms and ions, but the atomic orbitals that result from these solutions provide pictures of the behavior of electrons that can be extended to many-electron atoms and molecules (see Topics A3 and C3–C6).

Quantum numbers and nomenclature

The atomic orbitals of hydrogen are labeled by **quantum numbers**. Three integers are required for a complete specification.

- The **principal quantum number** n can take the values 1, 2, 3, ... It determines how far from the nucleus the electron is most likely to be found.
- The **angular momentum** (or **azimuthal**) **quantum number** l can take values from zero up to a maximum of $n - 1$. It determines the total angular momentum of the electron about the nucleus.
- The **magnetic quantum number** m can take positive and negative values from $-l$ to $+l$. It determines the direction of rotation of the electron. Sometimes m is written m_l to distinguish it from the spin quantum number m_s (see Topic A3).

Table 1 shows how these rules determine the allowed values of l and m for orbitals with $n = 1 - 4$. The values determine the structure of the periodic table of elements (see Section A4).

Atomic orbitals with $l = 0$ are called **s orbitals**, those with $l = 1, 2, 3$ are called ***p, d, f*** **orbitals**, respectively. It is normal to specify the value of n as well, so that, for example, 1*s* denotes the orbital with $n = 1, l = 0$, and 3*d* the orbitals with $n = 3, l = 2$. These labels are also shown in *Table 1*. For any type of orbital $2l + 1$ values of m are possible; thus there are always three *p* orbitals for any n, five *d* orbitals, and seven *f* orbitals.

Angular functions: 'shapes'

The mathematical functions for atomic orbitals may be written as a product of two factors: the **radial wavefunction** describes the behavior of the electron as a function of distance from the nucleus (see below); the **angular wavefunction** shows how it varies with the direction in space. Angular wavefunctions do not depend on n and are characteristic features of *s, p, d*, . . . orbitals.

Diagrammatic representations of angular functions for *s, p* and *d* orbitals are shown in *Fig. 1*. Mathematically, they are essentially **polar diagrams** showing how

Table 1. Atomic orbitals with n = 1–4

n	l	m	Name
1	0	0	1*s*
2	0	0	2*s*
2	1	−1, 0, +1	2*p*
3	0	0	3*s*
3	1	−1, 0, +1	3*p*
3	2	−2, −1, 0 +1, +2	3*d*
4	0	0	4*s*
4	1	−1, 0, +1	4*p*
4	2	−2, −1, 0, +1, +2	4*d*
4	3	−3, −2, −1, 0, +1, +2, +3	4*f*

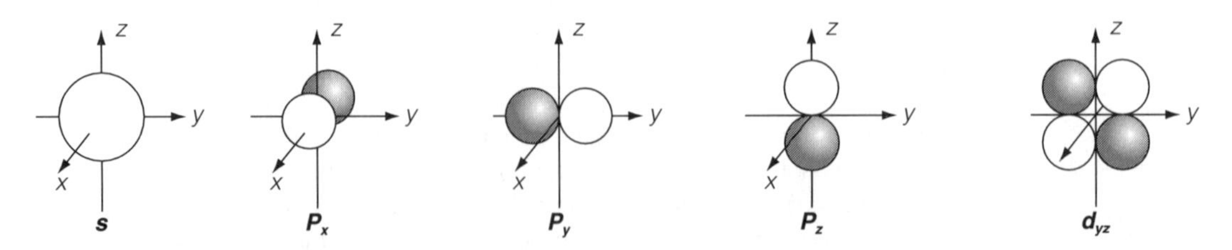

Fig. 1. The shapes of s, p and d orbitals. Shading shows negative values of the wavefunction. More d orbitals are shown in Topic H2, Fig. 1.

the angular wavefunction depends on the polar angles θ and ϕ. More informally, they can be regarded as **boundary surfaces** enclosing the region(s) of space where the electron is most likely to be found. An *s* orbital is represented by a sphere, as the wavefunction does not depend on angle, so that the probability is the same for all directions in space. Each *p* orbital has two lobes, with positive and negative values of the wavefunction either side of the nucleus, separated by a **nodal plane** where the wavefunction is zero. The three separate *p* orbitals corresponding to the allowed values of *m* are directed along different axes, and sometimes denoted p_x, p_y and p_z. The five different *d* orbitals (one of which is shown in *Fig. 1*) each have two nodal planes, separating two positive and two negative regions of wavefunction. The *f* orbitals (not shown) each have three nodal planes.

The shapes of atomic orbitals shown in *Fig. 1* are important in understanding the bonding properties of atoms (see Topics C3–C5 and H2).

Radial distributions

Radial wavefunctions depend on *n* and *l* but not on *m*; thus each of the three 2*p* orbitals has the same radial form. The wavefunctions may have positive or negative regions, but it is more instructive to look at how the **radial probability distributions** for the electron depend on the distance from the nucleus. They are shown in *Fig. 2* and have the following features.

- Radial distributions may have several peaks, the number being equal to $n-l$.
- The outermost peak is by far the largest, showing where the electron is most likely to be found. The distance of this peak from the nucleus is a measure of the radius of the orbital, and is roughly proportional to n^2 (although it depends slightly on *l* also).

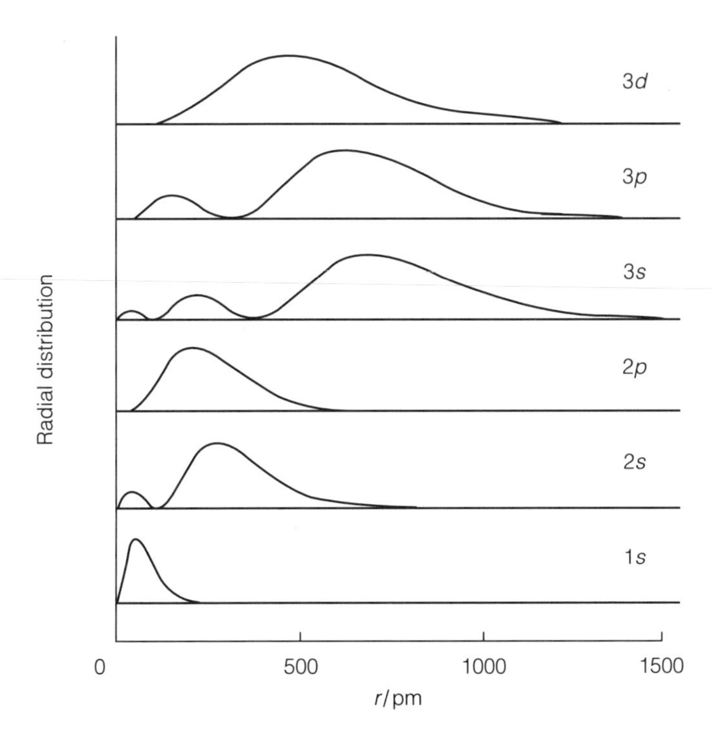

Fig. 2. Radial probability distributions for atomic orbitals with n = 1–3.

Radial distributions determine the energy of an electron in an atom. As the average distance from the nucleus increases, an electron becomes less tightly bound. The subsidiary maxima at smaller distances are not significant in hydrogen, but are important in understanding the energies in many-electron atoms (see Topic A3).

Energies in hydrogen

The energies of atomic orbitals in a hydrogen atom are given by the formula

$$E_n = -R/n^2 \quad (1)$$

We write E_n to show that the energy depends only on the principal quantum number n. Orbitals with the same n but different values of l and m have the same energy and are said to be **degenerate**. The negative value of energy is a reflection of the definition of energy zero, corresponding to $n = \infty$ which is the **ionization limit** where an electron has enough energy to escape from the atom. All orbitals with finite n represent bound electrons with lower energy. The **Rydberg constant** R has the value 2.179×10^{-18} J, but is often given in other units. Energies of individual atoms or molecules are often quoted in **electron volts** (eV), equal to about 1.602×10^{-19} J. Alternatively, multiplying the value in joules by the Avogadro constant gives the energy per mole of atoms. In these units

$$R = 13.595 \text{ eV per atom}$$
$$= 1\,312 \text{ kJ mol}^{-1}$$

The predicted energies may be compared with measured **atomic line spectra** in which light quanta (photons) are absorbed or emitted as an electron changes its energy level, and with the **ionization energy** required to remove an electron. For a hydrogen atom initially in its lowest-energy **ground state**, the ionization energy is the difference between E_n with $n = 1$ and ∞, and is simply R.

Hydrogenic ions

The exact solutions of Schrödinger's equation can be applied to **hydrogenic ions** with one electron: examples are He^+ and Li^{2+}. Orbital sizes and energies now depend on the atomic number Z, equal to the number of protons in the nucleus. The average radius $\langle r \rangle$ of an orbital is

$$\langle r \rangle \approx n^2 a_0/Z \quad (2)$$

where a_0 is the **Bohr radius** (59 pm), the average radius of a 1*s* orbital in hydrogen. Thus electron distributions are pulled in towards the nucleus by the increased electrostatic attraction with higher Z. The energy (see Equation 1) is

$$E_n = -Z^2R/n^2 \quad (3)$$

The factor Z^2 arises because the electron-nuclear attraction at a given distance has increased by Z, and the average distance has also decreased by Z. Thus the ionization energy of He^+ ($Z = 2$) is four times that of H, and that of Li^{2+} ($Z = 3$) nine times.

A3 Many-electron atoms

Key Notes

The orbital approximation

Putting electrons into orbitals similar to those in the hydrogen atom gives a useful way of approximating the wavefunction of a many-electron atom. The electron configuration specifies the occupancy of orbitals, each of which has an associated energy.

Electron spin

Electrons have an intrinsic rotation called spin, which may point in only two possible directions, specified by a quantum number m_s. Two electrons in the same orbital with opposite spin are paired. Unpaired electrons give rise to paramagnetism.

Pauli exclusion principle

When the spin quantum number m_s is included, no two electrons in an atom may have the same set of quantum numbers. Thus a maximum of two electrons can occupy any orbital.

Effective nuclear charge

The electrostatic repulsion between electrons weakens their binding in an atom; this is known as screening or shielding. The combined effect of attraction to the nucleus and repulsion from other electrons is incorporated into an effective nuclear charge.

Screening and penetration

An orbital is screened more effectively if its radial distribution does not penetrate those of other electrons. For a given *n*, *s* orbitals are least screened and have the lowest energy; *p*, *d*, . . . orbitals have successively higher energy.

Hund's first rule

When filling orbitals with $l > 0$, the lowest energy state is formed by putting electrons so far as possible in orbitals with different *m* values, and with parallel spin.

Related topics

Atomic orbitals (A2)

Molecular orbitals: homonuclear diatomics (C3)

The orbital approximation

Schrödinger's equation cannot be solved exactly for any atom with more than one electron. Numerical solutions using computers can be performed to a high degree of accuracy, and these show that the equation does work, at least for fairly light atoms where relativistic effects are negligible (see Topic A5). For most purposes it is an adequate approximation to represent the wavefunction of each electron by an atomic orbital similar to the solutions for the hydrogen atom. The limitation of the **orbital approximation** is that electron repulsion is included only approximately and the way in which electrons move to avoid each other, known as electron correlation, is neglected.

A state of an atom is represented by an **electron configuration** showing which orbitals are occupied by electrons. The ground state of hydrogen is written $(1s)^1$ with one electron in the 1*s* orbital; two excited states are $(2s)^1$ and $(2p)^1$. For helium with two electrons, the ground state is $(1s)^2$; $(1s)^1(2s)^1$ and $(1s)^1(2p)^1$ are excited states.

The energy required to excite or remove one electron is conveniently represented by an **orbital energy**, normally written with the Greek letter ε. The same convention is used as in hydrogen (see Topic A2), with zero being taken as the ionization limit, the energy of an electron removed from the atom. Thus energies of bound orbitals are negative. The ionization energy required to remove an electron from an orbital with energy ε_1 is then

$$I = -\varepsilon_1$$

which is commonly known as **Koopmans' theorem**, although it is better called Koopmans' approximation, as it depends on the limitations of the orbital approximation.

Electron spin

In addition to the quantum numbers *n*, *l* and *m*, which label its orbital, an electron is given an additional quantum number relating to an intrinsic property called **spin**, which is associated with an angular momentum about its own axis, and a magnetic moment. The rotation of planets about their axes is sometimes used as an analogy, but this can be misleading as spin is an essentially quantum phenomenon, which cannot be explained by classical physics. The direction of spin of an electron can take one of only two possible values, represented by the **quantum number** m_s, which can have the values $+1/2$ and $-1/2$. Often these two states are called **spin-up** and **spin-down** or denoted by the Greek letters α and β.

Electrons in the same orbital with different m_s values are said to be **paired**. Electrons with the same m_s value have **parallel spin**. Atoms, molecules and solids with unpaired electrons are attracted into a magnetic field, a property know as **paramagnetism**. The magnetic effects of paired electrons cancel out, and substances with no unpaired electrons are weakly **diamagnetic**, being repelled by magnetic fields.

Experimental evidence for spin comes from an analysis of **atomic line spectra**, which show that states with orbital angular momentum ($l > 0$) are split into two levels by a magnetic interaction known as **spin-orbit coupling**. It occurs in hydrogen but is very small there; spin-orbit coupling increases with nuclear charge (*Z*) approximately as Z^4 and so becomes more significant in heavy atoms. **Dirac's equation**, which incorporates the effects of relativity into quantum theory, provides a theoretical interpretation.

Pauli exclusion principle

Electron configurations are governed by a limitation known as the **Pauli exclusion principle**:

- no two electrons can have the same values for all four quantum numbers *n*, *l*, *m* and m_s.

An alternative statement is

- a maximum of **two** electrons is possible in any orbital.

Thus the three-electron lithium atom cannot have the electron configuration $(1s)^3$; the ground state is $(1s)^2(2s)^1$. When *p*, *d*, ... orbitals are occupied it is important to remember that 3, 5, ... *m* values are possible. A set of *p* orbitals with any *n* can be occupied by a maximum of six electrons, and a set of *d* orbitals by 10.

Effective nuclear charge

The electrostatic repulsion between negatively charged electrons has a large influence on the energies of orbitals. Thus the ionization energy of a neutral helium atom (two electrons) is 24.58 eV compared with 54.40 eV for that of He^+ (one electron). The effect of repulsion is described as **screening** or **shielding**. The combined effect of attraction to the nucleus and repulsion from other electrons gives an **effective nuclear charge** Z_{eff}, which is less than that (Z) of the 'bare' nucleus. One quantitative definition is from the orbital energy ε using the equation (cf. Equation 3, Topic A2):

$$\varepsilon = -Z_{eff}^2 R/n^2$$

where n is the principal quantum number and R the Rydberg constant. For example, applying this equation to He ($n = 1$) gives $Z_{eff} = 1.34$.

The difference between the 'bare' and the effective nuclear charge is the **screening constant** σ:

$$\sigma = Z - Z_{eff}$$

For example, $\sigma = 0.66$ in He, showing that the effect of repulsion from one electron on another has an effect equivalent to reducing the nuclear charge by 0.66 units.

Screening and penetration

The relative screening effect on different orbitals can be understood by looking at their radial probability distributions (see Topic A2, *Fig.* 2). Consider a lithium atom with two electrons in the lowest-energy 1*s* orbital. Which is the lowest-energy orbital available for the third electron? In hydrogen the orbitals 2*s* and 2*p* are **degenerate**, that is, they have the same energy. But their radial distributions are different. An electron in 2*p* will nearly always be outside the distribution of the 1*s* electrons, and will be well screened. The 2*s* radial distribution has more likelihood of **penetrating** the 1*s* distribution, and screening will not be so effective. Thus in lithium (and in all many-electron atoms) an electron has a higher effective nuclear charge, and so lower energy, in 2*s* than in 2*p*. The ground-state electron configuration for Li is $(1s)^2(2s)^1$, and the alternative $(1s)^2(2p)^1$ is an excited state, found by spectroscopy to be 1.9 eV higher.

In a similar way with $n = 3$, the 3*s* orbital has most penetration of any other occupied orbitals, 3*d* the least. Thus the energy order in any many-electron atom is $3s < 3p < 3d$.

Hund's first rule

For a given n and l the screening effect is identical for different m values, and so these orbitals remain degenerate in many electron atoms. In the ground state of boron $(1s)^2(2s)^2(2p)^1$ any one of the three m values $(-1, 0, +1)$ for the p electron has the same energy. But in carbon $(1s)^2(2s)^2(2p)^2$ the different alternative ways of placing two electrons in the three 2*p* orbitals do not have the same energy, as the electrons may repel each other to different extents. Putting two electrons in an orbital with the same m incurs more repulsion than having different m values. In the latter case, the exclusion principle makes no restriction on the spin direction (m_s values), but it is found that there is less repulsion if the electrons have **parallel** spin (same m_s). This is summarized in **Hund's first rule**:

- when electrons are placed in a set of degenerate orbitals, the ground state has as many electrons as possible in different orbitals, and with parallel spin.

The mathematical formulation of many-electron wavefunctions accounts for the rule by showing that electrons with parallel spin tend to avoid each other in a way that cannot be explained classically. The reduction of electron repulsion that results from this effect is called the **exchange energy**.

A4 THE PERIODIC TABLE

Key Notes

History

The periodic table – with elements arranged horizontally in periods and vertically in groups according to their chemical similarity – was developed in an empirical way in the 19th century. A more rigorous foundation came, first with the use of spectroscopy to determine atomic number, and, second with the development of the quantum theory of atomic structure.

Building up

The 'aufbau' or 'building up' principle gives a systematic method for determining the electron configurations of atoms and hence the structure of the periodic table. Elements in the same group have the same configuration of outer electrons. The way different orbitals are filled is controlled by their energies (and hence their different screening by other electrons) and by the Pauli exclusion principle.

Block structure

The table divides naturally into *s*, *p*, *d* and *f* blocks according to the outer electron configurations. *s* and *p* blocks form the main groups, the *d* block the transition elements, and the *f* block the lanthanides and actinides.

Group numbers and names

Modern group numbering runs from 1 to 18, with the *f* blocks being subsumed into group 3. Older (and contradictory) numbering systems are still found. Some groups of elements are conventionally given names, the most commonly used being alkali metals (group 1), alkaline earths (2), halogens (17) and noble gases (18).

Related topics

Many-electron atoms (A3)
Trends in atomic properties (A5)
Chemical periodicity (B2)

History

As more elements were discovered in the 19th century chemists started to note similarities in their properties. Early attempts to order the elements in a regular fashion were hampered by various difficulties, especially the fact (only later realized) that atomic masses do not increase regularly with atomic number. Mendeleev published the first satisfactory form of the periodic table in 1869, and although many details of layout have evolved since then, his basic idea has been retained, of ordering elements horizontally in **periods** so that they fall in vertical **groups** with similar chemical properties. Mendeleev was forced to leave some gaps for elements not yet discovered, and his ability to predict their properties vindicated his approach.

The first satisfactory determination of **atomic number** (as opposed to atomic mass) came from Moseley's studies of X-ray spectra in 1917. By determining the wavelength, and hence frequency, of X-rays emitted from different elements, Moseley observed different series of X-ray lines. In each series the frequency (ν) of each line varied with atomic number (Z) according to the formula

$$\nu = C(Z - \sigma)^2 \quad (1)$$

where C and σ are constants for a given series. **Moseley's law** can be understood from the quantum theory of many-electron atoms. X-rays are produced when atoms are bombarded with high-energy electrons. These knock out electrons from filled orbitals, thus providing 'vacancies' into which electrons can move from other orbitals. Different series of lines come from different vacancies; for example, the highest-energy K series is excited when a 1s electron is removed. Equation 1 then expresses the energy difference between two types of orbital, with C depending on the values of n involved, and σ on the screening constants (see Topic A3).

Using Moseley's law allowed the remaining uncertainties in the structure of the periodic table to be resolved. At about the same time the theoretical ideas of the quantum theory allowed the structure of the table to be understood. Bohr's **aufbau** (or **building up**) **principle** (see below) was developed before the final version of the theory was available; following Schrödinger's equation (1926) the understanding was complete. The periodic table with its theoretical background remains one of the principal conceptual frameworks of inorganic chemistry. A complete table is shown inside the front cover of this book.

Building up

According to the aufbau principle, the ground-state electron configuration of an atom can be found by putting electrons in orbitals, starting with that of lowest energy and moving progressively to higher energy. It is necessary to take into account both the exclusion principle and the modification of orbital energies by screening and penetration effects (see Topic A3). Thus following He $(1s)^2$, the electron configuration of Li is $(1s)^2(2s)^1$, as the 2s orbital is of lower energy than 2p. Following Be, the 2p orbitals are first occupied in B (see *Table 1*). A total of six electrons can be accommodated in these three orbitals, thus up to Ne.

Following completion of the $n = 2$ orbitals, 3s and then 3p shells are filled. The electron configurations of the elements Na–Ar thus parallel those of Li–Ne with only a change in the principal quantum number n. An abbreviated form of the configurations is often used, writing [He] for the filled configuration $(1s)^2$ and [Ne] for [He]$(2s)^2(2p)^6$. The **inner shell** orbitals denoted by these square brackets are too

Table 1. Electron configuration of ground-state atoms up to K (Z = 19)

H	$(1s)^1$
He	$(1s)^2$ = [He]
Li	[He]$(2s)^1$
Be	[He]$(2s)^2$
B	[He]$(2s)^2(2p)^1$
C	[He]$(2s)^2(2p)^2$
N	[He]$(2s)^2(2p)^3$
O	[He]$(2s)^2(2p)^4$
F	[He]$(2s)^2(2p)^5$
Ne	[He]$(2s)^2(2p)^6$ = [Ne]
Na	[Ne]$(3s)^1$
Mg	[Ne]$(3s)^2$
Al	[Ne]$(3s)^2(3p)^1$
Si	[Ne]$(3s)^2(3p)^2$
P	[Ne]$(3s)^2(3p)^3$
S	[Ne]$(3s)^2(3p)^4$
Cl	[Ne]$(3s)^2(3p)^5$
Ar	[Ne]$(3s)^2(3p)^6$ = [Ar]
K	[Ar]$(4s)^1$

tightly bound to be involved in chemical interactions: it is the **valence** or **outer electrons** that determine chemical properties. The group structure of the periodic table depends on the fact that similar outer electron configurations are reflected in similar chemical behaviour.

It might be expected that 3*d* orbitals would fill after 3*p*, but in fact this does not happen, because the extra penetration of *s* compared with *d* orbitals significantly lowers the energy of 4*s*. This fills first, so that following Ar the first two elements of the fourth period K ($[Ar](4s)^1$) and Ca ($[Ar](4s)^2$) have configurations parallel to Na and Mg, respectively. The 3*d* orbitals then fill, giving the 10 elements Sc-Zn, followed by 4*p*. The fifth period follows similarly, 5*s*, 4*d* then 5*p*. In the sixth period another change takes place, with filling of the 4*f* shell after 6*s* and before 5*d*. The seventh incomplete period begins with 7*s* followed by 5*f* and would be expected to continue in the same way, but these elements become increasingly radioactive and hard to make or study (see Topic I2).

The order of filling of shells is conveniently summarized in *Fig. 1*. It is important to note that it reflects the order of energies at the appropriate point, and that this order changes somewhat as more electrons are added. Thus following completion of the 3*d* shell, increasing atomic number stabilizes these orbitals rapidly so that they are no longer chemically active; in an element such as Ga ($[Ar](3d)^{10}(4s)^2(4p)^1$) the valence orbitals are effectively only the 4*s* and 4*p*, so that its chemistry is similar to that of Al ($[Ne](3s)^2(3p)^1$). The same is true following completion of each *d* and *f* shell.

Block structure

The filling of the table described above leads to a natural division of the periodic table into blocks according to the outer electron configurations of atoms (see *Fig. 2*). Elements of the **s block** all have configurations $(ns)^1$ or $(ns)^2$. In periods 2 and 3 these are followed immediately by the ***p* block** with configurations $(ns)^2(np)^x$. Lower *p* block elements are similar as the $(n-1)d$ orbitals are too tightly bound to be chemically important. The *s* and *p* blocks are collectively known as **main**

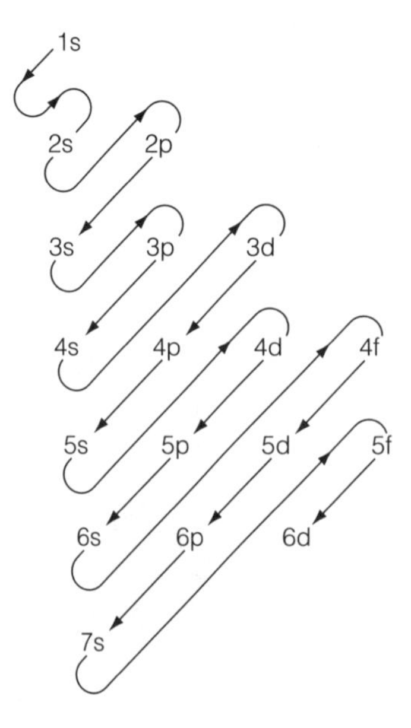

Fig. 1. Showing the order of filling of orbitals in the periodic table.

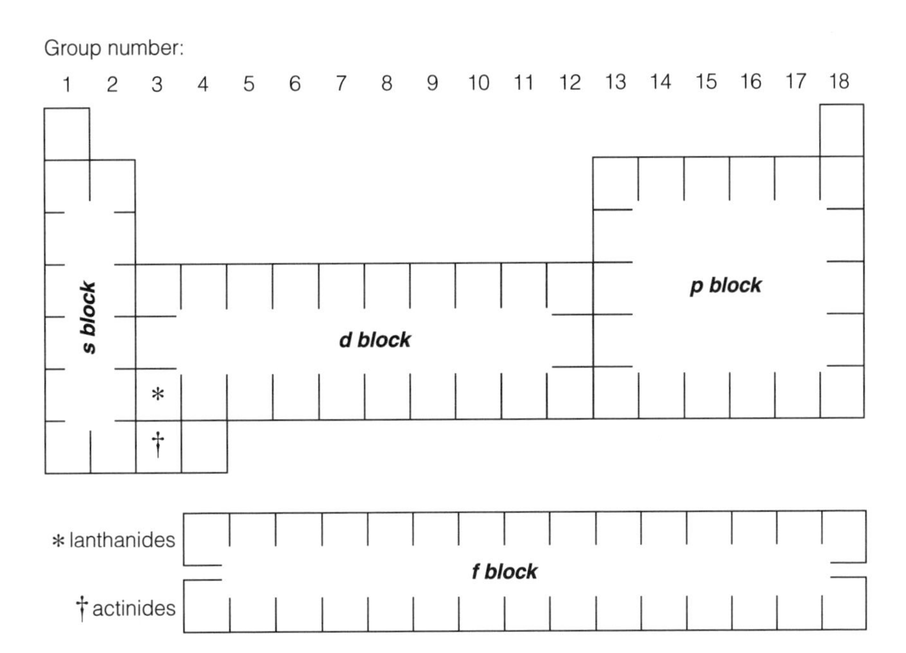

Fig. 2. Structure of the periodic table, showing the s, p, d and f blocks.

groups. ***d*-block** elements of periods 4, 5 and 6 have ns and $(n-1)d$ outer electrons, and are known as **transition elements**. Their configurations show some complexities as the s and d orbitals are similar in energy (see Topic H1). The ***f*-block** elements are known as the **lanthanides** ($4f$) and **actinides** ($5f$). For ease of presentation they are generally shown as separate blocks below the main table. In the case of the lanthanides, this procedure is chemically justified as the elements have very similar properties (see Topic I1).

Group numbers and names

The numbering of groups in the periodic table has a confused history reflecting developments in understanding and presenting the table itself. In the current nomenclature used in this book, groups are numbered 1–18, with the lanthanides and actinides all subsumed into group 3. Older numberings based on 1–8 are still found, with a division into A and B subgroups which unfortunately differs according to the continent. In the UK, the s- and early d-block elements are numbered 1A–8A (the last encompassing modern group numbers 8, 9 and 10), followed by numbers 1B (now 11) to 8B. In the USA, 1A–8A refer to main groups, with d-block elements numbered B. This confusion is resolved by the newer system.

Some groups of elements are conventionally given names. Group 1 elements (not hydrogen) are called **alkali metals** and those of group 2 **alkaline earths**. Groups 17 and 18 are the **halogens** and **noble gases**, respectively. Sometimes group 16 are called **chalcogens** although this normally excludes the first element oxygen: thus the term **chalcogenide** refers to compounds with sulfur, selenium and tellurium. Lanthanides were previously called **rare earths**; although the term is no longer used by chemists it is still common in geochemistry (where it often includes yttrium in group 3 in the previous period, not a lanthanide but chemically very similar).

A5 Trends in atomic properties

Key Notes

Energies and sizes — Trends in orbital energy and size reflect changes in the principal quantum number and effective nuclear charge. They are seen experimentally in trends in ionization energy (IE) and apparent radius of atoms.

Horizontal trends — Increasing nuclear charge causes a general increase of IE and a decrease of radius across any period. Breaks in the IE trend are found following the complete or half filling of any set of orbitals.

Vertical trends — A general increase of radius and decrease in IE down most groups is dominated by the increasing principal quantum number of outer orbitals. Effective nuclear charge also increases, and can give rise to irregularities in the IE trends.

States of ionization — IEs for positive ions always increase with the charge. Electron affinities are the IEs of negative ions and are always less than IEs for neutral atoms.

Relativistic effects — Deviations from the nonrelativistic predictions become significant for heavy atoms, and contribute to especially high IEs for later elements in the sixth period.

Related topics

Many-electron atoms (A3)
The periodic table (A4)
Chemical periodicity (B2)

Energies and sizes

The **first ionization energy (IE)** of an atom (M) is the energy required to form the positive ion M^+:

$$M \rightarrow M^+ + e^-$$

The IE value reflects the energy of the orbital from which the electron is removed, and so depends on the principal quantum number (n) and effective nuclear charge (Z_{eff}; see Topic A3):

$$IE = Z_{eff}^2 R/n^2 \qquad (1)$$

The average radius of an orbital depends on the same factors (see Topic A2):

$$\langle r \rangle \approx n^2 a_0 / Z_{eff} \qquad (2)$$

Smaller orbitals generally have more tightly bound electrons with higher ionization energies.

It is sometimes useful to assume that the distance between two neighboring atoms in a molecule or solid can be expressed as the sum of atomic or ionic radii. **Metallic, covalent** or **ionic radii** can be defined according to the type of bonding between atoms, and **van der Waals' radii** for atoms in contact but not bonded. Such empirically derived radii are all different and are not easily related to any simple predictions based on isolated atoms. They are, however, qualitatively related to orbital radii and all follow the general trends discussed below (see, e.g. Topic D4, *Table 1*, for ionic radii).

Horizontal trends

Increasing nuclear charge is accompanied by correspondingly more electrons in neutral atoms. Moving from left to right in the periodic table, the increase of nuclear charge has an effect that generally outweighs the screening from additional electrons. **Increasing Z_{eff} leads to an increase of IE across each period,** which is the most important single trend in the periodic table (see Topic B2). At the same time, the atoms become smaller.

As illustrated for the elements Li–Ne in *Fig. 1*, the IE trend across a period is not entirely regular. Irregularities can be understood from the electron configurations involved (see Topics A3 and A4). Ionization of boron removes an electron from a $2p$ orbital, which is less tightly bound than the $2s$ involved in lithium and beryllium. Thus the IE of B is slightly less than that of Be. Between nitrogen and oxygen, the factors involved in Hund's rule are important. Up to three $2p$ electrons can be accommodated in different orbitals with parallel spin so as to minimize their mutual repulsion. For O $(2p)^4$ and subsequent elements in the period some electrons are paired and repel more strongly, leading to IE values less than would be predicted by extrapolation from the previous three elements.

The trends shown in *Fig. 1* are sometimes cited as evidence for a 'special stability' of filled and half-filled shells. This is a misleading notion. The general increase of IE across a period is entirely caused by the increase of nuclear charge. Maxima in the plot at filled shells $(2s)^2$ and half-filled shells $(2p)^3$ occur only because of the decrease after these points. It is the exclusion principle that controls such details, by forcing the next electron either to occupy another orbital type (as in boron) or to pair up giving a doubly occupied orbital (as in oxygen).

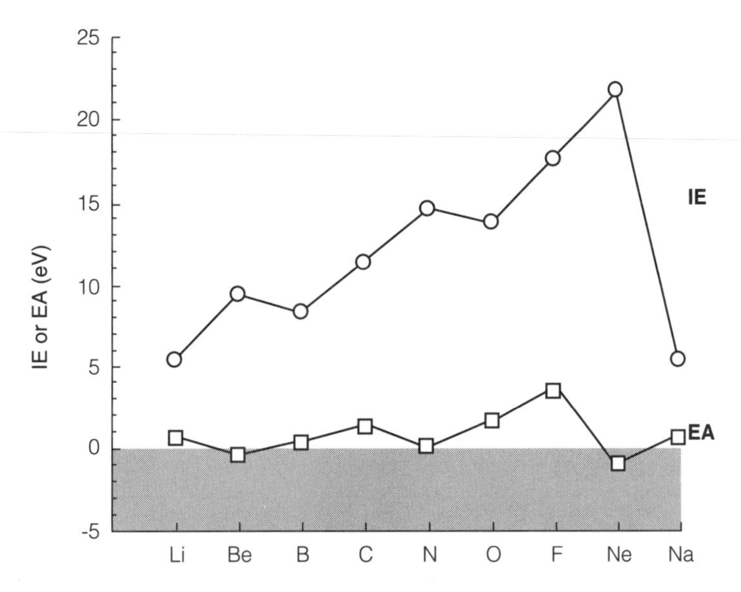

Fig. 1. Ionization energies (IE) and electron affinities (EA) for the elements Li–Na.

Vertical trends

The IE generally decreases down each group of elements. *Figure 2* shows this for hydrogen and the elements of group 1, all of which have the $(ns)^1$ outer electron configuration. The main influence here is the **increasing value of principal quantum number *n***. The fall in IE is, however, much less steep than the simple hydrogenic prediction ($1/n^2$; see Topic A2). There is a substantial increase of nuclear charge between each element, and although extra inner shells are occupied, they do not provide perfect shielding. Thus, contrary to what is sometimes stated, **effective nuclear charge increases down the group**. In the resulting balance between increasing n and increasing Z_{eff} (see Equation 1) the former generally dominates, as in group 1. There is, however, nothing inevitable about this, and there are occasions in later groups where Z_{eff} increases sufficiently to cause an increase of IE between an element and the one below it.

Figure 2 also shows the group 11 elements Cu, Ag and Au, where an *ns* electron is also being ionized. The increase of IE along period 4 between K ($Z = 19$) and Cu ($Z = 29$) is caused by the extra nuclear charge of 10 protons, partly shielded by the 10 added 3*d* electrons. A similar increase occurs between Rb and Ag in period 5. In period 6, however, the 4*f* shell intervenes (see Topic A4) giving 14 additional elements and leading to a total increase of Z of 24 between Cs and Au. There is a much more substantial increase of IE therefore, and Au has a higher IE than Ag. (Relativistic effects also contribute; see below.) Similarly irregular trends in IE may have some influence on the chemistry of *p*-block elements (see Topics F1 and G1).

Orbital radii also depend on n^2 and generally increase down each group. Because the radius depends on Z_{eff} and not on Z_{eff}^2 (see Equation 2) irregular

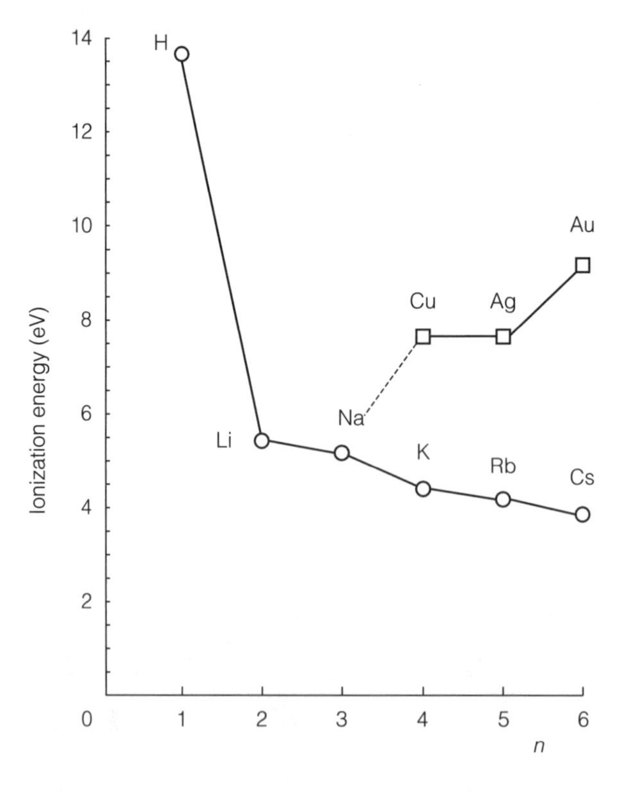

Fig. 2. Ionization energies for elements with $(ns)^1$ outer electron configurations.

1s¹ 1s²2s¹

changes in this quantity have less influence than they do on IEs. (See, however, transition metals, Topics H1 and H5).

There is another interesting feature of vertical trends, arising also from the way in which the periodic table is filled. For orbitals of a given *l* there is a more significant change, both in IE and size, between the first and second periods involved than in subsequent cases. *Figure 2* illustrates this for *s* orbitals, where the IE decreases much more from hydrogen (1*s*) to lithium (2*s*) than between the lower elements. Such a distinction is reflected in the chemical properties of group 1 elements, hydrogen being nonmetallic and the other elements metals (see Topic B2). Similar, although less dramatic, differences are found with 2*p* and 3*d*. Thus period 2 *p*-block elements are in many ways different from those lower in the *p* block, and 3*d* series elements distinct from those of the 4*d* and 5*d* series.

States of ionization

The successive energies required to create more highly charged ions, M^{2+}, M^{3+} . . . are the **second, third, . . . IEs**. The values always **increase with the degree of ionization**. When electrons are removed from the same shell, the main effect is that with each successive ionization there is one less electron left to repel the others. The magnitude of the change therefore depends on the size of the orbital, as electrons in smaller orbitals are on average closer together and have more repulsion. Thus with Be $(2s)^2$ the first two IEs are 9.3 and 18.2 eV, whereas with Ca $(4s)^2$ the values are 6.1 and 11.9 eV, not only smaller to start with (see above) but with a smaller difference. The third IE of both elements is very much higher (154 and 51 eV, respectively) because now the outer shell is exhausted and more tightly bound inner shells (1*s* and 3*p*, respectively) are being ionized. The trends are important in understanding the stable valence states of elements.

The **electron affinity** of an atom may be defined as the **ionization energy of the negative ion**, thus the energy input in the process:

$$M^- \rightarrow M + e^-$$

although some books use a definition with the opposite sign. Electron affinities are always less than ionization energies because of the extra electron repulsion involved (see *Fig. 1*). As with successive IEs, the difference depends on the orbital size. Some apparently anomalous trends can be understood in this way. For example, although the IE of F is greater than that of Cl (17.4 and 13.0 eV, respectively) the electron affinity of F is smaller (3.4 eV compared with 3.6 eV) partly because the smaller size of F^- provides more repulsion from the added electron.

Some atoms have negative electron affinities, meaning that the negative ion is not stable in the gas phase. Second and subsequent electron affinities are always negative because of the high degree of repulsion involved in forming a multiply charged negative ion. Thus the O^{2-} ion is not stable in isolation. This does not invalidate the ionic description of compounds such as MgO, as the O^{2-} ion is now surrounded by positive Mg^{2+} ions which produce a stabilizing effect (the lattice energy; see Topic D6).

As expected, ion sizes decrease with increasing positive charge, and negative ions are larger. In most ionic compounds, anions are larger than cations (see Topics D3 and D4).

Relativistic effects

Schrödinger's equation does not take into account effects that are important when particles travel at a speed comparable with that of light. There are two important aspects: moving charged particles experience magnetic as well as electric fields; and the **special theory of relativity** predicts effects such an enhancement of the mass of fast-moving particles. These effects were incorporated into the quantum mechanical wave theory by **Dirac's equation** (1928). One remarkable prediction is the existence of **electron spin** (see Topic A3) and the occurrence of spin-orbit splitting in atomic spectra. The energies of orbitals are also altered, especially for electrons close to highly charged nuclei, as it is then that they are travelling fast. Inner shells are most affected but they are not important in chemistry. For very heavy elements even outer shells show an influence of relativity. This is true for the 6*s* shell in gold and mercury, and the 6*p* shell in subsequent elements of period 6. Relativistic effects increase the binding energy of these electrons. They thus contribute to the irregularities in group trends, and make an appreciable contribution to the high IEs and hence chemical inertness of some heavy elements.

B1 Electronegativity and Bond Type

Key Notes

Definitions

Electronegativity is the power of an atom to attract electrons to itself in a chemical bond. Different numerical estimates agree on qualitative trends: electronjegativity increases from left to right along a period, and generally decreases down groups in the periodic table. Elements of low electronegativity are called electropositive.

The bonding triangle

Electropositive elements form metallic solids. Electronegative elements form molecules or polymeric solids with covalent bonds. Elements of very different electronegativity combine to form solids that can be described by the ionic model.

Bond polarity

The polarity of a bond arises from the unequal sharing of electrons between atoms with different electronegativities. There is no sharp dividing line between polar covalent and ionic substances.

Related topics

Trends in atomic properties (A5)
Electron pair bonds (C1)
Introduction to solids (D1)

Definitions

Electronegativity may be defined as the **power of an atom to attract electrons to itself in a chemical bond**. It is the most important chemical parameter in determining the type of chemical bonds formed between atoms. It is hard to quantify in a satisfactory way, especially as electronegativity is not strictly a property of atoms on their own, but depends to some extent on their state of chemical combination. Nevertheless several scales have been devised.

- **Pauling electronegativity** is based on bond energies (see Topic C7), using the empirical observation that bonds between atoms with a large electronegativity difference tend to be stronger than those where the difference is small. This scale was historically the first to be devised and although it lacks a firm theoretical justification is still widely used.
- **Mulliken electronegativity** is the average of the first ionization energy and the electron affinity of an atom (see Topic A5), reflecting the importance of two possibilities in bond formation, losing an electron or gaining one. The scale has the advantage that electronegativity values can be estimated not only for the ground states of atoms, but for other electron configurations and even for polyatomic fragments.
- **Allred–Rochow electronegativity** is proportional to Z_{eff}/r^2, where Z_{eff} is the effective nuclear charge of valence orbitals (see Topic A3), and r the covalent radius of the atom. The value is proportional to the effective electrostatic attraction on valence electrons by the nucleus, screened by inner shell electrons.

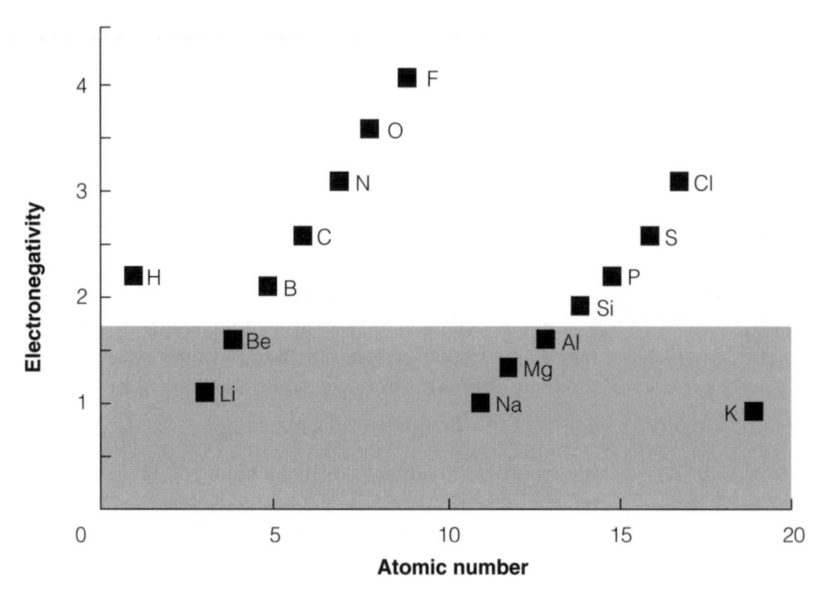

Fig. 1. Pauling electronegativity values for the elements H–K. Elements in the shaded region are metallic (see Topic B2).

Each scale produces different numbers and they should not be mixed. The broad general trends do, however, agree: electronegativity **increases towards the right and decreases towards the bottom in the periodic table**. It thus follows the same trend as atomic ionization energies (see Topic A5). Elements in early groups have low values and are called **electropositive**. *Figure 1* shows the Pauling electronegativities of elements up to potassium. Elements of group 18 in early periods do not form any stable compounds, and so the most electronegative element is fluorine.

The bonding triangle

The bonding triangle (see *Fig. 2*) is a useful way of showing how the electronegtivities of two elements A and B (which may be the same) determine the type of bond formed between them. The horizontal and vertical scales show the Pauling electronegativities of the two elements. (Other scales would do equally well at this qualitative level.) Pure elements (A = B) appear on the diagonal, and various compounds are shown within the triangle. Three basic regions are distinguished.

- When A and B are both electropositive they form a **metallic solid**, characterized by high electrical conductivity and a structure where each atom is surrounded by many others (often 12; see Topic D2). Metallic bonding involves the **delocalization of electrons** throughout the solid. The electrons are shared between atoms as in covalent bonding (see below), but in a less specific way and without the directional character of covalent bonds.
- When A and B are both electronegative they form **covalent compounds**. These may consist of individual **molecules** (O_2, H_2O, etc.) or of **giant covalent lattices** (**polymeric solids**) with a continuous network of bonds. Although the dividing line between these types is not sharp, very highly electronegative atoms (F, O, Cl, etc.) have more tendency to molecular behavior in both their elements and their compounds. Covalent solids do not conduct electricity well. The most important feature of this bonding, whether in molecules and solids, is its highly directional and specific nature. Thus the neighbors to any atom are limited in number (e.g. four in the case of elemental silicon, three for phosphorus, two for

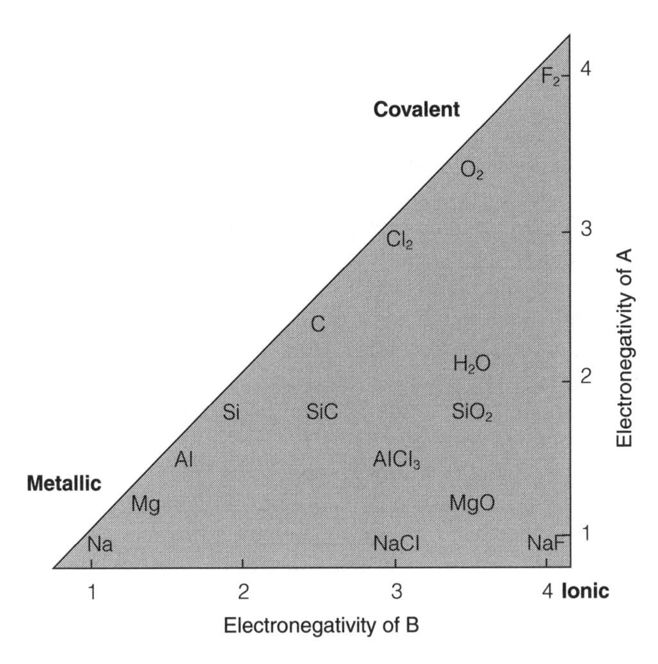

Fig. 2. The bonding triangle, showing a selection of elements and compounds plotted against the Pauling electronegativities.

sulfur, one for chlorine), and are generally found in specific geometrical arrangements. The simplest view of covalent bonding involves the **sharing of electrons** in specific, **localized bonds** between atoms (see Topic C1).

- When one atom is very electropositive and the other very electronegative, a solid compound is formed that is often regarded as **ionic**. In this picture there is a complete transfer of one or more electrons, giving **cations** of the electropositive element and **anions** of the electronegative one, which are then held together by electrostatic attraction (see Topics D3, D4 and D6). Solids are formed rather than molecules because the force is not directional, and greatest stability is achieved by packing several anions around each cation and vice versa.

Bond polarity

A covalent bond between two atoms of the same element is described as **homopolar**, one between different elements as **heteropolar**; the general term **bond polarity** describes the unequal sharing of electrons between two atoms, and is a feature of heteropolar bonds when the two elements concerned have a different electronegativity. The more electronegative atom draws electrons and thus acquires a partial negative charge, with the other atom becoming correspondingly positive. One manifestation of such polarity is the formation of an **electric dipole moment**, the magnitude of which is equal to the product of the charges and their average separation. The dipole moments decrease in a series of molecules such as HF > HCl > HBr > HI as might be expected from the falling difference in electronegativities. Dipole moments are, however, not always easy to interpret, as they can be influenced by other factors, such as the relative orientation of bonds in polyatomic molecules and the distribution of nonbonding electrons. Dipole moments are an important source of intermolecular forces (see Topic C9).

Polar covalent bonds can be regarded as having some degree of ionic character, and the distinction between 'ionic' and 'covalent' bond types is sometimes hard to make. Some compounds have clear examples of both types of bonding simultaneously. Thus $CaCO_3$ has well-defined carbonate ions (CO_3^{2-}) with C and O covalently bonded together; the complex ion also interacts ionically with Ca^{2+}. Such **complex ions** need not be discrete entities but can form polymeric covalent networks with a net charge, with ionic bonds to cations (e.g. silicates; see Topics D6 and F4). Even when only two elements are present, however, bonding may be hard to describe in simple terms.

When a compound is molecular under normal conditions it is usual to regard it as covalent (although 'ionic molecules' such as NaCl(g) can at be made by vaporizing the solid compounds at high temperatures). When two elements of different electronegativity form a solid compound alternative descriptions may be possible. Consider the compounds BeO and BN. Both form structures in which every atom is surrounded tetrahedrally by four of the other kind (BN also has an alternative structure similar to that of graphite). For BeO this is a plausible structure on ionic grounds, given that the Be^{2+} ion must be much smaller than O^{2-} (see Topic D4). On the other hand, many of the structures and properties of beryllium compounds are suggestive of some degree of covalent bonding (see Topic G3). Thus one can think of BeO as predominantly ionic, but with the oxide ion **polarized** by the very small Be^{2+} ion so that electron transfer and ionic character are not complete. For BN the electronegativity difference between elements is much less, and it would be more natural to think of polar covalent bonding. The tetrahedral structure of BN can be understood from its similarity to diamond, where each carbon atom is covalently bonded to four others. The difference between two descriptions 'polarized ionic' and 'polar covalent' is not absolute but only one of degree. Which starting point is better cannot be laid down by rigid rules but is partly a matter of convenience.

One should beware of using oversimplified criteria of bond type based on physical properties. It is sometimes stated that 'typical' ionic compounds have high melting points and dissolve well in polar solvents such as water, whereas covalent compounds have low melting points and dissolve well in nonpolar solvents. This can be very misleading. Diamond, a purely covalent substance, has one of highest melting points known and is insoluble in any solvent. Some compounds well described by the ionic model have fairly low melting points; others are very insoluble in water on grounds that can be explained perfectly satisfactorily in terms of ions (see Topic E4).

B2 CHEMICAL PERIODICITY

Key Notes

Introduction

Major chemical trends, horizontally and vertically in the periodic table, can be understood in terms of changing atomic properties. This procedure has its limitations and many details of the chemistry of individual elements cannot be predicted by simple interpolation from their neighbors.

Metallic and non-metallic elements

Metallic elements are electropositive, form electrically conducting solids and have cationic chemistry. Non-metallic elements, found in the upper right-hand portion of the periodic table, have predominantly covalent and anionic chemistry. The chemical trend is continuous and elements on the borderline show intermediate characteristics.

Horizontal trends

Moving to the right in the periodic table, bonding character changes as electronegativity increases. The increasing number of electrons in the valence shell also gives rise to changes in the stoichiometry and structure of compounds. Similar trends operate in the *d* block.

Vertical trends

The increased size of atoms in lower periods is manifested in structural trends. For each block, changes in chemistry between the first and second rows concerned are often more marked than those between lower periods.

Related topics

The periodic table (A4)
Trends in atomic properties (A5)
Introduction to nonmetals (F1)
Introduction to nontransition metals (G1)
Introduction to transition metals (H1)

Introduction

The periodic table was devised by Mendeleev in response to observed regularities in the chemistry of the elements before there was any understanding of their electronic basis (see Topic A4). His procedure was vindicated by his ability to predict the properties and simple chemistry of the then unknown elements gallium and germanium by simple interpolation between known elements in neighboring positions. Chemical periodicity was thus seen to be a powerful tool in the interpretation and even prediction of the chemical properties of elements.

Since Mendeleev the range of chemical compounds known has expanded enormously and it has become apparent that such simple interpolation procedures have many limitations. In a few groups (especially the *s* block) the chemistry is fairly similar, and most of the observed trends in the group can be interpreted straightforwardly from changes of atomic properties such as radius. In the *p* and *d* blocks, however, this is not so easy. Complications arise partly from the fact that atomic trends are themselves less regular (because of the way in which the periodic table is filled), and partly from the greater complexities in chemical bonding, which respond in a more subtle way to changes in orbital size and energy. The periodic table remains the most important framework for understanding the comparative chemistry of elements, and many major trends can be understood from the atomic trends described in Topic A5. Most elements have peculiarities, however, which although they can be rationalized in terms of periodic trends, would probably not have been predicted if they were not known.

Metallic and non-metallic elements

The most important classification of elements is that of metallic versus non-metallic. **Metallic elements** form solids that are good conductors of electricity, and have structures with many near neighbors and where bonding is not strongly directional. **Non-metallic elements** form molecules or covalent solids, which are generally poor conductors of electricity and where bonding is markedly directional in character. This distinction on the basis of physical properties is fairly clear-cut and is shown in the periodic table in *Fig. 1*. All elements of the *s*, *d* and *f* blocks are metallic (except hydrogen), non-metallic ones being confined to the upper right-hand part of the *p* block. The most obvious atomic parameter that determines this behavior is **electronegativity** (see Topic B1, especially *Fig. 1*).

Different types of chemical behavior are associated with the two kinds of element.

- Typical metallic elements are good reducing agents (for example, reacting with water to produce dihydrogen) and form hydrated **cations** in aqueous solution (Na^+, Mg^{2+}, etc.). They have solid halides and oxides, which are well described by the ionic model. The oxides are **basic** and either react with water to produce hydroxide ions (OH^-) or, if insoluble under neutral conditions, dissolve in acidic solutions. Their hydrides are solids with some ionic (H^-) character.
- Typical non-metallic elements form ionic compounds with electropositive metals. They form **anions** in water, either monatomic (e.g. Cl^-) or **oxoanions** (e.g. NO^{3-}, SO_4^{2-}). They have molecular hydrides and halides. Their oxides are either molecular or polymeric covalent in structure, and are **acidic**, reacting with water (as do halides) to produce **oxoacids** (H_2CO_3, H_2SO_4, etc.)

It must be recognized that this classification has many limitations, and borderline behavior is common. In addition to their typical cationic behavior, most metallic elements form some compounds where bonding is predominantly

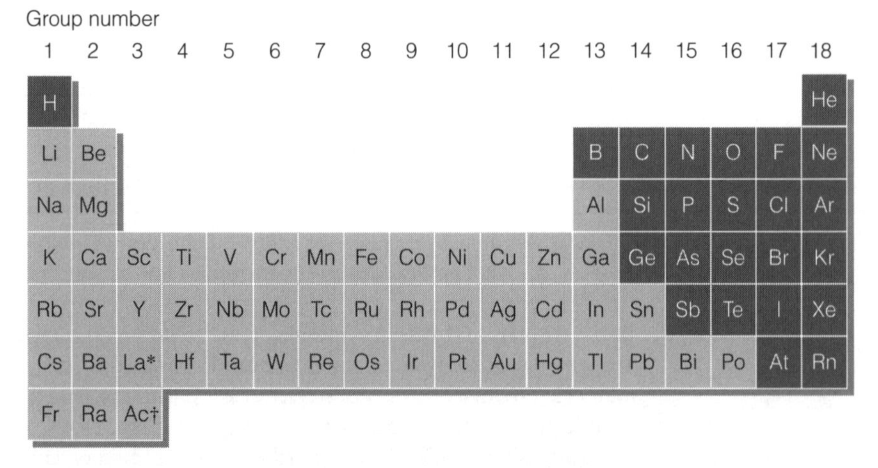

Fig. 1. Periodic table showing metallic and (heavily shaded) non-metallic elements.

covalent (see, e.g. Topic H10). Some form anionic species such as MnO^{4-} or even Na^- (see Topic G2). Many metals in later groups are much less electropositive than the typical definition would suggest, and the metal-nonmetal borderline in the *p* block involves a continuous gradation in chemical behavior rather than a discontinuous boundary (see Topic G6). Non-metallic elements close to the metallic borderline (Si, Ge, As, Sb, Se, Te) show less tendency to anionic behavior and are sometimes called **metalloids**.

Horizontal trends

The major horizontal trends towards the right in any block are a general increase of ionization energy (which is reflected in an increase in electronegativity), a contraction in size, and an increase in the number of electrons in the valence shell. In main groups, the effect of changing electronegativity is obvious in determining the metal-nonmetal borderline. The number of valence electrons has a clear influence on the stoichiometry of compounds formed (NaF, MgF_2, AlF_3, etc.).

Main group elements commonly form ions with **closed shell** configurations: hence cations (Na^+, Mg^{2+}, Al^{3+}) in which all electrons have been lost from the valence shell, and anions (F^-, O^{2-}) in which the valence shell has been filled. This observation suggests some 'special stability' of filled shells, but, as in atomic structure (see Topic A5), such an interpretation is misleading. The stoichiometry of stable ionic compounds depends on the balance between the energy required to form ions and the **lattice energy**, which provides the bonding (see Topic D6). Such an approach provides a better understanding not only of why closed-shell ions are often found, but also of cases where they are not, as happens frequently in the *d* block (see Topics H1 and H3).

In covalent compounds some regularities in stoichiometry can also be understood from the increasing number of valence electrons. Thus the simple hydrides of groups 14, 15, 16 and 17 elements have the formulae EH_4, EH_3, EH_2 and EH, respectively, reflecting the **octet rule**. Filling the valence shell creates progressively more nonbonding electrons and limits the capacity for bonding. Such **non-bonding electrons** also influence the geometrical structures of the molecules (see Topics C1 and C2).

The general increase of electronegativity (or decline in electropositive character) and contraction in size is apparent also in *d*-block chemistry. The formation of closed-shell ions (Sc^{3+}, Ti^{4+}, etc.) is a feature of only the early groups. As ionization energies increase more electrons are prevented from involvement in bonding. Non-bonding *d* electrons also influence the structures and stabilities of compounds, but because of the different directional properties of *d* orbitals compared with *p*, these effects are best understood by a different approach, that of ligand field theory (see Section H2).

Vertical trends

The general decrease of ionization energy down a group is reflected in the trend towards metallic elements in the *p* block. Another change is the general increase in radius of atoms down a group, which allows a **higher coordination number**. Sometimes this is reflected in the changing stoichiometry of stable compounds: thus ClF_3, BrF_5 and IF_7 are the highest fluorides known for elements of group 17. In other groups the stoichiometry is fixed but the structure changes: thus the coordination of the metallic element by fluorine is four in BeF_2, six in MgF_2 and eight in CaF_2. Although exceptions occur (see Topics G4 and H5) this is a common trend irrespective of different modes of bonding.

One further general feature of vertical trends is important, and reflects the analogous trends in atomic properties mentioned in Topic A5. For each block (*s*, *p*, *d*) the first series involved has somewhat distinct chemistry compared with subsequent ones. Hydrogen (1*s*) is non-metallic and very different from the other *s*-block elements. The 2*p*-series elements (B–F) have some peculiarities not shared with the rest of the *p* block (e.g. a limitation in the number of valence-shell electrons in molecules, and the frequent formation of multiple bonds; see Topic F1). In the *d* block, the elements of the 3*d* series also show characteristic differences from the 4*d* and 5*d* series (e.g. forming many more compounds with unpaired electrons; see Topics H1–H5).

B3 STABILITY AND REACTIVITY

Key Notes

Introduction

Stability and reactivity can be controlled by thermodynamic factors (depending only on the initial and final states and not on the reaction pathway) or kinetic ones (very dependent on the reaction pathway). Both factors depend on the conditions, and on the possibility of different routes to decomposition or reaction.

Enthalpy and Hess' Law

Enthalpy change (ΔH) is the heat input to a reaction, a useful measure of the energy change involved. As ΔH does not depend on the reaction pathway (Hess' Law) it is often possible to construct thermodynamic cycles that allow values to be estimated for processes that are not experimentally accessible. Overall ΔH values for reactions can be calculated from tabulated enthalpies of formation.

Entropy and free energy

Entropy is a measure of molecular disorder. Entropy changes (ΔS) can be combined with ΔH in the Gibbs free energy change (ΔG), which determines the overall thermodynamic feasibility of a reaction. As with ΔH, ΔG can be estimated from thermodynamic cycles and tabulated values, the latter always referring to standard conditions of pressure or concentration.

Equilibrium constants

The equilibrium constant of reaction is related to the standard Gibbs free energy change. Equilibrium constants change with temperature in a way that depends on ΔH for the reaction.

Reaction rates

Reaction rates depend on the concentrations of reagents, and on a rate constant that itself depends on the energy barrier for the reaction. Reaction rates generally increase with rise in temperature. Catalysts provide alternative reaction pathways of lower energy.

Related topics

Bond strengths (C7)
Lattice energies (D6)
Industrial chemistry: catalysts (J5)

Introduction

We tend to say that substances are 'stable' or 'unstable', 'reactive' or 'unreactive' but these terms are relative and may depend on many factors. Is important to specify the conditions of temperature and pressure, and what other substances are present or could act as potential routes to decomposition. Thermodynamic and kinetic factors can also be important.

Thermodynamics deals with overall energy and entropy changes, and their relation to the direction of reaction and the position of equilibrium. Such quantities depend only on the initial and final states, and not at all on the reaction pathway. It is often possible to assess the thermodynamic feasibility of a reaction without any knowledge of the mechanism. On the other hand, the rate of a reaction does depend on the pathway; this is the subject of **chemical kinetics**, and thermodynamic considerations alone cannot predict how fast a reaction will take place.

Many known substances are **thermodynamically stable**, but others are only **kinetically stable**. For example, the hydrides B_2H_6 and SiH_4 are thermodynamically unstable with respect to their elements, but in the absence of heat or a catalyst (and of atmospheric oxygen and moisture) the rate of decomposition is extremely slow. To assess why some substances are unknown, it is important to consider different possible routes to decomposition. For example, the unknown CaF(s) is probably thermodynamically stable with respect to the elements themselves, but certainly unstable (thermodynamically and kinetically) with respect to the reaction

$$2CaF(s) \rightarrow Ca(s) + CaF_2(s)$$

Thermodynamic and kinetic factors depend on temperature and other conditions. For example, CaF(g) can be formed as a gas-phase molecule at high temperatures and low pressures.

Enthalpy and Hess' Law

The enthalpy change (ΔH) in a reaction is equal to the heat input under conditions of constant temperature and pressure. It is not exactly equal to the total energy change, as work may be done by expansion against the external pressure. The corrections are generally small, and enthalpy is commonly used as a measure of the energies involved in chemical reactions. **Endothermic reactions** (positive ΔH) are ones requiring a heat input, and **exothermic reactions** (negative ΔH) give a heat output.

Hess' Law states that ΔH does not depend on the pathway taken between initial and final states, and is a consequence of the **First Law of Thermodynamics**, which asserts the conservation of total energy. *Figure 1* shows a schematic thermodynamic cycle where the overall ΔH can be expressed as the sum of the values for individual steps:

$$\Delta H = \Delta H_1 + \Delta H_2 + \ldots \quad (1)$$

It is important that they need not represent any feasible mechanism for the reaction but can be any steps for which ΔH values are available from experiment or theory. Hess' Law is frequently used to estimate ΔH values that are not directly accessible, for example, in connection with **lattice energy** and **bond energy** calculations (see Topics D6 and C7).

Enthalpy change does depend on conditions of temperature, pressure and concentration of the initial and final states, and it is important to specify these. **Standard states** are defined as pure substances at standard pressure (1 bar), and

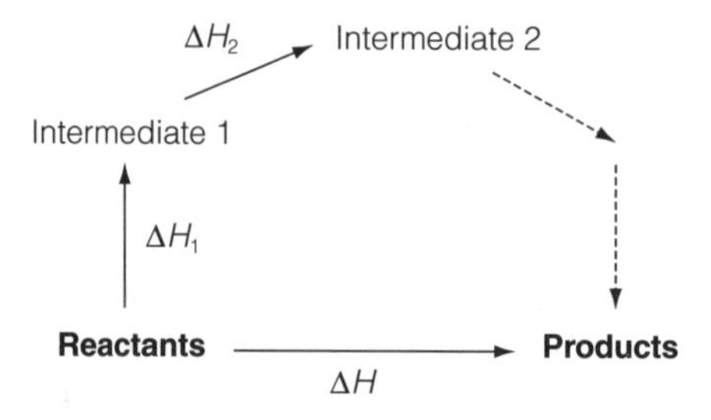

Fig. 1. Schematic thermodynamic cycle illustrating the use of Hess' Law (see Equation 1).

the temperature must be additionally specified, although 298 K is normally used. Corrections must be applied for any other conditions. The **standard enthalpy of formation** ($\Delta H_f^\ominus$) of any compound refers to formation from its elements, all in standard states. Tabulated values allow the standard enthalpy change $\Delta H^\ominus$ in any reaction to be calculated from

$$\Delta H^\ominus_{\text{reaction}} = \sum_{\text{products}} \Delta H_f^\ominus - \sum_{\text{reactants}} \Delta H_f^\ominus \qquad (2)$$

which follows from Hess' Law. By definition, $\Delta H_f^\ominus$ is zero for any element in its stable (standard) state.

Entropy and free energy

Entropy (S) is a measure of molecular 'disorder', or more precisely 'the number of microscopic arrangements of energy possible in a macroscopic sample'. Entropy increases with rise in temperature and depends strongly on the state. Entropy changes (ΔS) are invariably positive for reactions that generate gas molecules. The **Second Law of Thermodynamics** asserts that the total entropy always increases in a spontaneous process, and reaches a maximum value at equilibrium. To apply this to chemical reactions it is necessary to include entropy changes in the surroundings caused by heat input or output. Both internal and external changes are taken account of by defining the **Gibbs free energy change** (ΔG): for a reaction taking place at constant temperature (T, in kelvin)

$$\Delta G = \Delta H - T\Delta S \qquad (3)$$

From the Second Law it can be shown that ΔG is always negative for a feasible reaction at constant temperature and pressure (and without any external driving force such as electrical energy) and is zero at equilibrium.

As with enthalpies, ΔS and ΔG for reactions do not depend on the reaction pathway taken and so can be estimated from thermodynamic cycles like that of *Fig. 1*. They depend even more strongly than ΔH on concentration and pressure. Tabulated standard entropies may be used to estimate changes in a reaction from

$$\Delta S^\ominus_{\text{reaction}} = \sum_{\text{products}} \Delta S^\ominus - \sum_{\text{reactants}} \Delta S^\ominus$$

which is analogous to Equation 2 except that $S^\ominus$ values are not zero for elements. The direct analogy to Equation 2 may also be used to calculate $\Delta G^\ominus$ for any reaction where the standard free energies of formation $\Delta G_f^\ominus$ are known.

Equilibrium constants

For a general reaction such as

$$a\text{A} + b\text{B} \rightleftharpoons c\text{C} + d\text{D}$$

the **equilibrium constant** is

$$K = \frac{[\text{C}]^c[\text{D}]^d}{[\text{A}]^a[\text{B}]^b}$$

where the terms [A], [B], . . . strictly represent **activities** but are frequently approximated as concentrations or partial pressures. (This assumes **ideal thermodynamic behavior** and is a much better approximation for gases than in solution.) Pure liquids and solids are not included in an equilibrium constant as they are present in their standard state. A very large value ($\gg 1$) of K indicates a strong thermodynamic tendency to react, so that very little of the reactants (A and B) will remain at equilibrium. Conversely, a very small value ($\ll 1$) indicates very little tendency to react: in this case the reverse reaction (C and D going to A and B) will be very favorable.

For any reaction K may be related to the **standard Gibbs free energy change** ($\Delta G^\ominus$) according to

$$\Delta G^\ominus = -RT \ln K \quad (4)$$

where R is the gas constant (= 8.314 J K^{-1} mol^{-1}) and T the absolute temperature (in K). Thus equilibrium constants can be estimated from tabulated values of $\Delta G_f^\ominus$, and trends may often be interpreted in terms of changes in $\Delta H^\ominus$ and $\Delta S^\ominus$ (see Equation 3).

Equilibrium constants change with temperature in a way that depends on $\Delta H^\ominus$ for the reaction. In accordance with **Le Chatelier's principle**, K increases with rise in temperature for an endothermic reaction, and decreases for an exothermic one.

Reaction rates

The rate of reaction generally depends on the concentration of reactants, often according to a power law such as

$$\text{Rate} = k\,[A]^n[B]^m$$

where k is the **rate constant** and n and m are the **orders of reaction** with respect to reactants A and B. Orders of reaction depend on the mechanism and are **not** necessarily equal to the stoichiometric coefficients a and b. The rate constant depends on the mechanism and especially on the **energy barrier** or **activation energy** associated with the reaction pathway. High activation energies (E_a) give low rate constants because only a small fraction of molecules have sufficient energy to react. This proportion may be increased by raising the temperature, and rate constants approximately follow the **Arrhenius equation**:

$$k = Ae^{-E_a/RT}$$

Large activation energies arise in reactions where covalent bonds must be broken before new ones are formed, or where atoms must move through solids. Reactions involving free radicals, or ions in solution, often have small (sometimes zero) activation energies.

Reactions may be accelerated by the presence of a **catalyst**, which acts by providing an alternative pathway with lower activation energy. A true catalyst by definition can be recovered unchanged after the reaction, and so does not alter the thermodynamics or the position of equilibrium (see Topic J5).

B4 OXIDATION AND REDUCTION

Key Notes

Definitions

Oxidation means combination with a more electronegative element or the removal of electrons. Reduction means combination with a less electronegative element or the addition of electrons. A complete redox reaction involves both processes.

Oxidation states

Oxidation states of atoms in a compound are calculated by assigning electrons in a bond to the more electronegative element. In simple ionic compounds they are the same as the ionic charges. In any redox reaction the oxidation states of some elements change.

Balancing redox reactions

In complete redox reactions the overall changes in oxidation state must balance. When reactions involve ions in water it is convenient to split the overall reaction into two half reactions. To balance these it may also be necessary to provide water and H^+ or OH^-.

Extraction of the elements

Redox reactions are used in the extraction of nearly all elements from naturally occurring compounds. Carbon is used to reduce some metal oxides, but many elements require stronger reducing agents, or the use of electrolysis.

Related topic

Electrode potentials (E5)

Definitions

Oxidation originally meant 'combination with oxygen' and reduction 'removal of oxygen'. These definitions have been greatly expanded. **Oxidation** implies combination with a more electronegative element, the removal of a less electronegative one, or simply the **removal of electrons**. **Reduction** is the reverse of oxidation and in general implies **addition of electrons**. In any reaction where one species is oxidized, another must be reduced: the term redox reaction is used to express this. Two examples are: the reaction of zinc in aqueous acid,

$$Zn(s) + 2H^+(aq) \rightarrow Zn^{2+}(aq) + H_2(g) \quad (1)$$

where zinc metal is oxidized to Zn^{2+}, and hydrogen reduced from H^+ to H_2; and the reduction of zinc oxide by carbon,

$$ZnO(s) + C(s) \rightarrow Zn(s) + CO(g) \quad (2)$$

where zinc is reduced from ZnO to the metal, elemental carbon is oxidized to CO, and oxygen, combined with a less electronegative element on both sides, is not oxidized or reduced.

A strong **oxidizing agent** is a substance capable of oxidizing many others, and is thus itself easily reduced; conversely, a strong **reducing agent** is itself easily oxidized; these terms usually imply thermodynamic reaction tendency although kinetics may also be important (see Section B3). Atmospheric dioxygen is a good oxidizing agent, but many substances (e.g. organic compounds) are kinetically stable in air. Strong reducing agents include electropositive metals, especially those of group 1 (see Section G2).

Oxidation states

The **oxidation state** (or **oxidation number**) is a number applied to each atom in a compound in such as way as to keep track of changes occurring in redox reactions. For simple ionic compounds the oxidation state is equal to the charge on the ions. Roman numbers should be used to distinguish oxidation states from ionic charges, e.g. Na^{I} and Cl^{-I} in NaCl. In polar covalent bonds the electrons are assigned to the more electronegative atoms **as if the bonding were ionic**. Oxidation states are calculated by the following rules.

(i) Bonds between the same element are not counted. Elements have oxidation state zero. In an ion such as peroxide O_2^{2-} the electrons in the O–O bond are distributed equally, making O^{-I}.
(ii) Except in cases such as O_2^{2-} the most electronegative and electropositive elements in a compound have an oxidation state equal to their normal ionic charge: K^{I}, Ca^{II}, F^{-I}, O^{-II}.
(iii) The sum of the oxidation states must equal the charge on the species, and is therefore zero in a neutral compound. Using this rule and (ii) above, we have H^{I} in H_2O, H^{-I} in CaH_2 and Mn^{VII} in MnO_4^-.
(iv) Complex formation, and donor–acceptor interaction in general (see Topic C8) do not alter the oxidation state. Both $[Ni(NH_3)_6]^{2+}$ and $[Ni(CN)_4]^{2-}$ have Ni^{II}, complexed by NH_3 and CN^- respectively.

A **redox reaction** is any reaction involving changes of oxidation state. In Equation 1 the changes are from Zn^0 to Zn^{II} and from H^{I} to H^0. The reaction

$$ZnO(s) + 2H^+(aq) \rightarrow Zn^{2+}(aq) + H_2O$$

is **not** a redox reaction as no change of oxidation state takes place.

Specifying the oxidation state of an element can be a useful way of naming compounds, especially when variable stoichiometries are possible (see Topic B5). Thus we have iron(II) chloride ($FeCl_2$) and iron(III) chloride ($FeCl_3$). The older names 'ferrous' and 'ferric' respectively are still encountered for such compounds but are potentially confusing. In current terminology the -ous and -ic suffixes (referring to a lower and a higher oxidation state, respectively) are only used for some oxoacids (e.g. $H_2S^{IV}O_3$, sulfurous acid, and $H_2S^{VI}O_4$, sulfuric acid; see Topic F7).

Balancing redox reactions

In any complete redox reaction the **changes in oxidation state must balance** so that the totals on the two sides are the same. Difficulties can arise with ions in solution, as the ionic charges may not be the same as the oxidation states. Consider the unbalanced redox reaction in acidified aqueous solution:

$$MnO_4^- + Fe^{2+} \rightarrow Mn^{2+} + Fe^{3+}$$

It is easiest to balance the redox changes by first splitting this into two **half reactions**, one involving oxidation, and the other reduction. The oxidation step is

$$Fe^{2+} \rightarrow Fe^{3+} + e^- \quad (3)$$

with electrons (e^-) being removed. The conversion of MnO_4^- to Mn^{2+} involves a change of oxidation state from Mn^{VII} to Mn^{II} and so is a reduction requiring five electrons. To balance the half reaction

$$MnO_4^- + 5e^- \rightarrow Mn^{2+}$$

four oxygen atoms are required on the right-hand side, which (in aqueous solution) will be in the form of H_2O. The reaction

$$MnO_4^- + 5e^- + 8H^+ \rightarrow Mn^{2+} + 4H_2O \qquad (4)$$

is then completed by balancing hydrogen with $8H^+$ on the left-hand side, as this reaction takes place in acid. The overall redox reaction is now written by combining the two half reactions in such a way that the free electrons are eliminated. This requires 5 moles of Equation 3 to every 1 mole of Equation 4, giving

$$MnO_4^- + 5Fe^{2+} + 8H^+ \rightarrow Mn^{2+} + 5Fe^{3+} + 4H_2O$$

In alkaline solution it is more appropriate to use OH^- rather than H^+ (see Topic E2). The other species present may also be different from those in acid, as many metal cations form insoluble hydroxides or even oxoanions (see Topic E4). As an example, consider the reaction of aluminum metal with water to form $[Al^{III}(OH)_4]^-$ and H_2. The balanced half reactions are

$$Al(s) + 4OH^- \rightarrow [Al(OH)_4]^- + 3e^-$$

and

$$2H_2O + 2e^- \rightarrow 2OH^- + H_2$$

which may be combined in the appropriate proportions (two to three) to give

$$2Al(s) + 2OH^- + 6H_2O \rightarrow 2[Al(OH)_4]^- + 3H_2$$

A particular advantage of the half-reaction approach is that it leads naturally to the discussion of the thermodynamics of redox reactions in terms of **electrode potentials** (see Topic E5).

Extraction of the elements

Very few elements occur naturally in uncombined form (see Topic J2). Most are found in compounds where they are in positive or (less often) negative oxidation states (e.g. Ti^{IV}, Zn^{II} and Cl^{-I} in TiO_2, ZnS and NaCl, respectively). Extraction of these elements therefore requires redox chemistry, using appropriate reducing or oxidizing agents. Thermodynamic considerations are very important (see Topic B3).

Iron is produced in greater quantities than any other metal, by reduction of Fe_2O_3 with carbon (coke). The overall reaction approximates to

$$Fe_2O_3 + \tfrac{3}{2}C \rightarrow 2Fe + \tfrac{3}{2}CO_2$$

At 25°C, $\Delta G^{\ominus}$ for this reaction is +151 kJ mol^{-1} so that it is not thermodynamically feasible at room temperature. However, it is strongly endothermic ($\Delta H^{\ominus}$ = +234 kJ mol^{-1}) and so by Le Chatelier's principle the equilibrium is shifted in favor of the products at higher temperatures. In a blast furnace it takes place above 1000°C, heat being provided from the combustion of carbon in air, which is blown through the reaction mixture.

Carbon is a convenient and cheap reducing agent for metal oxides, but for many elements it cannot be used. With some highly electropositive metals (e.g. Al) the oxide is too stable (i.e. its $\Delta G_f^{\ominus}$ is too negative), and the temperature required for reduction by carbon is too high to be technically or economically viable. Some elements (e.g. Ti) react with carbon to form a carbide. In these cases other redox processes are necessary. *Table 1* summarizes the common methods. Hydrogen can be used to reduce oxides or halides, or a very strongly reducing metal such as sodium or calcium to reduce halides.

In **electrolysis** a redox process with positive ΔG is induced by providing electrical energy. Reduction takes place at the **cathode** (the negative electrode, which provides electrons), and oxidation at the **anode** (the positive electrode). For example, electrolysis of molten NaCl gives elemental Na at the cathode and Cl_2 at the anode. Many very electropositive elements (e.g. Na, Ca, Al) and a few very electronegative ones (F, Cl) are obtained by this method.

Table 1. Extraction of elements from their compounds

Method of extraction	Elements
Reduction of oxide with C	Si, P, Mn, Fe, Sn
Conversion of sulfide to oxide, then reduction with C	Co, Zn, Pb, Bi
Reaction of sulfide with O_2	Cu, Hg
Electrolysis of solution or molten salt	Li, Be, B, F, Na, Ca, Al, Cl, Ni, Cu, Ga, Sr, In, Ba, La–Lu, Tl
Reduction of halide with Na or other electropositive metal	Be, Mg, Si, K, Ti, V, Cr, Rb, Zr, Cs, La–Lu, Hf, U
Reduction of halide or oxide with H_2	B, Ni, Ge, Mo, Ru, W, Re
Oxidation of anion with Cl_2	Br, I

B5 DESCRIBING INORGANIC COMPOUNDS

Key Notes

Formulae

Stoichiometric (empirical) formulae describe only the relative numbers of atoms present. Molecular formulae and/or representations giving structural information should be used when they are appropriate. The physical state of a substance is often specified.

Names

Systematic nomenclature can be based on three systems, binary, substitutive (similar to that in organic chemistry) or coordination. Many nonsystematic or trivial names are used.

Structure and bonding

The coordination number and geometry of an atom describe the number of bonded atoms and their arrangement in space. Oxidation states rather than valencies are generally used for describing different possible stoichiometries.

Related topics

Hydrogen (F2)
Oxygen (F7)
Complexes: structure and isomerism (H6)

Formulae

It is important to distinguish the **stoichiometric** or **empirical formula** of a molecular substance from its **molecular formula**. The former expresses only the relative numbers of atoms present, in the simplest possible ratio. For example, the compound of stoichiometry P_2O_5 contains P_4O_{10} molecules. Molecular formulae should be used when they are known. On the other hand, in a solid where clear molecular or other units do not exist the empirical formula is generally used. For example, NaCl is an ionic substance and the formula does not imply that molecules are present.

When solids contain identifiable groups such as molecules or **complex ions** the formula is written to indicate this: for example, NH_4NO_3 is much more informative for ammonium nitrate than the empirical formula $N_2H_4O_3$. This is often used in molecular formulae, for example, in NH_2OH (**1**) and $Ni(CO)_4$, which are intended to show the groupings of atoms present. For **coordination compounds** formed by transition metals formulae are written with square brackets as in $[Ni(NH_3)_6]Br_2$, which indicates that six NH_3 are attached to directly to Ni, but not the two Br. Complex ions formed by main-group elements can be written in a similar way, for example, $[PCl_4]^+$ and $[BF_4]^-$, although usage is not very systematic.

```
H
 \
  N—O
 /   \
H     H
```

1

When a metallic and a nonmetallic element are present, the metallic one is always written first, as in NaCl and PbO_2. For compounds between two or more nonmetals they are listed conventionally in the following order, based roughly on a sequence of increasing electronegativity:

Xe, Kr, B, Ge, Si, C, Sb, As, P, N, H, Te, Se, S, I, Br, Cl, O, F.

For example, we have OF_2 and ClO_2, which are therefore called oxygen difluoride and chlorine dioxide, respectively (see below).

When the **physical state** of a substance is important it is specified as in NaCl(s), H_2O(l) and HCl(g) for solids, liquids and gases, respectively. (l) is assumed to be a pure liquid or the major component (solvent) in a solution. For substances dissolved in water the designation (aq) (for 'aqueous') is used. Thus solid sodium chloride dissolving in water is expressed:

$$NaCl(s) \rightarrow Na^+(aq) + Cl^-(aq)$$

NaCl(aq) means dissolved NaCl molecules and is incorrect for this substance.

Names

The systematic naming of chemical substances is not easy, and the authoritative guide to inorganic nomenclature fills a book of 299 pages. Very many **trivial names** such as water (H_2O) will always remain in use. Systematic nomenclature is based on three systems.

Binary names

Simple examples are **sodium chloride** (NaCl), **phosphorus trichloride** (PCl_3) and **dinitrogen tetroxide** (N_2O_4). The oxidation state may be given as an alternative to the stoichiometry, as in **manganese (IV) oxide**, or **manganese dioxide**, MnO_2 (see Topic B4). This is unnecessary when only one possibility is known, as in **magnesium bromide** ($MgBr_2$).

Elements are named in the same order as they appear in the formula (see above). Although there is no implication that the compound is ionic, the names ending in -ide are the same as those used for anions (e.g. **fluoride**, F^-). For the elements listed in *Table 1*, anion names are derived not from English but from the Latin name which gives the chemical symbol. For example, CsAu is **cesium auride**.

Binary naming may be extended to include complex ions with recognized names as in **ammonium chloride** (NH_4Cl), **sodium cyanide** (NaCN) and **magnesium sulfate** (Mg_2SO_4). Some common oxoanions are listed elsewhere (Topic F7, *Table 1*). Many other complex ions can be named systematically as discussed below.

Table 1. Anion names derived from Latin roots; the -ate form is used for complex anions

Element	Symbol	Anion name
Copper	Cu	Cupride/cuprate
Gold	Au	Auride/aurate
Iron	Fe	Ferride/ferrate
Lead	Pb	Plumbide/plumbate
Silver	Ag	Argentide/argentate
Tin	Sn	Stannide/stannate

Substitutive names

This is the system used in organic chemistry, as in **dichloromethane**, CH_2Cl_2, which can be regarded as derived from methane CH_4 by replacing two hydrogens with chlorine. (There is no assumption that this is a chemically feasible route for preparation.) It may be extended to inorganic molecules using the appropriate hydride names (see Topic F2, *Table 1*). Thus from **silane** (SiH_4) we obtain names such as **chlorosilane** (SiH_3Cl) and **tetrachlorosilane** ($SiCl_4$), the latter being also called silicon tetrachloride. For nitrogen compounds derived from ammonia (NH_3) the root **amine** is used, as, for example, in **hydroxylamine** (NH_2OH).

Coordination names

This system is used in compounds that can be regarded as complexes formed by the coordination of atoms or groups to a central atom. Examples from transition metal chemistry are **tetraamminecopper(2+) ion** or **tetraamminecopper(II) ion**, $[Cu(NH_3)_4]^{2+}$, and **tetrachlorocuprate(2−)** or **tetrachlorocuprate(II)**, $[CuCl_4]^{2-}$ (see Topic H6 for further examples, and nomenclature for isomers). Either the overall charge on the complex ion or the oxidation state of the central atom is given (the latter always with Roman numerals). Anion names end in -ate and use the Latin roots for elements shown in *Table 1*. Coordination names are also widely used for complex ions with main-group elements, for example, **tetrahydroborate**, $[BH_4]^-$; in this case the charge or oxidation state may be omitted as only one possibility is known.

Examples of the use of coordination names in the binary system are the solids **hexaamminenickel dibromide**, $[Ni(NH_3)_6]Br_2$, and **potassium hexafluorophosphate(V)**, $K[PF_6]$.

Structure and bonding

The complete description of a **chemical structure** involves specifying the relative coordinates of the atoms present, or alternatively giving all **bond lengths** and **bond angles**. A simple example is shown in **2**. Less complete information is satisfactory for most descriptive purposes. The **coordination number** (CN) of an atom is the number of bonded atoms, irrespective of the type (ionicity, multiplicity, etc.) of bond involved. For very simple molecular compounds this is obvious from the formula (e.g. O in H_2O and C in CO_2 (**3**) both have CN = 2). However, polymeric and ionic solids have greater CN values (e.g. 4, 6 and 8, respectively, for Si in SiO_2, Ti in TiO_2 and U in UO_2), and it should not generally be assumed that the CN is given directly by the stoichiometry.

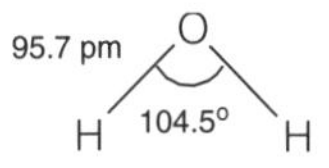

95.7 pm O
H 104.5° H

2

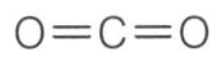

O=C=O

3

The geometrical arrangement around an atom is sometimes described as its **coordination sphere**. Different geometrical arrangements may be described by simple informal terms (e.g. H_2O (**2**) is **bent** and CO_2 (**3**) **linear**), or by the names of polyhedra, such as tetrahedra and octahedra (see Topics C2 and D3).

Describing bonding in a consistent way is much harder. The term **valency**, meaning the number of bonds formed by an atom, is useful in simple molecular substances. Stoichiometries such as CH_4, CO_2 and H_2O can be rationalized by assuming the valencies C(4), H(1) and O(2). One can extend the idea by recognizing the possibility of variable valency; for example, three for phosphorus in PCl_3

and P_2O_3, and five in PCl_5 and P_2O_5. Unfortunately, the simple valence idea has serious limitations and can be misleading outside a narrow area. For example:

- Given the 'normal' valencies of C and O, how can one account for the stability of CO, and the fact that it apparently has a triple bond (see Topic C1)?
- PCl_5 in its solid form contains $[PCl_4]^+$ and $[PCl_6]^-$ ions. What is the valency of P here?

Much more serious problems arise with transition metal compounds. For these and other reasons the word valency has been largely abandoned by inorganic chemists. When it is necessary to distinguish different stoichiometries such as PCl_3 and PCl_5 the **oxidation state** is more frequently used. This is defined according to clearer rules than valency, but as they depend on the electronegativity difference of atoms, the oxidation state can be very uninformative about bonding. For example, every sulfur atom forms two covalent bonds in the compounds H_2S, H_2S_2 (**4**), S_2Cl_2 and SCl_2, and yet the oxidation state of sulfur is respectively −2, −1, +1 and +2.

S—S
H H

4

As a final example, consider phosphorus acid H_3PO_3 (**5**). The oxidation state of phosphorus is +3, its coordination number 4, and its valency 5. All these numbers give useful information, but they must not be confused.

O
‖
H—P—OH
|
OH

5

C1 ELECTRON PAIR BONDS

Key Notes

Lewis and valence structures

A Lewis structure shows the valence electrons in a molecule. Two shared electrons form a single bond, with correspondingly more for multiple bonds. Some atoms may also have nonbonding electrons (lone-pairs). Valence structures show the bonds simply as lines.

Octets and 'hypervalence'

In most stable molecules and ions of the elements C–F, each of these atoms has eight electrons (an octet) in its valence shell. Expansion of the octet and increased valency is possible with elements in periods 3 and below.

Resonance

When several alternative valence structures are possible, the bonding may be described in terms of resonance between them.

The range of antigens

Formal charges are assigned by apportioning bonding electrons equally between the two atoms involved. They can be useful to rationalize apparent anomalies in bonding, and to assess the likely stability of a proposed valence structure.

Limitations

Many covalent molecules and ions cannot be understood in terms of electron pair bonds between two atoms. They include electron-deficient boron hydrides and transition metal compounds.

Related topics

Electronegativity and bond type (B1)

Molecular shapes: VSEPR (C2)

Lewis and valence structures

A single covalent bond is formed when two atoms share a pair of electrons Double and triple bonds can be formed when two or three such pairs are shared. A **Lewis structure** is a representation of a molecule or complex ion that shows the disposition of **valence electrons** (inner shells are not drawn) around each atom. **1–4** show Lewis structures of CH_4, H_2O, O_2 and N_2, the last two molecules having a double and triple bond, respectively. These representations are entirely equivalent to the **valence structures (1′–4′)** in which each bonding pair of electrons is represented by a line.

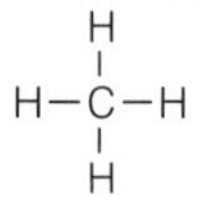

1' 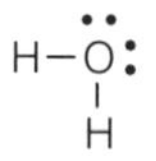 2' 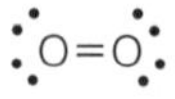 3'  4'

A molecule such as H_2O has **nonbonding** or **lone-pair** electrons localized on one atom rather than shared. The presence of these has important consequences for both the shape of a molecule and its chemical properties (see Topics C2 and C8).

Simple complex ions such as ammonium (NH_4^+ **5**) and tetrahydroborate (BH_4^- **6**) can be drawn in a similar way; the valence structures shown are essentially identical to those for CH_4 as the total number of valence electrons is the same in all examples. The **isoelectronic principle** suggests that molecules or ions having the same number of valence electrons should have similar valence structures, although this idea has limitations.

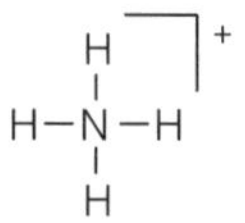

5 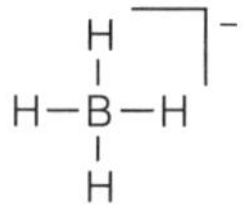 6

Octets and 'hypervalence'

A great majority of simple molecules containing the elements C–F of the second period can be represented by Lewis structures with **eight** electrons around each of these atoms, including all shared electrons and lone-pairs. The **octet rule** provides a systematization of the normal **valencies** of these elements: for example, a nitrogen atom has five electrons in its valence shell and so must share three more to achieve an octet, thus forming three bonds. Hydrogen is limited to **two** electrons in its valence shell, and these differences may be understood from the valence atomic orbitals available for electrons, 1*s* only for H, 2*s* and 2*p* in the second period; the exclusion principle then limits the number of electrons that can be accommodated (see Topics A3 and A4).

Some molecules containing boron (e.g. BF_3 **7**) have an **incomplete octet** and this has implications for their chemical reactivity (see Topics C8 and F3). Generally, however, structures with complete octets are preferred. Thus the triple-bonded representation for carbon monoxide (**8**) is better than the double-bonded one (**8'**) where carbon only has six valence-shell electrons.

7  8 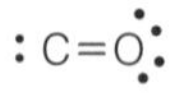 8'

Nonmetallic elements of the third and subsequent periods form some compounds entirely analogous to those of the same group in period 2. Thus we have H_2S, H_2Se and H_2Te similar to H_2O, all with octets. These heavier elements, however, are capable of **octet expansion** or **hypervalence**, the latter term implying a valency higher than 'normal'. Examples are SF_4 and SF_6 (**9, 10**) where sulfur has respectively 10 and 12 electrons in its valence shell. Hypervalence is sometimes considered to be a consequence of the availability of further orbitals for bonding (e.g. *3d* in addition to *3s* and *3p* for sulfur). Although this may play a part, it is generally thought that other differences between the periods are equally important, especially size and electronegativity (see Topic F1).

9 **10**

Resonance

Sometimes more than one valence structure is possible and there appears to be no unique assignment. A familiar organic example is in the disposition of double and single C—C bonds in benzene (see Topic C6, Structure 6). In the carbonate ion (CO_3^{2-} **11**) the three structures shown are equivalent by symmetry, and experimentally all three C—O bonds have equal length. We describe this situation as **resonance** between the different structures, and represent it by the double-headed arrows shown in **11**. The term is misleading as it suggests a rapid oscillation between different structures, which certainly does not happen. It is better to think of a wavefunction that is formed by combining the structures, none of which on their own describe the bonding correctly.

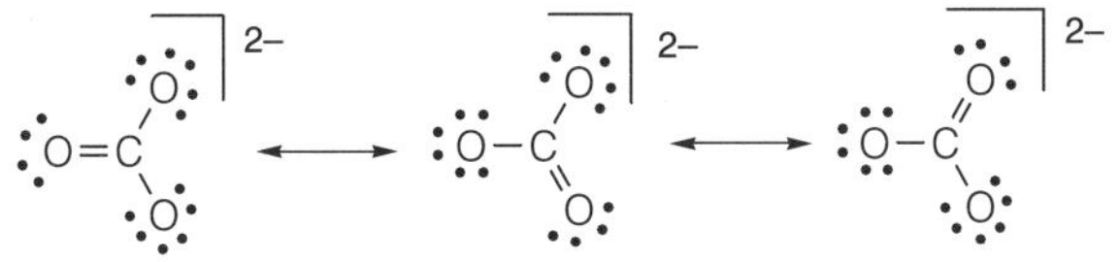

11

Resonance may also be appropriate with different valence structures that are not equivalent but look equally plausible, as in nitrous oxide (N_2O **12**).

12

Formal charges

Atoms are often found in bonding situations that do not correspond to their 'normal' valency. Such cases can be rationalized by the concept of **formal charge**. A formal charge on an atom is essentially the charge that would remain if all covalent bonds were broken, with the electrons being assigned equally to the atoms involved. More mathematically, it is defined as

$$\begin{aligned}\text{formal charge} = {} & (\text{no. of valence electrons in neutral atom}) \\ & - (\text{no. of nonbonding electrons}) \\ & - (1/2)\,(\text{no. of electrons in bonds formed})\end{aligned}$$

The formal charges in CO and in the two valence structures for N_2O are shown in **13** and **14**. The isoelectronic principle allows us to understand these structures by analogy. Thus C^- and O^+ are both isoelectronic to neutral N and can similarly form three bonds. The N_2O structures can be understood with the isoelectronic relations N^- and O (two bonds expected), N^+ and C (four bonds) and O^- and F (one bond).

$:\overset{-}{C}\equiv\overset{+}{O}:$

13

$\overset{-}{N}=\overset{+}{N}=O \longleftrightarrow :N\equiv\overset{+}{N}-\overset{-}{O}$

14

Formal charges are frequently drawn in organic structures; for example, 'trivalent' carbon can occur as a carbocation (C^+ isoelectronic to B, and with an incomplete octet) or a carbanion (C^- isoelectronic to N, with a nonbonding pair). They are not always written on inorganic valence structures, but the idea is useful in judging the viability of a proposed structure. Some general principles are:

- structures without formal charges are preferred if possible;
- structures with formal charges outside the range −1 to +1 are generally unfavorable;
- negative formal charges should preferably be assigned to more electronegative atoms, positive charges to more electropositive atoms.

Thus in N_2O (**14**), the structure with O^- is probably more significant than that with N^-. The BF molecule (**15**) is isoelectronic with CO but the corresponding triple-bonded structure appears very unlikely because it requires formal charges B^{2-} and F^{2+}. The single-bonded form without charges may best describe the bonding.

$:\overset{2-}{B}\equiv\overset{2+}{F}:$ $\quad$ $:B-F:$

15

Formal charge is very different from **oxidation state**, which is assigned by apportioning electrons in a bond to the more electronegative atom rather than equally (Topic B4). Both are artificial assignments, useful in their respective ways, but neither is intended as a realistic judgment of the charges on atoms.

Limitations

The model described in this section can be justified theoretically using the quantum mechanical **valence bond theory**. Nevertheless, there are many molecules where bonding cannot be described simply in terms of electron pairs localized between two atoms. Diborane is an example. The structure (**16**) as often drawn appears to have eight bonds and would therefore seem to need 16 valence electrons. In fact, there are only 12 and the molecule is sometimes described as **electron deficient**. Two pairs of electrons form **three-center bonds** each linking two boron atoms and a bridging hydrogen, as illustrated in the preferable way of drawing the valence structure in **16′**. Transition metal chemistry (see Section H) is another area where bonding often cannot be described in terms of localized electron pairs.

The resonance concept is one way of overcoming some of the limitations of the localized electron pair model, but such cases are treated more naturally by **molecular orbital theory**, which is not limited to bonds involving two atoms (see Topics C5 and C6).

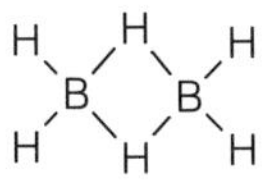

16

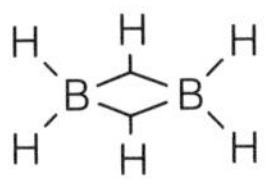

16'

C2 MOLECULAR SHAPES: VSEPR

Key Notes

VSEPR principles

The valence shell electron pair repulsion (VSEPR) model is based on the observation that the geometrical arrangement of bonds around an atom is influenced by nonbonding electrons present. It is assumed that electron pairs – whether bonding or nonbonding – repel each other and adopt a geometrical arrangement that maximizes the distances between them.

Using VSEPR

It is first necessary to decide which atoms are bonded together. Drawing a valence structure gives the total number of electron pairs around an atom, sometimes known as its steric number. The basic VSEPR geometry is then used to assign positions for bonding and nonbonding electrons.

Extensions, difficulties and exceptions

In spite of a lack of firm theoretical foundation the VSEPR model is widely applicable to molecular geometries and even to some solids. Occasionally it fails to predict the correct structure.

Related topics Electron pair bonds (C1) Molecular orbitals: polyatomics (C5)

VSEPR principles

Stereochemical arrangement becomes an issue whenever an atom is bonded to two or more others. Thus triatomic species AB_2 can be linear (e.g. CO_2, ICl_2^-) or bent (e.g. H_2O **1**, NO_2^-). It is observed that when a central atom has no nonbonding electrons, the surrounding atoms are usually arranged in a regular way that spaces them as far apart as possible. When nonbonding electron pairs are present in the valence structure, however, less regular arrangements of bonds are often found. The **valence shell electron pair repulsion** (VSEPR) model is based on the idea that both bonding and nonbonding electron pairs in the valence shell of an atom 'repel' each other. This idea is useful but can be misleading if taken too literally. Detailed calculations show that the shape of a molecule is determined by a combination of factors, of which the electrostatic repulsion between electrons is not the most important. Furthermore, the real electron distribution in a molecule is much more evenly spread out than the localized pictures used in VSEPR (**1**, **2**, . . .) suggest. It is best to think of 'repulsion' as coming primarily from the **exclusion principle** (see Topic A3), which forces electron pairs to occupy orbitals in different regions of space.

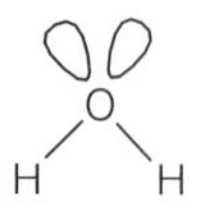

O
H H

1

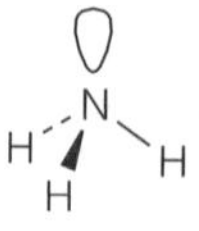

N
H H
H

2

The basic principles of the model are as follows.

(i) Valence electron pairs round an atom (whether bonding or nonbonding) adopt a geometry that maximizes the distance between them. The basic geometries usually observed with 2–7 pairs are shown in *Fig. 1.*

(ii) Nonbonding electron pairs are closer to the central atom than bonding pairs and have larger repulsions: in fact, the order of interactions is

nonbonding–nonbonding > nonbonding–bonding > bonding–bonding.

(iii) If double (or triple) bonds are present the four (or six) electrons involved behave as if they were a single pair, although they exert more repulsion than do the two electrons of a single bond

(iv) As the terminal atoms become more electronegative relative to the central one, bonding electron pairs are drawn away from the central atom and so repel less.

Using VSEPR

Before applying VSEPR to a molecule or complex ion it is necessary to know the **connectivity**, that is, which atoms are bonded together. With a species of formula AX_n this often gives no problem, especially if X is a monovalent atom or group (e.g. H, F, CH_3). Sometimes it is not so obvious, and a useful rule (which does not apply to hydrides or organic groups) is that **the central atom is usually the least electronegative one**. For example, in N_2O one of the nitrogen atoms is central (see Topic C1, Structure 12). Drawing a valence structure including nonbonding pairs on the central atom then gives the total number of 'pairs' (multiple bonds counted as a single 'pair'). This is sometimes called the steric number (SN) of the central atom A, equal to the number (n) of bonded atoms, plus the number of nonbonding electron pairs. It may generally be assumed that one electron from A is used in each bond formed to X. Thus in SF_4 four electrons are used in single S—F bonds, leaving two electrons (i.e. one pair) nonbonding, so that SN = 5. In $XeOF_4$ the Xe=O double bond uses two electrons from xenon, and again there is one nonbonding pair making SN = 6. In complex ions account must be taken of the charge. Thus in ClO_3^- we can include the charge on the ion and assign eight

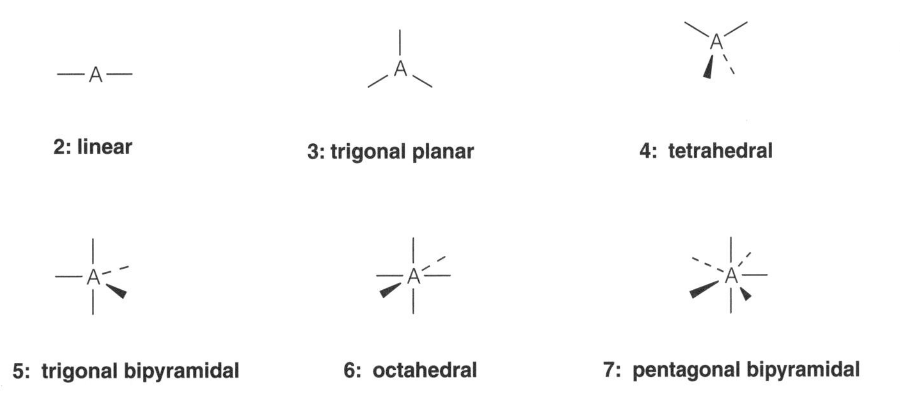

Fig. 1. Basic VSEPR geometries with 2–7 electron pairs.

valence electrons to chlorine. Six are involved in bonding, so that there is one nonbonding pair, and SN = 4.

Steric numbers 2–4

The shapes shown in *Fig. 1* are simple and the rules generally easy to apply. Examples without nonbonding electrons are:

- **linear species** (SN = 2): BeH_2 (gas phase only; see Topic F3), $HgCl_2$, CO_2 and ions isoelectronic to it such as NO_2^+, N_3^- and NCO^-;
- **trigonal planar species** (SN = 3): BF_3, CO_3^{2-} and NO_3^-;
- **tetrahedral species** (SN = 4): CH_4, NH_4^+, $SiCl_4$, $POCl_3$ and SO_4^{2-}.

AX_2 species with SN = 3 or 4 are **bent**, with the nonbonding pairs occupying positions of the trigonal plane or tetrahedron, respectively (e.g. water, **1**). As predicted by rules (ii) and (iv) the XAX bond angles are less than the ideal values of 120° (SN = 3) or 109.5° (SN = 4), and tend to decrease as the electronegativity difference between A and X increases. Some examples with their bond angles are:

- SN = 3: NO_2^- (115°), ClNO (113°);
- SN = 4: H_2O (104.5°), H_2S (92°), F_2O (102°).

In AX_3 with SN = 4 the nonbonding pair forces the bonds to be **pyramidal** (see ammonia, **2**). Examples with their bond angles are:

NH_3 (107°), PH_3 (93°), ClO_3^- (106°).

Steric number 5

The normal shape adopted by five groups is the **trigonal bipyramid**, as with PF_5. There are now two inequivalent types of position, two **axial** (top and bottom in *Fig. 1*) and three **equatorial**. It appears that the equatorial positions allow more space than axial ones. Thus bulkier groups (e.g. Cl in PF_4Cl) tend to be found in these positions, as do nonbonding pairs when these are present. With successively one, two and three nonbonding pairs, the molecular shapes are as follows.

- AX_4 is often described as a **'see-saw'** with two axial and two equatorial X positions, the former being slightly bent out of the ideal linear configuration by the lone-pair repulsion. Examples are SF_4 (**3**) and XeO_2F_2 (where O in preference to F occupies the equatorial position; see rule iii).
- AX_3 gives a **T-shape**, as in ClF_3 (**4**).
- AX_2 is **linear** as the bonded atoms are axial. Examples are XeF_2 (**5**) and I_3^-.

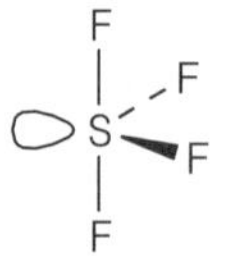

F, S, F, F, F — **3**

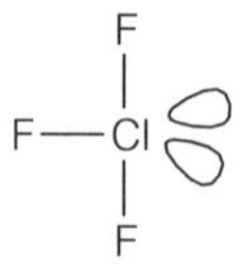

F, F—Cl, F — **4**

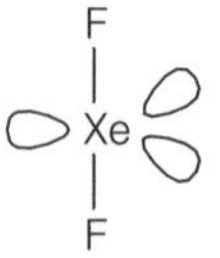

F, Xe, F — **5**

Steric number 6

The basic shape is octahedral and is found with SF_6 and PF_6^-. All positions are equivalent and with one nonbonding pair AX_5 adopts a **square pyramidal** structure (e.g. BrF_5 **6**, and $XeOF_4$, where repulsion between the double bond and the lone-pair is minimized by putting these *trans* to each other). When two non-nonbonding pairs are present they minimize their repulsion (rule ii) by adopting the *trans* configuration, giving a **square planar** molecule (e.g. XeF_4, **7** and ICl_4^-).

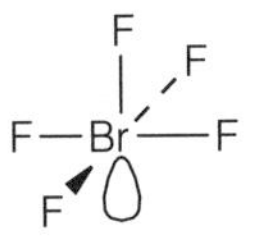

6 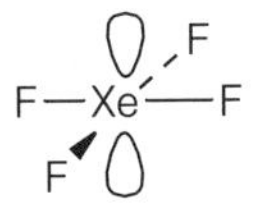 **7**

Steric number 7

The only simple examples are the **pentagonal bipyramidal** IF_7 (see *Fig. 1*) and the ion XeF_5^-, which is pentagonal planar, having two lone-pairs occupying the axial positions (for XeF_6 see below).

Extensions, difficulties and exceptions

One of the problems with VSEPR is that its rules appear somewhat arbitrary and hard to justify in a rigorous quantum-mechanical formulation. The interpretation of small variations in bond angle is often considered to be particularly dubious. In spite of this (and of the exceptions noted later) the model is surprisingly useful. Although the discussion has concentrated on cases where single atoms are bound to a central one, VSEPR should be able to predict the geometry around any atom in a complex molecule, where main-group atoms are involved. (It cannot be generally applied to transition metals; see Topics H2 and H6.) For example, in hydroxylamine, H_2NOH, the bonds around the nitrogen are pyramidal, those around the oxygen bent as expected. The model is even useful in interpreting solid-state structures containing ions such as Sn^{2+} where nonbonding electrons appear to have a stereochemical influence (see Topic G6).

One type of exception to VSEPR arises when apparently nonbonding electrons are really involved to some extent in bonding. For example, the geometry around nitrogen is planar when bonded to carbonyl groups in the peptide linkage (-NH-CO-) in proteins, and in trisilylamine, $(SiH_3)_3N$ (**8**, only one of three equivalent resonance structures shown). In both cases the 'nonbonding' pair on nitrogen is used to form partial double bonds. In **8** this requires valence expansion by the silicon, and contrasts with pyramidal trimethylamine $(CH_3)_3N$, where the carbon cannot accommodate extra electrons (see Topics C1 and F4).

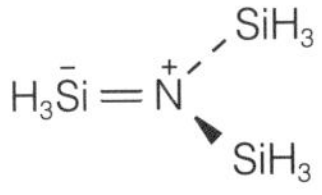

8

AX_5 species with no lone-pairs are occasionally square pyramidal rather than the normal trigonal pyramid of *Fig. 1* (see, e.g. Topic F6, Structure 2). Other difficulties arise with AX_6 where there is one nonbonding pair. This is the case with XeF_6, which, as predicted, is not regularly octahedral. A unique shape cannot be determined in the gas phase, however, as the molecule appears to be highly **fluxional** and converts rapidly between different distorted configurations. By contrast, the ions $[SeCl_6]^{2-}$ and $[TeCl_6]^{2-}$ are regularly octahedral in spite of having a nonbonding pair. There is no simple explanation, although the comparatively large size of the chloride ion could be a factor.

Other notable exceptions are some of the group 2 dihalides such as BaF_2, which in the gas phase are bent, not linear as VSEPR predicts. (In their normal solid-state forms they have different structures; see Topics D3 and G3.) Two factors that are thought to contribute are (i) the use of valence *s* and *d* orbitals for bonding (rather than *s* and *p* as is normal in later main groups, and (ii) the possibility that core polarization could lower the energy of the bent form.

C3 MOLECULAR ORBITALS: HOMONUCLEAR DIATOMICS

Key Notes

Bonding and antibonding orbitals

Molecular orbitals (MOs) are wavefunctions for electrons in molecules, often formed by the linear combination of atomic orbitals (LCAO) approximation. Overlapping atomic orbitals can give bonding and antibonding MOs. Electrons in bonding MOs have an increased probability of being in the region between the nuclei.

MO diagrams

An MO diagram is a representation of the energies of bonding and antibonding MOs formed from atomic orbitals. Electrons are assigned to MOs in accordance with the exclusion principle, leading to the same building-up procedure as for the periodic table of elements. The bond order is defined as half the number of net bonding electrons.

Second period diatomics

2*p* atomic orbitals can give rise to both σ and π MOs, the former overlapping along the molecular axis and the latter perpendicular to it. Multiple bonding in O_2 and N_2 arises from π as well as σ bonding. Trends in bond strengths and lengths follow predicted bond orders. The O_2 molecule has two unpaired electrons, a fact not predicted by simple electron-pair bonding models.

Related topics Atomic orbitals (A2) Many-electron atoms (A3)

Bonding and antibonding orbitals

Just as an atomic orbital (AO) is the wavefunction for an electron in an atom (Topic A2) so a **molecular orbital** (MO) is that for a molecule; an MO may extend over two or more atoms. Exact MOs may be obtained by solution of Schrödinger's equation for the one-electron ion H_2^+, but as in atoms the extension of the orbital approach to many-electron systems involves approximations (see Topic A3). For MOs a further approximation is often made: rather than using the exact but very complicated wavefunction for H_2^+ it is convenient to express each MO as a **linear combination of AOs**, the so-called **LCAO approximation**. This approximation needs to be refined considerably for quantitative calculations on computers, but it is adequate for qualitative purposes and provides very useful pictures of how chemical bonds form, and how they are related to the valence AOs of the atoms involved.

Figure 1 shows how the LCAO method works for H_2. The diagrams show how the value of a wavefunction varies along the molecular axis. *Figure 1a* shows wavefunctions ϕ_1 and ϕ_2 for the 1*s* valence AOs on the two H atoms. An MO wavefunction ψ is constructed by writing

$$\psi = c_1\phi_1 + c_2\phi \qquad (1)$$

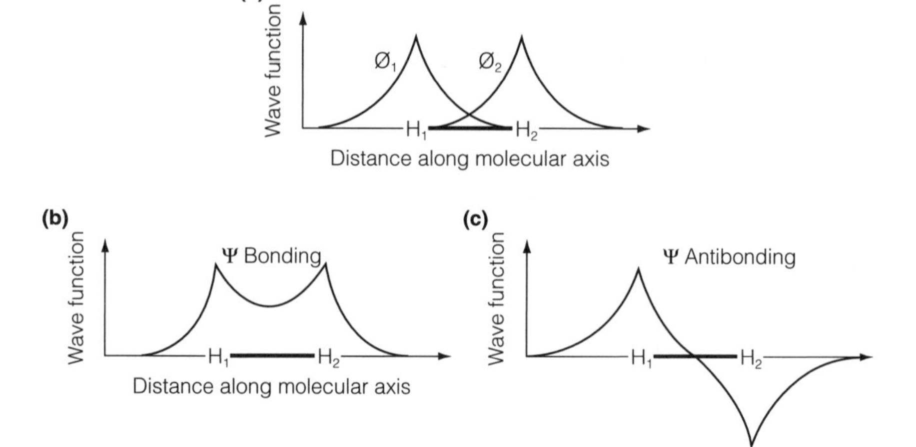

Fig. 1. Formation of a bonding and antibonding MO from the overlap of 1s AOs in H_2 (see text)

where c_1 and c_2 are numerical coefficients. The square of ψ gives the electron probability distribution, and in a **homonuclear diatomic** such as H_2 (where each atom is the same) ψ^2 must have the same value on each atom. Thus we have $c_1^2 = c_2^2$, from which either $c_1 = c_2$ or $c_1 = -c_2$. The former combination is the **bonding MO**, and the latter the **antibonding MO**, shown in *Fig. 1b* and *c*, respectively. Near each atomic nucleus both MOs resemble the 1*s* AO (apart from sign, which does not affect the probability distribution). An electron close to one nucleus is hardly affected by the presence of the other one. The important difference occurs in the region between the nuclei, where the bonding MO predicts an **increase of electron density** compared with the isolated atoms. This may be seen from

$$\psi^2 = (\phi_1 + \phi_2)^2$$
$$= \phi_1^2 + \phi_2^2 + 2\phi_1\phi_2$$

The first two terms give the same electron density as found in the two 1*s* AOs, whereas $2\phi_1\phi_2$ represents an increase of density in the region between the two atoms where the orbitals **overlap**. An electron in the bonding MO has lower energy than in the AO of an isolated atom, as it has enhanced probability of being close to both nuclei simultaneously. The electrostatic repulsion between the nuclei is effectively shielded, leading to a stable bond. Conversely, in an antibonding MO an electron has higher energy, with a reduced probability of being in the internuclear region, and internuclear repulsion being 'deshielded'.

In the ground state of H_2 two electrons occupy the bonding MO, the antibonding one not being used. The simple MO model thus shows how bonding arises from an increase of electron density in the internuclear region, and how **overlap of AOs** is essential for this to happen.

MO diagrams

Figure 2 shows an **MO diagram** for H_2. The AO energies of the two isolated atoms are displayed on the left- and right-hand sides, and in the center are shown the bonding and antibonding MOs resulting from their overlap. They are labeled $1s\sigma_g$ and $1s\sigma_u$, respectively, the σ designation referring to their symmetric nature about

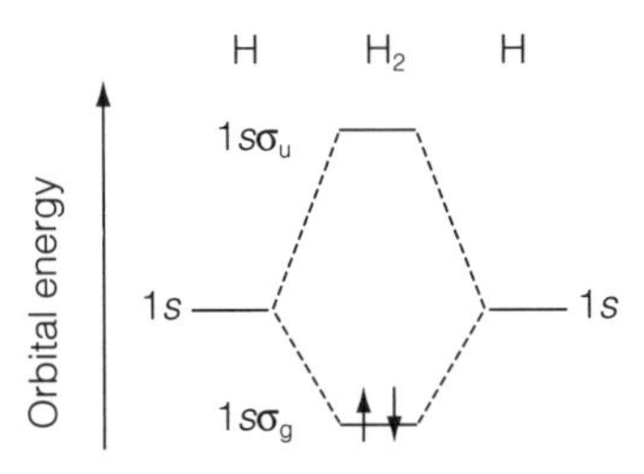

Fig. 2. MO diagram for H_2 showing the bonding MO occupied by two electrons with paired spins.

the molecular axis (see below) and *g* and *u* (from the German *gerade* and *ungerade*, respectively) to their **even** or **odd** behavior under inversion through the center of symmetry of molecule (the mid-point of the bond).

Occupation of MOs is governed by the **Pauli exclusion principle** (see Topic A3). In *Fig. 2* two electrons are shown with opposite spin (one upward and one downward pointing arrow) corresponding to the **electron configuration** $(1s\sigma_g)^2$. When more than two electrons are present some must occupy an orbital of higher energy. Thus He_2 (four electrons) would have the electron configuration $(1s\sigma_g)^2(1s\sigma_u)^2$, which could be represented on a similar MO diagram. Calculations show that the increase of energy (relative to isolated AOs) in forming the antibonding MO more than compensates for the stabilization of the bonding MO. He_2 is therefore an unstable molecule with higher energy than two individual He atoms. In MO theory the 'repulsion' between closed shells (the He core in this example) is seen to be a consequence of the exclusion principle, which forces occupation of antibonding as well as bonding orbitals.

The definition of **bond order (BO)** in MO theory recognizes that a 'normal' single bond is formed by two electrons (see Topic C1). We define

$$\text{BO} = (1/2)[(\text{no. of electrons in bonding MOs}) - (\text{no. of electrons in antibonding MOs})]$$

In the above examples, H_2 and He_2 have bond orders of one and zero, respectively. Fractional values are possible, as in the molecular ions H_2^+ and He_2^+, which are both bonded but more weakly than H_2, as indicated by the electron configurations and BO shown below:

H_2^+ $(1s\sigma_g)^1$ BO = (1 − 0)/2 = 1/2
He_2^+ $(1s\sigma_g)^2(1s\sigma_u)^1$ BO = (2 − 1)/2 = 1/2

Second period diatomics

In extending the MO theory to second period elements two additional principles need to be considered. First, only **valence-shell orbitals** are shown in MO diagrams, as inner shells are too tightly bound to be involved in bonding, and do not overlap significantly. Second, both 2*s* and 2*p* valence orbitals need to be included (see Topics A2–A4). The three AOs in a *p* shell differ only in their direction in space, and in an atom are **degenerate** (i.e. have the same energy). In diatomic molecules, however, they are distinguished by the way they overlap (see *Fig. 3*): *p*σ orbitals point along the direction of the bond, and *p*π orbitals (of which there are two equivalent and degenerate ones) are perpendicular to it. Each type of orbital can combine to form MOs in the same way as in H_2. Two fundamental rules in the LCAO MO model are important:

- the number of MOs formed is equal to thc total number of starting AOs;
- only AOs of the same symmetry type combine to make MOs.

Figure 3 shows that the $p\sigma$ AOs, which point towards each other, overlap more strongly than do $p\pi$ MOs, so that both the bonding stabilization and the antibonding destabilization is greater for $p\sigma$. The order of MO energies formed from the $2p$ AO on the two atoms is therefore:

σ bonding < π bonding < π antibonding < σ antibonding
$2p\sigma_g$ $2p\pi_u$ $2p\pi_g$ $2p\sigma_u$

The symmetry designation of each MO is given and it should be noted that the *g* or *u* labeling for π MOs is the reverse of that for σ MOs.

Figure 4 shows the MO diagram for dioxygen O_2. As in *Fig. 2* the energies of $2s$ and $2p$ AOs of the separate atoms are shown and between them the MOs formed from overlap. Lowest in energy are $2s\sigma_g$ and $2s\sigma_u$, respectively the bonding and antibonding combinations of $2s$ AOs. At higher energies are MOs resulting from $2p$ AOs. *Figure 4* shows occupation of MOs in O_2 (12 valence electrons). The BO calculated as above is two, as expected from the normal valence structure (see Topic C1). However, the degenerate $2p\pi_g$ MOs are occupied by two electrons, and as in atoms, **Hund's first rule** shows that the ground state will be formed by putting one electron in each orbital with parallel spin (see Topic A3). Thus O_2 is correctly predicted to have **two unpaired electrons** and as a consequence is paramagnetic (see Topic F7).

One of the advantages of MO theory is that the same diagram can be used for molecules with different numbers of electrons. The electron configurations and BOs for some molecules and ions, which can be derived using *Fig. 4*, are shown in *Table 1* together with the experimental dissociation energies and bond lengths. Higher predicted BO values correspond to stronger and shorter bonds.

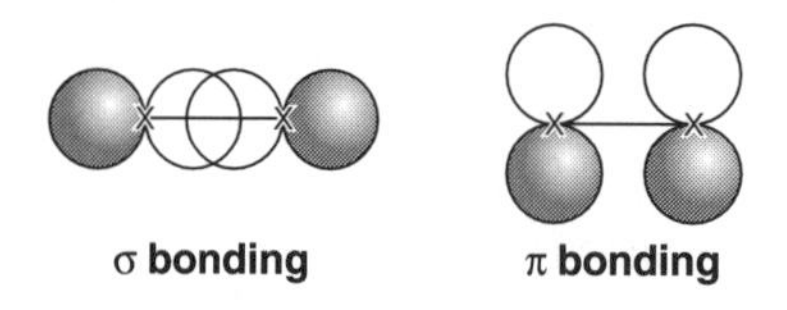

Fig. 3. Bonding MOs formed by σ and π overlap of 2p AOs. Negative regions of the wavefunction are shaded.

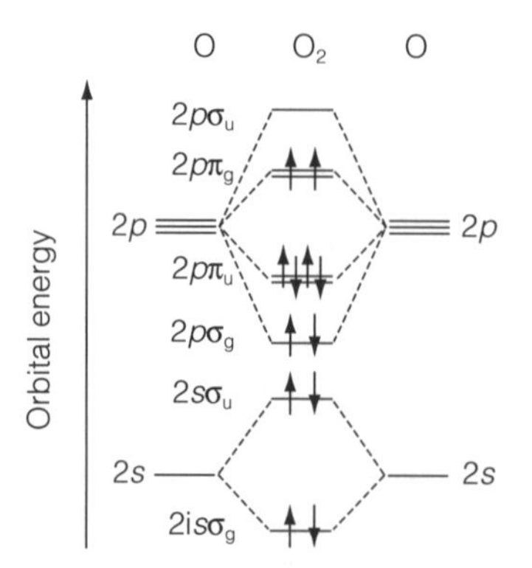

Fig. 4. MO diagram for O_2.

Table 1. Electron configurations, bond orders (BO), dissociation energies (D) and bond lengths (R) for some homonuclear diatomics

Electron configuration		BO	D (kJ mol^{-1})	R (pm)
N_2	$(2s\sigma_g)^2(2s\sigma_u)^2(2p\sigma_g)^2(2p\pi_u)^4$	3	945	110
O_2^+	$(2s\sigma_g)^2(2s\sigma_u)^2(2p\sigma_g)^2(2p\pi_u)^4(2p\pi_g)^1$	2.5	630	112
O_2	$(2s\sigma_g)^2(2s\sigma_u)^2(2p\sigma_g)^2(2p\pi_u)^4(2p\pi_g)^2$	2	498	121
O_2^-	$(2s\sigma_g)^2(2s\sigma_u)^2(2p\sigma_g)^2(2p\pi_u)^4(2p\pi_g)^3$	1.5	–	128
F_2	$(2s\sigma_g)^2(2s\sigma_u)^2(2p\sigma_g)^2(2p\pi_u)^4(2p\pi_g)^4$	1	158	142

For more sophisticated purposes (e.g. interpretation of molecular spectra) some refinements need to made to this simple picture. In particular, the possibility of overlap between a $2s$ orbital on one atom and the $2p\sigma$ on the other can change the order of MO energies. This does not alter the electron configurations of any species shown in *Table 1*, but is important for some molecules such as C_2 (known at high temperatures, e.g. in flames) and for heteronuclear molecules (see Topic C4).

C4 Molecular orbitals: heteronuclear diatomics

Key Notes

Basic principles

Orbitals from atoms of different elements overlap to give unsymmetrical MOs, the bonding MO being more concentrated on the more electronegative atom. The greater the electronegativity difference, the greater the degree of localization.

HF and BH

In HF the F 2*s* orbital is too low in energy to be involved significantly in bonding, but in BH both 2*s* and 2*p* orbitals on B can contribute to MOs with H 1*s*. This situation is described as *sp* hybridization, and leads to bonding and nonbonding MOs with a spatial localization similar to that assumed in VSEPR theory.

CO

When *sp* hybridization occurs on both atoms in a diatomic the order of MOs can be hard to predict. In CO the highest occupied MO (HOMO) is a nonbonding σ orbital resembling a lone-pair on C; the lowest unoccupied MO (LUMO) is a π antibonding orbital.

Related topics

Electronegativity and bond type (B1)

Molecular orbitals: homonuclear diatomics (C3)

Basic principles

In a **heteronuclear molecule** a bond is formed between different atoms, and the most important difference from the homonuclear case (Topic C3) is that molecular orbitals (MOs) are no longer shared equally between atoms. Consider a molecule where each atom has just one valence atomic orbital (AO): an example would be gas-phase LiH with 2*s* on Li and 1*s* on H. When MOs are constructed using the LCAO approximation

$$\psi = c_1\phi_1 + c_2\phi_2 \qquad (1)$$

the coefficients c_1 and c_2 are **no longer equal**. In LiH the two AOs differ greatly in energy, as H has a higher ionization energy and higher electronegativity than Li. If ϕ_2 is the AO of lower energy (i.e. of higher ionization energy or greater electronegativity; see Topics A5 and B1), then the bonding MO has $c_2 > c_1$. The square of each coefficient gives the electron density in the appropriate AO, and so the bonding MO has more electron density on the more electronegative atom. Bonding is provided by a combination of two effects: some increase of density between the atoms as in a homonuclear molecule, together with some electron transfer giving a partially ionic distribution of the form $Li^{\delta+}H^{\delta-}$. As the electronegativity difference between atoms increases, so does the localization of the MO, making the charge distribution more ionic.

Figure 1 shows the MO diagram appropriate to this case. The antibonding MO is also shown; the electron distribution here is localized in the reverse direction to that in the bonding MO, but this orbital is not occupied when only two electrons are present.

As well as providing a description of the transition between purely covalent and purely ionic bonding, the model described above has a consequence that is important in more complex cases. AOs of very different energy do not mix significantly; the resulting MOs are hardly different from the AOs themselves, and it is a good approximation to neglect their interaction.

HF and BH

The two molecules HF and BH illustrate cases where more orbitals are involved (see *Fig. 2*). In HF the fluorine is more electronegative, and hence its AOs are lower in the diagram than that of H. It may be assumed that the 2*s* AO on F is too far removed in energy from the 1*s* on H to interact significantly. The bonding and antibonding MOs are formed from combinations of H 1*s* with F 2*p*. The two $p\pi$ AOs on F have no corresponding AO on H to interact with (as 1*s* is of σ symmetry; see Topic C3) and so remain **nonbonding**. The orbital occupancy shown corresponds to a bond order (BO) of one, because the F 2*s* orbital is also nonbonding. The bonding orbital is more localized on F and the charge distribution is $H^{\delta+}F^{\delta-}$.

In BH the electronegativity differences are reversed, and bonding orbitals will be more localized on H. However, 2*s* and $2p\sigma$ AOs on boron are of comparable energy and both can contribute to the bonding. As the number of MOs formed is equal to the number of starting AOs (see Topic C3), the H 1*s*, and B 2*s* and $2p\sigma$ AOs combine to form three MOs, of which one is bonding, one approximately nonbonding, and one antibonding. The MOs shown in *Fig. 2* may be understood in terms of ***sp* hybrid AOs** formed on boron by mixing 2*s* with $2p\sigma$. Two such hybrids are formed, one pointing towards hydrogen and one away. The former hybrid combines with H 1*s* giving bonding and antibonding combinations, whereas the other does not overlap much and is nonbonding. The four valence electrons in BH thus make a bonding pair and a nonbonding pair oriented in opposite directions as predicted in the VSEPR model (Topic C2). The bond order is one, with the charge distribution $B^{\delta+}H^{\delta-}$ as predicted on electronegativity grounds.

Labeling of MOs in the heteronuclear case follows the same σ or π classification as for homonuclear diatomics but the subscripts *g* and *u* are **not** given, as there is

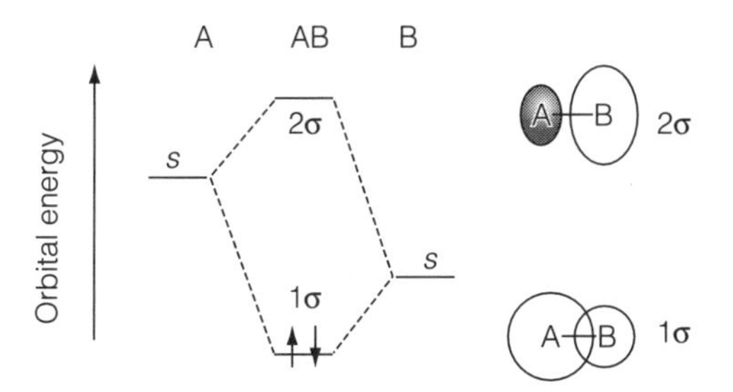

Fig. 1. MO diagram for a heteronuclear molecule with one valence s orbital per atom. The form of the MOs is also shown, with shading indicating negative regions of wavefunction.

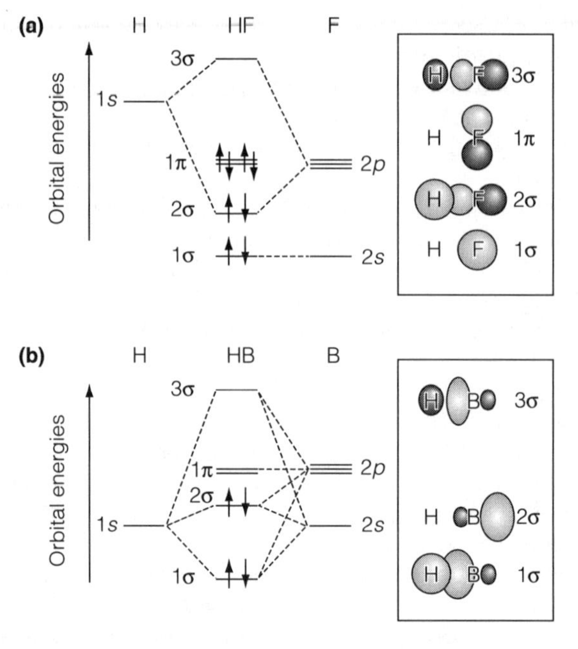

Fig. 2. MO diagrams with the approximate forms of orbitals shown for (a) HF and (b) BH (negative regions shaded).

no center of inversion symmetry. Different σ and π MOs are labeled 1, 2, 3, . . . in order of increasing energy. Normally only valence-shell orbitals are included but occasionally the labeling includes inner shell orbitals as well. Labeling of the MOs in *Fig. 2* follows the normal convention with the MO derived from the inner shell 1*s* AO on B or F not included. Sometimes the designation σ* or π* is used to distinguish antibonding MOs from the σ or π bonding MOs.

CO

The MO diagram for CO shown in *Fig. 3* illustrates a more complex example, useful for understanding the bonding in carbonyl compounds such as $Ni(CO)_4$ (see Topic H9). Formation of the π MOs is straightforward to understand as there is

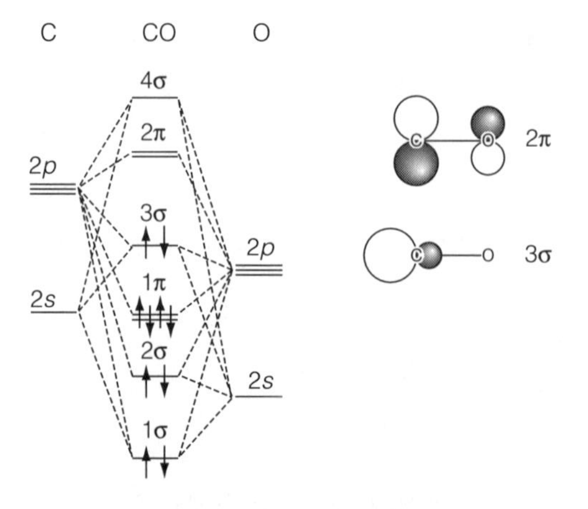

Fig. 3. MO diagram for CO, showing the form of the frontier orbitals 3σ and 2π.

only one pair of equivalent AOs on each atom. They combine to form the bonding 1π (concentrated more on oxygen) and the antibonding 2π MOs. The σ MOs are more complex as there are four AOs involved, the 2*s* and 2*p*σ on each atom. They give four MOs, labeled 1σ–4σ in *Fig. 3*. Their forms and energies can be understood by imagining the preliminary formation of two *sp* hybrids on each atom. The hybrids that point towards each other overlap strongly and combine to form the strongly bonding 1σ and strongly antibonding 4σ MOs. The other two hybrids each point away from the other atom and so do not overlap strongly; they remain almost nonbonding and form 'lone-pair' type orbitals, with 2σ localized on oxygen and 3σ on carbon. The 10 valence electrons occupy the MOs as shown. The **highest occupied MO (HOMO)** is the 'carbon lone-pair' 3σ, and the **lowest unoccupied MO (LUMO)** the antibonding 2π. The HOMO and LUMO are called the **frontier orbitals** and can be used to understand the interaction of a molecule with other species (see Topics C8 and H9). Their approximate form in CO is shown in *Fig. 3*. The bonding in CO comes primarily from two electrons in 1σ and four in 1π, giving a bond order of three as predicted by the valence structure (Topic C1, Structure 8).

C5 MOLECULAR ORBITALS: POLYATOMICS

Key Notes

Localized and delocalized orbitals

Alternative bonding descriptions are often possible in polyatomic molecules, involving either localized (two center) or delocalized (three or more center) molecular orbitals. The overall electron distributions predicted may be the same in both models.

Directed valence

Directed valence theory uses two-center bonding orbitals, with hybrid combinations of atomic orbitals. Some of the features of VSEPR are reproduced, but detailed interpretations of bond angles are different.

Multiple bonding

Multiple bonds are provided by π-type molecular orbitals as with diatomics.

Three-center bonding

Three-center bonds are necessary for the description of some molecules. Bridge bonds in diborane are of the three-center two-electron type, whereas three-center four-electron bonding provides an explanation of some hypervalent molecules.

Related topics

Molecular shapes: VSEPR (C2)

Molecular orbitals: heteronuclear diatomics (C4)

Localized and delocalized orbitals

When the molecular orbital (MO) theory is applied to polyatomic molecules alternative descriptions are possible, as shown in *Figure. 1* for the linear gas-phase molecule BeH_2. There is no reason why an MO must be confined to just two atoms. In *Fig. 1a* the two orbitals shown are formed respectively from a 2*s* and a 2*p* atomic orbital (AO) on beryllium, combined with hydrogen 1*s* AOs of appropriate sign to give a bonding MO. It is also possible to form antibonding combinations (not shown). The four valence electrons in the ground state of BeH_2 can be regarded as occupying the two **three-center** (3c) or **delocalized MOs** shown. Bonding stabilization is provided, as in the diatomic case, by a combination of increased electron density in the overlap regions between atoms, and a transfer of electrons to the more electronegative hydrogen atoms.

The alternative picture in *Fig. 1b* is based on *sp* hybrid orbitals on the central atom, pointing in opposite directions as in the MO description of BH (see Topic C4). Each hybrid is combined to form a **two-center** (2c) or **localized MO** with the appropriate hydrogen AO; again, antibonding MOs can be made but are not shown. In this description of BeH_2 two electrons occupy each of the 2c MOs, giving a picture similar to that assumed in VSEPR theory where two electron pairs around an atom adopted a linear configuration (see Topic C2).

The 3c and 2c bonding descriptions look different, but so long as both orbitals

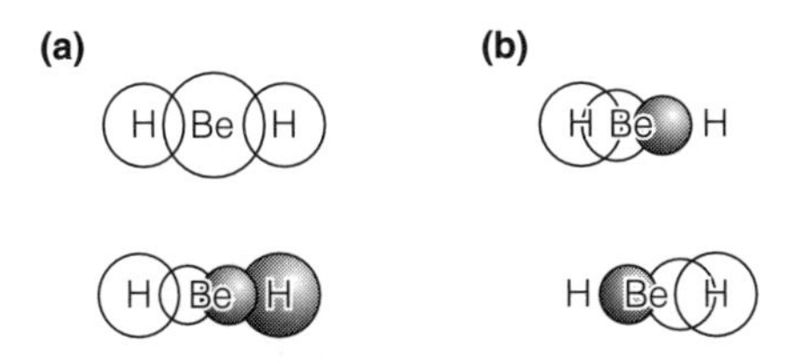

Fig. 1. Bonding MOs for BeH_2. (a) 3-centre, (b) 2-centre representations. In each case both MOs are doubly occupied.

are doubly occupied in each case, they are in fact **equivalent**. In the orbital approximation any set of occupied orbitals may be replaced by a linear combination of them without changing the overall many-electron wavefunction. The two 2c MOs of *Fig. 1b* can be formed by making linear combinations of the 3c MOs in *Fig. 1a*, and conversely the 3c MOs could be reconstructed by combining the 2c MOs. The two pictures show different ways of 'dissecting' the total electron distribution into contributions from individual pairs, but as electrons are completely indistinguishable such dissections are arbitrary and do not predict any observable differences.

The two MO approaches to polyatomic systems, localized and delocalized, are useful in different circumstances. When localized descriptions are possible, they correspond more closely to the simple chemical pictures of electron-pair bonds provided by the Lewis and VSEPR models. Such descriptions are not always possible, and 3c or other delocalized models provide an alternative to the resonance approach (see below and Topic C6). Delocalized MO theory is also more useful for interpreting electronic spectra of molecules.

Directed valence

The localized 2c MO picture depends on hybrid AOs that point towards other atoms and provide **directed valence**. Combining *s* with one *p* orbital in a valence shell gives two ***sp* hybrids** directed at 180° apart. Two *p* orbitals with *s* make **sp^2 hybrids** directed at 120° in a plane. These can be used to describe a trigonal planar molecule such as BF_3. Combining *s* with all three *p* orbitals gives **sp^3 hybrids** directed towards the corners of a tetrahedron. These are the geometrical arrangements assumed by VSEPR for two, three and four electron pairs, respectively (see Section C2). In the 2c MO description of methane CH_4, each of the sp^3 hybrids on carbon is combined to make a bonding MO with one hydrogen 1*s* orbital. The four equivalent bonding MOs are occupied by two electrons each.

Nonbonding electron pairs can also be assumed to occupy hybrids on the central atom. Thus in ammonia NH_3, three hybrids on nitrogen are directed towards hydrogen atoms and form bonding combinations. The fourth does not overlap with a hydrogen atom and remains nonbonding. In water H_2O there are two bonding MOs and two nonbonding. The bond angles in these molecules (107° in NH_3, 104.5° in H_2O compared with the ideal tetrahedral angle of 109.5° found in CH_4) suggest that the bybrids used for bonding and nonbonding MOs are not quite equivalent. A smaller bond angle corresponds to more *p* character and less *s* in the hybrid. (The angle between pure *p* orbitals is 90°; see Topic A2.) Valence *s* orbitals are more tightly bound to an individual atom than are *p* orbitals and so do not contribute as much to bonding MOs (see the discussion of HF in Topic C4). On the other hand, hybrid AOs with some *s* character are more strongly directed than are pure *p* orbitals and so can overlap more strongly with neighboring atoms.

The degree of hybridization therefore depends on a balance of factors. NH_3 and H_2O have angles fairly close to the ideal sp^3 prediction, although the bonding orbitals have slightly more p character and the nonbonding MOs will have correspondingly more s. In PH_3 and H_2S the angles are closer to 90°, showing that the balance has changed and that bonding MOs are constructed mostly with valence p orbitals with s remaining largely nonbonding. This trend can be attributed to the weaker bond strengths (compared with s–p energy separations) for elements lower in a group (see Topic C7). The explanation of bond angles provided by VSEPR is very different.

Multiple bonding

As with diatomics (see Topic C3) multiple bonds are provided by the overlap of **$p\pi$ orbitals** perpendicular to the direction of the bond, in contrast to the **σ orbitals**, which point in the bond direction. A simple example is ethene, C_2H_4, *Fig. 2a*, where the planar structure of the molecule results from sp^2 bonding with each carbon forming two σ bonds to hydrogens, and one to the other carbon. The p orbitals not involved in the hybrids are directed perpendicular to the molecule, and can overlap to form the π bonding MO shown, which is occupied by two electrons. The combination of $\sigma + \pi$ MOs gives a double C=C bond. Maximum bonding overlap of the π orbitals depends on the coplanar arrangement of atoms, and there is a significant barrier to rotation about double bonds, unlike single bonds where groups can rotate fairly freely. Triple bonds (e.g. in C_2H_2) are provided by the overlap of two sets of perpendicular $p\pi$ orbitals, as in diatomics such as N_2 and CO (see Topics C3 and C4).

In some cases where a localized description of σ bonding is possible this is not so for the π bonds. An example is the carbonate ion CO_3^{2-}, where a resonance picture is necessary in simple models (see Topic C1, Structure 11). *Figure 2b* shows the planar framework, with sp^2 bonding in the central atom. The $p\pi$ AOs of the four atoms can overlap together to form a delocalized MO as shown. Out of the three orbital combinations possible for the three oxygen πAOs only one can overlap and bond with carbon in this way. There are two others (not shown), which remain nonbonding on the oxygen. Thus one π bonding MO is distributed over three C—O bonds, with nonbonding charge density corresponding to two MOs distributed over the three oxygen atoms. This is essentially similar to the resonance picture.

Three-center bonding

An example of where the 2c bonding picture is not possible is in diborane B_2H_6 (see Topic C1, Structure 16). The terminal B—H bonds can be described in simple 2c terms, but the number of electrons available suggests that each bridging hydrogen forms part of a 3c bond involving the two boron atoms. The MO method provides a simple interpretation (**1**). Four H atoms are disposed roughly tetrahedrally

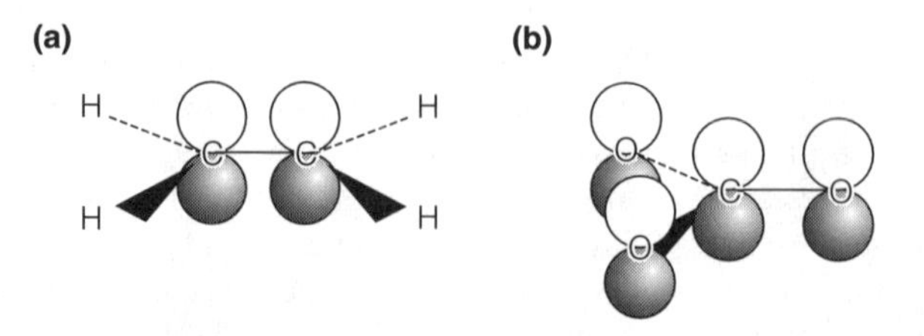

Fig. 2. π bonding MOs in (a) C_2H_4, (b) CO_3^{2-}.

around each boron; this arrangement shows that sp^3 hybrids are used. Two such hybrids form normal 2c bonds by overlap with the 1*s* AO on the terminal hydrogens. The others are combined as in **1** to form two 3c bridge bonds (only one shown). In addition to B—H overlap there is some direct overlap between the boron hybrids, which provides some B—B bonding as well. The result is known as a **three-center two-electron (3c2e) bond**. 3c2e bonds with bridging hydrogen occur in other circumstances, for example the normal form of BeH_2, which has a polymeric chain structure with all H atoms in bridging positions (see Topic G3). Other groups such as methyl CH_3 can do this, for example in dimeric aluminum methyl, $Al_2(CH_3)_6$, which has a structure essentially similar to B_2H_6 with CH_3 in place of H (see Topic G4).

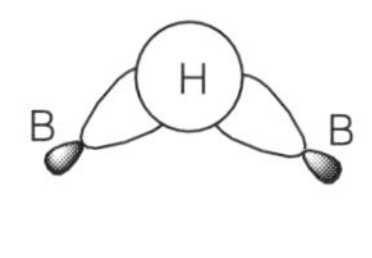

1

Another type of bridging hydrogen occurs in the symmetrical ion $[FHF]^-$ formed by hydrogen bonding between F^- and HF (see Topic F2). To understand this, first count electrons and orbitals as follows: F 2*s* AOs have two electrons each, too tightly bound for bonding (as in HF); the 2$p\pi$ AOs on each F are too far apart to overlap, thus forming nonbonding orbitals holding a total of eight electrons. This leaves four electrons (out of a total valence count of 16) to occupy MOs formed from the two F $p\sigma$ and the H 1*s* AO. The two occupied MOs are shown in *Fig. 3*. There is a 3c bonding MO where H 1*s* is combined with both F $p\sigma$ AOs, and also a nonbonding MO formed from a fluorine combination that has the wrong symmetry to interact with hydrogen. The four electrons in the these MOs give rise to a **three-center four-electron (3c4e) bond**. Effectively each F—H bond is only 'half' a covalent bond as in the 3c2e case, but unlike that situation there are also two electrons localized on the terminal atoms, giving a negative charge there. The result is equivalent to the resonance formulation **2**.

$$F^- \quad H{-}F \quad \longleftrightarrow \quad F{-}H \quad F^-$$

2

3c4e bonding models are an alternative to the use of *d* orbitals in hypervalent compounds with octet expansion. One interpretation of a molecule such as XeF_2 with five electron pairs around a central atom would use sp^3d hybrids, which include *d* orbitals in the valence shell of Xe. Calculations show that this picture greatly overestimates the contribution of *d* orbitals to the bonding. An alternative approach considers 3c4e bonds that use only *p* orbitals on the central atom. The result corresponds to the resonance picture **3**, which requires only eight electrons in Xe valence shell (see Topics F1 and F10).

$$F^- \quad \overset{+}{Xe}{-}F \quad \longleftrightarrow \quad F{-}\overset{+}{Xe} \quad F^-$$

3

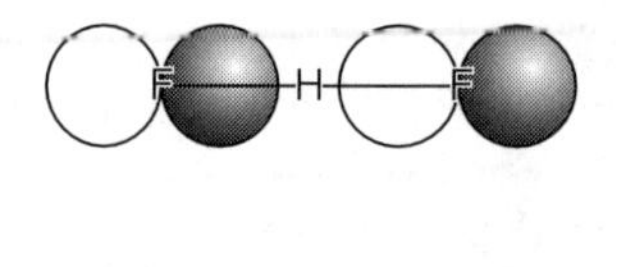

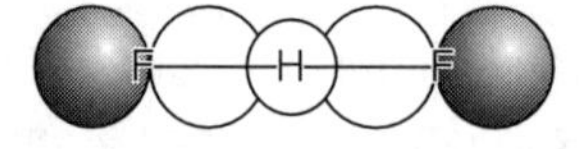

Fig. 3. Occupied MOs in the 3c4e descriptionof $[FHF]^-$.

C6 Rings and clusters

Key Notes

Introduction

Ring and cluster molecules and ions arise in many areas of chemistry. Some structures can be understood using simple two-center bonding models; in others it is necessary to used delocalized MO models.

Aromatic rings

The Hückel MO model predicts that rings will have aromatic stability if they have $4n + 2$ delocalized π electrons, where n is a whole number. Inorganic applications include S_2N_2, which has six π electrons.

Wade's rules

Borane clusters can be classified as *closo*, *nido* or *arachno* with successively more open structures, and respectively $2n + 2$, $2n + 4$ and $2n + 6$ skeletal bonding electrons, where n is the number of boron atoms. The rules may be applied to 'naked' clusters formed by *p*-block metals, and extended to transition metal compounds.

Related topics

Molecular orbitals: polyatomics (C5)

Boron (F3).

Introduction

Ring and cluster compounds arise in many areas of chemistry. **Rings** are most often formed by nonmetallic elements with directional covalent bonding. They include homoelement rings such as S_8 (**1**) and benzene C_6H_6, and ones with heteroelement bonding such as S_2N_2 (**2**), borazine $B_3N_3H_6$ (see Topic F3) and the silicate ion $[Si_3O_9]^{3-}$ (see Topic D5). **Clusters** are polyhedral arrangements of atoms found very widely in the periodic table: nonmetals (e.g. P_4 **3** and boranes discussed below), main-group metals (e.g. $[Pb_5]^{2-}$ **4**) and transition metals (often with ligands such as CO; see Topic H9). Heteroelement bonding is also possible as in the A_4B_4 structure **5**, adopted by S_4N_4 (where A=S and B=N) and As_4S_4 (where A=As and B=S).

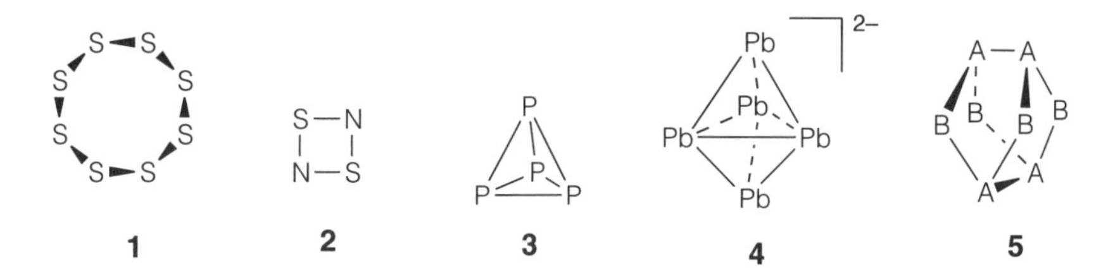

Complex molecular structures do not necessarily require complex bonding models, and indeed much of organic chemistry can be understood using rather elementary ideas. Some inorganic ring and cluster compounds such as S_8 and P_4 can be understood in terms of elementary electron pair bonds (Topic C1); the octets are achieved in each case with two nonbonding electron pairs for each S atom, and one for each P. Similarly in As_4S_4, each As forms three bonds and one lone-pair, each S atom two bonds and two lone-pairs. The reverse arrangement of S_4N_4 is slightly harder to understand but may still be accommodated within simple ideas by placing a formal negative charge on each two-bonded N, and a formal positive charge on each three-bonded S; as expected for bonds between formal S^+ entities, the S—S bonds (A—A in **5**) are abnormally long and weak.

Many rings and clusters, however, cannot be understood within the two-center two-electron bond framework. This is sometimes extended by assuming resonance, for example between the two Kekulé-type structures for benzene (**6**). A more natural approach is to extend the molecular orbital (MO) approach (see Topic C5) to many atoms. The so-called **Hückel theory** of ring systems makes important predictions relevant to inorganic molecules such as S_2N_2. Clusters such as boranes also need a delocalized MO approach. **Wade's rules** provide a useful systematization of the principles involved, and can be extended to other systems.

6

Aromatic rings

The **Hückel MO approach** treats the π electrons of rings such as benzene. We imagine a framework of σ bonds formed by sp^2 hybrids on each carbon atom (see Topic C5). The six remaining $2p$ π orbitals overlap to form six delocalized MOs. *Figure 1a* shows the pattern of orbital energies predicted. The lowest energy MO is formed by combining all $2p$ orbitals with positive overlap to give full bonding; higher energy MOs are progressively less bonding and more antibonding. *Figure 1* shows the assignment of six π electrons as in the ground state of benzene. **Aromatic stability** arises because the electrons are collectively more stable in these MOs than they would be in three separate double bonds.

The arrangement of MO energies for benzene is paralleled with other ring sizes: in each case there is a single orbital of lowest energy followed by pairs of equal energy. *Figure 1b* shows the energies for a four-membered ring. Assignment of four π electrons does **not** lead to a closed-shell ground state where every MO is either filled or empty, and indeed the four-π-electron molecule cyclobutadiene C_4H_4 is very unstable. This type of argument leads to the **Hückel 4n + 2 rule**: irrespective of the ring size, aromatic stability requires $4n + 2$ π electrons, where n is a whole number. Possible values are 2, 6, 10, . . . but not 4, 8, ... One consequence

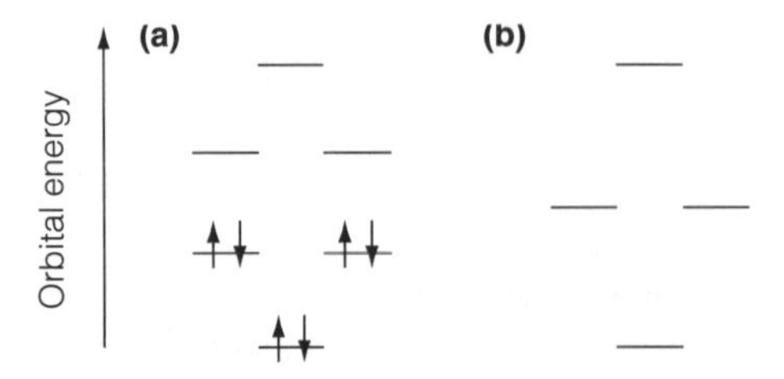

Fig. 1. Energies of π MOs in (a) benzene, (b) a four-membered ring compound.

is that the cyclopentadienyl fragment C_5H_5 is stable as a 6-π-electron anion $[C_5H_5]^-$, an important ligand for organometallic compounds (Topic H10).

There are examples of inorganic rings that conform to the Hückel rule. The heteroatom molecule $B_3N_3H_6$ is isoelectronic with benzene although much more reactive because of the polarity in the B—N bonds (see Topic F3). The S_2N_2 ring (**2**) is an example of a six-π-electron system although the ring is four-membered. Electrons can be counted by assigning two each to four localized S—N σ bonds, and two electrons to a lone-pair on each atom. Out of 22 valence electrons, six remain for the delocalized π system. The ions $[S_4]^{2+}$ and $[Se_4]^{2+}$ have the same valence electron count as S_2N_2 and are also square planar (see Topic F8).

Wade's rules

The apparently bewildering variety of structures adopted by boron-hydrogen compounds (**boranes**) can be rationalized by recognizing some major families, illustrated by the series in *Figure. 2*.

- ***Closo*** boranes with n boron atoms adopt closed polyhedral structures based on triangular faces such as the trigonal bipyramid (five vertices), octahedron (six) and icosahedron (12); such polyhedra are called **deltahedra**. The simplest examples are the ions $[B_nH_n]^{2-}$ such as $[B_6H_6]^{2-}$ illustrated.
- In ***nido*** ('nest-like') boranes n boron atoms are found roughly at the positions of the vertices of an n + 1-vertex deltahedron, with one vertex missing. The simplest general formula type is B_nH_{n+4}; for example, B_5H_9, where the boron atoms are placed at five of the corners of an octahedron.
- ***Arachno*** ('web-like') boranes are still more open and can be imagined as deltahedra with two vertices missing. They form a general series of formula B_nH_{n+6} (e.g. B_4H_{10}).

Wade's rules provide an electronic rationalization of the regularities, based on the MO prediction that an n atom deltahedron, with s and p valence orbitals, should have **n + 1 skeletal bonding MOs**. For example, in $[B_6H_6]^{2-}$ there are seven such MOs, and electrons may be counted as follows: there are 26 valence electrons; 12 are assigned to 'normal' two-center B-H bonds, leaving 14 for skeletal bonding. There is no simple way of assigning these 14 electrons to localized two-center or

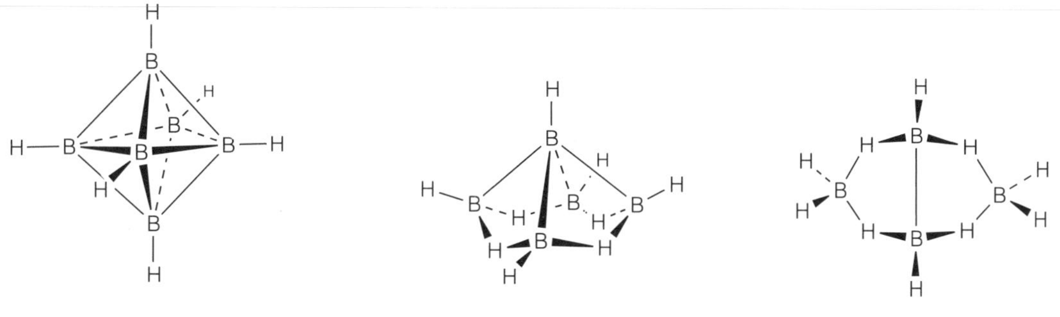

$[B_6H_6]^{2-}$ hexahydrohexaborate (2−)

B_5H_9 pentaborane (9)

B_4H_{10} tetraborane (10)

Fig. 2. Three boranes illustrating the closo/nido/arachno relationship (see text).

even three-center bonds. In the general case, we see that ***closo*** **boranes with *n* atoms should have 2*n* + 2 skeletal bonding electrons**. Isoelectronic replacements of atoms should preserve the structure; for example, $B_{10}C_2H_{12}$ is based on the same icosahedron as $[B_{12}H_{12}]^{2-}$.

Starting with the *closo* ion $[B_nH_n]^{2-}$ we can imagine removing one $[BH]^{2+}$ unit and adding 4 H^+ to give the $n-1$ *nido* borane $B_{n-1}H_{n+3}$ (e.g. $[B_6H_6]^{2-}$ gives B_5H_9). Neither of these operations should alter the number of skeletal bonding electrons, so B_5H_9 has 14, the same number as $[B_6H_6]^{2-}$, and in general ***nido*** **boranes with *n* atoms should have 2*n*+4 skeletal bonding electrons**. The argument may be repeated, starting from *nido* B_nH_{n+4}, removing BH^{2+} and adding 2 H^+, leading to the further conclusion that ***arachno*** **boranes with *n* atoms should have 2*n*+6 skeletal bonding electrons**.

Wade's rules may be applied to 'naked' clusters formed by *p*-block elements if it is assumed that each atom has one localized nonbonding electron pair. So in $[Pb_5]^{2-}$ there are 22 valence electrons, 10 used in lone-pairs, hence 12 for skeletal bonding: a *closo* structure is expected, as found (**4**). In $[Sn_9]^{4-}$ a similar count gives 22 skeletal bonding electrons, corresponding to $2n + 4$ and hence the *nido* structure observed (**7**). It should be noted that there are exceptions. Extension to transition metal clusters needs to accommodate the *d* bonding electrons, and leads to the **Wade–Mingos rules**.

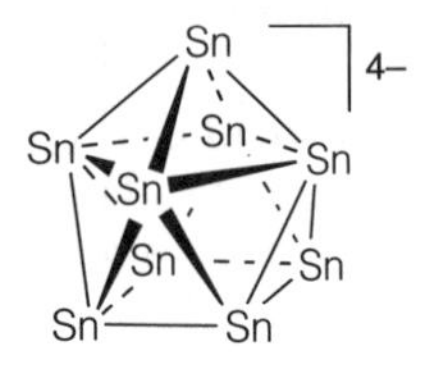

7

C7 Bond strengths

Key Notes

Bond enthalpies

Mean bond enthalpies are defined as the enthalpy changes involved in breaking bonds in molecules. They may be determined from thermochemical cycles using Hess' Law, although assumptions of transferability are sometimes required.

Major trends

Stronger bonds are generally formed with lighter elements in a group, when multiple bonding is present, and when there is a large electronegativity difference between the two elements. The 'anomalous' weakness of single bonds involving N, O and F is often attributed to repulsion between nonbonding electrons. C, N and O form especially strong multiple bonds.

Pauling electronegativity

The Pauling electronegativity scale is derived using an empirical relationship from bond enthalpies. For elements forming covalent bonds, it correlates fairly well with other scales.

Uses and limitations

Quantitative uses of bond energies are of very limited accuracy, but qualitative comparisons can be useful in interpreting trends in the structures and stability of covalent substances.

Other measures

Bond lengths, and spectroscopically measured stretching frequencies, are also useful comparative measures of bond strength.

Related topics

Electronegativity and bond type (B1)

Introduction to nonmetals (F1)

Bond enthalpies

The most straightforward measure of the strength of a bond is the energy required to break it. Estimates of such **bond energies** are normally obtained from thermochemical cycles using Hess' Law (see Topic B3) and are called **bond enthalpies**. A **bond dissociation enthalpy** is the enthalpy change involved in breaking the bond to one atom in a molecule, and in a diatomic is by definition equal to the bond enthalpy. Thus the enthalpy of dissociation of O_2 gives directly the (double) bond enthalpy $B(\mathrm{O{=}O})$.

When a molecule contains several equal bonds, the enthalpy required to dissociate them successively is not the same. Instead of dealing with individual bond dissociation energies, it is normal to define the **mean bond enthalpy**. Thus $B(\mathrm{O{-}H})$ is defined as half the enthalpy change in the process

$$H_2O(g) \rightarrow 2H + O$$

When several types of bonds are involved it is necessary to make assumptions about the energies of some of them. For example, it is normal to assume that the value of $B(O—H)$ obtained from H_2O can also be applied to H_2O_2. Then for the process

$$H_2O_2(g) \rightarrow 2H + 2O$$

we have

$$\Delta H = 2B(O—H) + B(O—O)$$

from which the (single) bond enthalpy $B(O—O)$ can be obtained. This quantity is not the same as ΔH for the dissociation of H_2O_2 into 2OH, as it is argued that the bonding in the hydroxyl radical OH has changed from the 'normal' situation where oxygen forms two bonds. The assumption of **transferability** involved in this method of determining bond enthalpies is, however, open to question (see below).

Major trends

A selection of single bond enthalpies is shown in *Table 1*. Some important trends are summarized below.

(i) Bond energies often become smaller on descending a main group (e.g. C—H > Si—H > Ge—H). This is expected as electrons in the overlap region of a bond are less strongly attracted to larger atoms. Some important exceptions are noted in (v) and (vi) below, and the reverse trend is generally found in transition metal groups (see Topic H1).

(ii) Bond energies increase with bond order, although the extent to which $B(A{=}B)$ is larger than $B(A—B)$ depends greatly on A and B, the largest differences occurring with elements from the set C, N, O (see *Table 2*). Strong multiple bonding involving these elements may be attributed to the very efficient overlap of $2p\pi$ orbitals compared with that of larger orbitals in lower periods.

(iii) In compounds AB_n with the same elements but different n values, $B(A—B)$ decreases as n increases (e.g. in the sequence ClF > ClF_3 > ClF_5). The differences are generally less for larger A, and more electronegative B.

(iv) Bonds are stronger between elements with a large electronegativity difference. This forms the basis for the **Pauling electronegativity scale** (see below).

(v) Single A—B bonds where A and B are both from the set N, O, F are weaker than expected from group comparisons. This is often attributed to a **repulsion between nonbonding electrons**, although as in other cases of 'electron repulsion' the effect may be attributed to the Pauli exclusion principle more than to electrostatic repulsion (see Topic C2).

Table 1. A selection of single-bond AB enthalpies (kJ mol^{-1})

A=	B=H	C	O	F	Cl
H	436	413	464	568	432
C	413	347	358	467	346
Si	318	307	466	597	400
Ge	285	-	385	471	340
N	391	286	214	278	190
P	321	-	360	490	322
As	297	-	326	487	309
O	464	358	144	214	206
S	364	289	–	–	271
Se	313	–	–	–	251

Table 2. Variation in bond enthalpy (kJ mol^{-1}) with bond order

A–B	Single	Double	Triple
C–C	347	612	838
C–O	358	805	1077
Si–O	466	638	–
N–N	167	247	942
N–O	214	587	–
O–O	144	498	–
S–S	266	429	–

(vi) Other exceptions to rule (i) above occur with A–O and A–X bonds (X being a halogen), which generally increase in strength between periods 2 and 3 (e.g. C–O < Si–O). This may be partly due to the increased electronegativity difference when A is period 3 (see (iv) above), but repulsion between lone-pairs electrons on nonbonded atoms may also play a role (e.g. F–F repulsion in CF_4, where the atoms are closer together than in SiF_4).

Pauling electronegativity

Pauling noted that B(A–B) is nearly always larger than the mean of the homonuclear A–A and B–B bond energies, and attributed this to the possibility of **ionic–covalent resonance** involving valence structures such as A^+B^- when B is the more electronegative atom. He related the bond strengths to the **electronegativities** x_A and x_B of the two elements according to the formula

$$B(\text{A–B}) = \sqrt{[B(\text{A–A})B(\text{B–B})]} + C(x_A - x_B)^2$$

where the constant C takes the value 96.5 if B values are in kJ mol^{-1}. As this formula depends only on the difference of electronegativities, it is necessary to choose one value to start the scale; Pauling chose 4.0 for the electronegativity of fluorine.

Pauling's formula should be regarded as purely empirical and without any rigorous theoretical foundation. Nevertheless, the electronegativity scale is widely used, and shows the same trends as ones based more directly on atomic quantities (see Topic B1). Pauling's formula provides a useful rationalization of some bond-strength trends, and can be used as a semiquantitative guide for estimating unknown bond enthalpies. It should not be used for solids with a high degree of ionic character, as these are best interpreted using lattice energies (see Topic D6).

Uses and limitations

Tabulated values of bond enthalpies can be used to estimate the enthalpy of formation of hypothetical compounds. Such estimates should be regarded as rough and not quantitatively reliable, as the assumptions of additivity and transferability that underlie these calculations are not accurate.

Trends in stability or structure of related compounds can often be usefully rationalized from bond strength trends. The decline in B(E–H) as a main group is descended leads to reduced thermodynamic stability of hydrides EH_n (see Topic F2). Double-bonded structures are much commoner with the elements C, N and O than with others in the same group. The stability of O_2 (double bonds) versus S_8 (single bonds) can be rationalized from the fact that B(O=O) is more than twice as large as B(O–O) but the same is not true of sulfur. In a similar way we have CO_2 (molecular with C=O) and SiO_2 (polymeric with single Si–O). The formation of multiple bonds is one of the main factors leading to differences in chemistry between $2p$ series elements and those in lower periods (see Topic F1).

Changes with valence state are important in understanding the stability of 'hypervalent' compounds. Thus SH_4 and SH_6 are unknown, whereas they would be thermodynamically stable compounds if their S–H bonds were as strong as in H_2S. The common formation of fluorides in high valency states (e.g. SF_6, IF_7) can be understood from a combination of factors. The F–F bond is itself rather weak, E–F bonds are generally strong, and they decline less rapidly with increasing n in EF_n molecules than in other compounds.

Other measures

Thermochemical bond energies may be hard to determine, either for experimental reasons or because of the limitations in the assumption of transferability that is often required. Alternative measures of comparative bond strength that are often useful include the following:

- the **bond length**, which for a given pair of elements decreases with increasing strength (e.g. with increasing bond order, as in the sequence N–N 145 pm, N=N 125 pm, N≡N 110 pm); bond length measurements are often useful for showing the existence of metal-metal bonds in transition metal compounds (see Topic H5);
- the **bond stretching frequency** measured by vibrational spectroscopy (e.g. IR) is related to the **stretching force constant** and increases with bond strength; IR measurements have been particularly useful in the study of CO as a ligand in transition metal carbonyl compounds (Topic H9).

C8 LEWIS ACIDS AND BASES

Key Notes

Definition and scope

A Lewis acid (or acceptor) can accept an electron pair from a Lewis base (or donor) to form a donor–acceptor complex. The formation of solvated ions, complexes in solution and coordination compounds are examples of this type of interaction.

Models of interaction

Contributions to the donor–acceptor interaction may come from electrostatic forces, and from the overlap between the highest occupied MO (HOMO) of the donor and the lowest unoccupied MO (LUMO) of the acceptor.

Hard–soft classification

Hard donors interact more strongly with hard acceptors, soft donors with soft acceptors. Harder acids tend to be more electropositive, and harder bases more electronegative. Softer donor and acceptor atoms tend to be larger and more polarizable.

Polymerization

Formation of dimers and polymeric structures is a manifestation of donor–acceptor interaction between molecules of the same kind.

Related topics

Solvent types and properties (E1)

Brønsted acids and bases (E2)

Complex formation (E3)

Definition and scope

A **Lewis acid** is any species capable of accepting a pair of electrons, and a **Lewis base** is a species with a pair of electrons available for donation. The terms **acceptor** and **donor** are also commonly used. Lewis acids include H^+ and metal cations, molecules such as BF_3 with incomplete octets, and ones such as SiF_4 where octet expansion is possible (see Topic C1). Any species with nonbonding electrons is potentially a Lewis base, including molecules such as NH_3 and anions such as F^-. The Lewis acid–base definition should not be confused with the Brønsted one (see Topic E2): Brønsted bases are also Lewis bases, and H^+ is a Lewis acid, but Brønsted acids such as HCl are **not** Lewis acids.

Lewis acids and bases may interact to give a **donor–acceptor complex**; for example,

$BF_3 + (CH_3)_3N\!: \rightarrow (CH_3)_3N\!: BF_3$ (**1**)

$SiF_4 + 2F^- \rightarrow [SiF_6]^{2-}$

The bond formed is sometimes denoted by an arrow (as in **1**) and called a **dative bond** but it is not really different from any other polar covalent bond. Thus the complex $[SiF_6]^{2-}$ has a regular octahedral structure where the two 'new' Si—F bonds are indistinguishable from the other four. (It is isoelectronic with SF_6; see Topic C2.)

1

The scope of the donor–acceptor concept is extremely broad and encompasses many types of chemical interaction, including the solvation and complexation of metal ions and the formation of coordination compounds by transition metals (see Topics E1, E3, H2 and H9). Many chemical reactions also depend on donor–acceptor interactions. For example, the hydrolysis of $SiCl_4$ to give $Si(OH)_4$ in water begins with a step such as

$$SiCl_4 + H_2O \rightarrow SiCl_4(OH_2)$$

where H_2O is acting as a donor to the $SiCl_4$ acceptor.

Models of interaction

Interaction between a Lewis acid and a base may have an **electrostatic contribution** as donor atoms are often electronegative and possess some partial negative change, whereas acceptor atoms may be positively charged. There is also an **orbital interaction**, which can be represented by the simple **molecular orbital (MO) diagram** of *Figure 1* (see Topic C4). On the left and right are represented respectively the lowest unoccupied MO (**LUMO**) of the acceptor A and the highest occupied MO (**HOMO**) of the donor D. The levels in the center show the formation of a more stable bonding MO and a destabilized antibonding MO in the complex. The electron pair from the donor occupies the bonding MO, which is partially shared between the two species.

Interaction between the orbitals in *Fig. 1* will be strongest when the energy difference between the acceptor LUMO and the donor HOMO is least. In this model the best acceptors will have empty orbitals at low energies, the best donors filled orbitals at high energies. By contrast, the strongest electrostatic interactions will take place between the smallest and most highly charged (positive) acceptor and (negative) donor atoms.

Hard–soft classification

It is found that the relative strength of donors depends on the nature of the acceptor and vice versa. The **hard and soft acid–base (HSAB)** classification is often used to rationalize some of the differences. When two acids (A_1 and A_2) are in competition for two bases (B_1 and B_2) the equilibrium

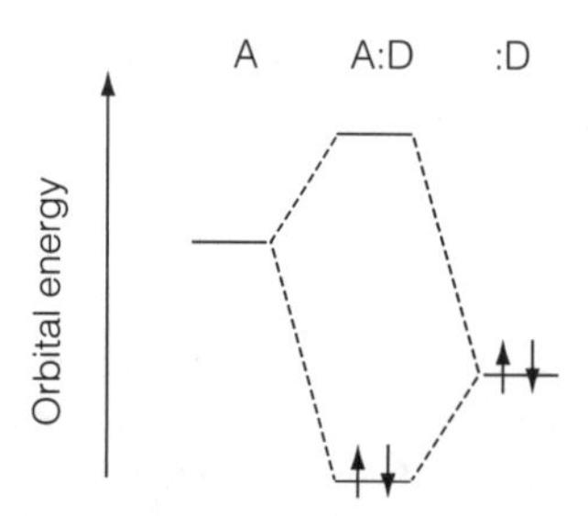

Fig. 1. Molecular orbital interaction between a donor (:D) and an acceptor (A).

$$A_1{:}B_1 + A_2{:}B_2 \rightleftharpoons A_1{:}B_2 + A_2{:}B_1$$

will lie in the direction where the harder of the two acids is in combination with the harder base, and the softer acid with the softer base. As a standard for comparison the prototype hard acid H^+ and soft acid $[(CH_3)Hg]^+$ are often used. Thus the equilibrium

$$[B{:}H]^+ + [(CH_3)Hg]^+ \rightleftharpoons H^+ + [(CH_3)Hg{:}B]^+$$

will lie to the left or right according to the degree of hardness of the base B.

Examples of hard acids are H^+, cations of very electropositive metals such as Mg^{2+}, and nonmetal fluorides such as BF_3. Soft acids include cations of late transition and post-transition metals such as Cu^+, Pd^{2+} and Hg^{2+} (see Topics G4, G6, H3 and H5). The hardness of bases increases with the group number of the donor atom (e.g. $NH_3 < H_2O < F^-$) and decreases down any group (e.g. $NH_3 > PH_3$, and $F^- > Cl^- > Br^- > I^-$).

Although the hard–soft classification provides a useful systematization of many trends it does not by itself provide an explanation of the different behavior. Generally it is considered that hard–hard interactions have a greater electrostatic component and soft–soft ones depend more on orbital interactions, but many other factors may be involved. Soft acceptor and donor atoms are often large and van der Waals' forces may contribute to the bonding (see Topic C9); some soft bases such as CO also show π-acceptor behavior (see Topic H9). It is also important to remember that hard and soft behavior is defined in a competitive situation. When reactions are studied in solution some competition with solvation is always present (see Topics E1 and E3).

Polymerization

The tendency of many molecules to aggregate and form **dimers** (e.g. Al_2Cl_6 **2**), larger oligomers, or extended **polymeric structures** can be regarded as a donor–acceptor interaction. Thus in the reaction

$$2AlCl_3 \rightarrow Al_2Cl_6$$

a chlorine atom bound to one $AlCl_3$ uses nonbonding electrons to complex with the other aluminum atom; as in most other examples of this type the bridging atoms are symmetrically disposed with identical bonds to each aluminum. Polymerization of AX_n molecules is more likely to occur when n is small, and when the atom A has vacant orbitals and is large enough to increase its coordination number. Many oxides and halides of stoichiometry AB_2 and AB_3 form structures that may be regarded as polymeric, although the distinction between this (polar covalent) description and an ionic one is not clear-cut (see Topics B1, D4 and F7).

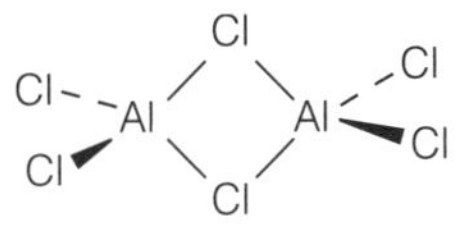

Cl Cl Cl Al Al Cl Cl Cl

2

Hydrogen bonding (see Topic F2) can also be regarded as a donor–acceptor interaction in which the acceptor LUMO is the (unoccupied) antibonding orbital of hydrogen bonded to an electronegative element.

C9 MOLECULES IN CONDENSED PHASES

Key Notes

Molecular solids and liquids

Intermolecular forces cause molecular substances to condense to form solids and liquids. Trouton's rule provides an approximate relationship between the normal boiling point of a liquid and the strength of intermolecular forces.

Intermolecular forces

Polar molecules have forces between permanent dipoles. With nonpolar molecules London dispersion (or van der Waals') forces arise between fluctuating dipoles; their magnitude is related to molecular polarizability, which generally increases with size. Molecules may also have more specific donor–acceptor interactions including hydrogen bonding.

Molecular polarity

The polarity of a molecule arises from charge separation caused by electronegativity differences in bonds, although contributions from lone-pairs and the consequences of molecular symmetry are also important. High polarity gives strong intermolecular forces, and also provides a major contribution to the dielectric constant.

Related topics

Electronegativity and bond type (B1)

Solvent types and properties (E1)

Molecular solids and liquids

The condensation of molecular substances into liquid and solid forms is a manifestation of **intermolecular forces**. The **enthalpies of fusion** (i.e. **melting) and vaporization** provide a direct measure of the energy required to overcome such forces.

We speak of **molecular solids** when molecules retain their identity, with geometries similar to those in the gas phase. The structures of molecular solids sometimes resemble those formed by close-packing of spheres (see Topic D2), although with highly unsymmetrical and polar molecules the directional nature of intermolecular forces may play a role. Molecular liquids are more disorganized, but the structural changes between solid and liquid can be subtle and the **melting point** of a molecular solid is not in general a good guide to the strength of intermolecular forces. A better correlation is found with the **normal boiling point**, as molecules become isolated in the vapor and the influence of intermolecular interactions is lost.

The enthalpy of vaporization ΔH_{vap} divided by the normal boiling point in kelvin (T_{b}) gives the **standard entropy of vaporization** (see Topic B3)

$$\Delta s_{vap}^{\ominus} = \Delta H_{\mathrm{vap}}/T_{\mathrm{b}}$$

and according to **Trouton's rule** its magnitude is normally around 90 J K^{-1} mol^{-1}. Trouton's rule is not quantitatively reliable, and breaks down when molecules have an unusual degree of organization in either the liquid or vapor phase (e.g.

because of hydrogen bonding); it does, however, express a useful qualitative relationship between the boiling point and the strength of intermolecular forces. *Figure 1* shows the normal boiling points for noble gas elements and some molecular hydrides.

Intermolecular forces

Between charged ions (whether simple or complex) the Coulomb attraction is the dominant force, as discussed in Topic D6. Even with neutral molecules, intermolecular forces are essentially electrostatic in origin. With **polar molecules** the **force between permanent electric dipoles** is the dominant one (see below). When polarity is absent the force arises from the interaction between instantaneous (fluctuating) dipoles, and is known as the **London dispersion** or **van der Waals' force**. Its strength is related to the **polarizability** of the molecules concerned. Polarizability generally increases with the size of atoms, and the sequence of boiling points He < Ne < Ar < Kr shown in *Fig. 1* reflects this. The boiling point increases down the group in most series of nonpolar molecules, for example, $CH_4 < SiH_4 < GeH_4$ (also in *Fig. 1*), the diatomic halogens $F_2 < Cl_2 < \ldots$, and the order $CF_4 < CCl_4 < CBr_4 < CI_4$ found with other molecular halides. (Ionic halides tend to show the reverse order, reflecting the decrease in lattice energy expected as the size of ions increases; see Topic D6.)

In addition to forces of a strictly nonbonding nature, molecules may have chemical interactions that contribute to the apparent intermolecular forces. Donor and acceptor centers on different parts of a molecule can lead to self-association and polymerization, as discussed in Topic C7. **Hydrogen bonding** is one manifestation of this type of interaction (see Topic F2), which is especially important in polar hydrides of period 2 elements, NH_3, H_2O and HF. The extent to which the boiling points of these compounds are out of line as a consequence can be seen in *Fig. 1*. Hydrogen bonding can also have an important influence on the structure of

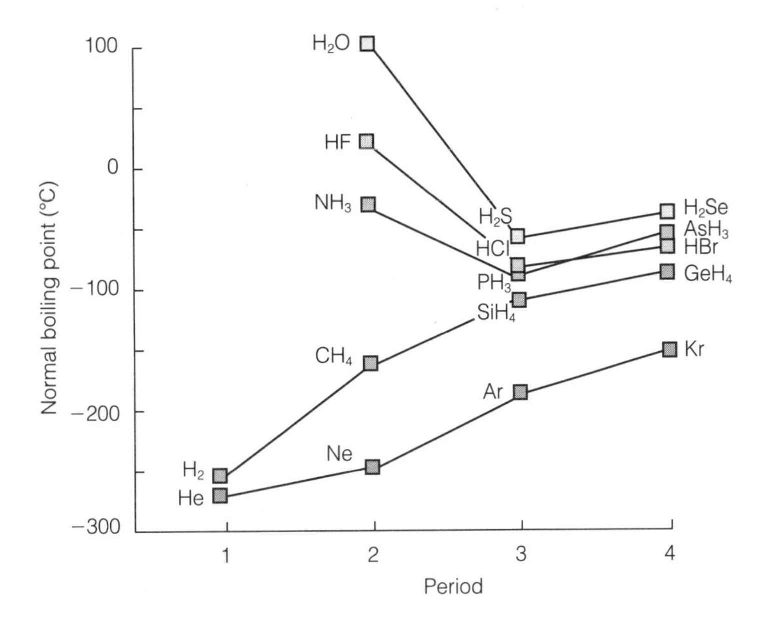

Fig. 1. Normal boiling points of some molecular hydrides, with noble gas elements for comparison.

liquids and solids: thus ice has structures in which each water molecule is hydrogen bonded to four others.

Molecular polarity

The polarity of a molecule is measured by its **dipole moment** μ: imagine charges $+q$ and $-q$ separated by a distance d, then, by definition, $\mu = qd$. A practical unit for μ at the molecular level is the **Debye** (D), equal to 3.336×10^{-30} C m. Polarity is a measure of charge separation in bonds and can often be related to the electronegativity difference between the atoms, as can be seen in the following series of μ/D values: HF (6.4); HCl (3.6); HBr (2.7); HI (1.4). There are, however, reasons why a unique correlation of dipole moments with electronegativity differences is impossible, as follows.

- Lone-pair electrons also have dipole moments. Sometimes these may reinforce the moment resulting from electronegativity difference (as in H_2O and NH_3, which both have very large net dipoles); in other cases the two contributions may oppose each other, as happens in CO where the net dipole is very small.
- In polyatomic molecules the dipoles associated with each bond add vectorially to give a resultant that depends on the bond angles. In highly symmetrical molecules such as BF_3, CF_4, PF_5 or SF_6 the net dipole moment is zero even though the individual bonds may be strongly polar.

Larger dipoles lead to stronger intermolecular forces although other factors (e.g. hydrogen bonding) can also be involved. For molecular species such as LiF(g) the dipole moment is very nearly that predicted for an ionic Li^+F^- charge distribution. Such ionic species do not, however, condense to form molecular solids or liquids, but rather ones with typical ionic structures where each anion is surrounded by several cations and vice versa (see Topic D3).

Application of an electric field to a substance causes a partial alignment of polar molecules; thus molecular dipole moments contribute to the **dielectric constant** of a liquid or solid, one of the most important physical properties determining the behavior of a solvent (see Topic E1). In nonpolar substances the dielectric constant arises from the molecular polarizability, and is generally much smaller than with polar molecules.

D1 INTRODUCTION TO SOLIDS

Key Notes

Crystals and glasses

A crystalline solid is characterized by a unit cell containing an arrangement of atoms repeated indefinitely; noncrystalline or glassy solids do not have a unit cell. Short-range order resulting from the local bonding of atoms may, however, be similar in crystals and glasses.

Looking at unit cells

Different representations of unit cells are possible. It is important to understand how to use them to determine the stoichiometry of the compound and the coordination of each atom.

Non-stoichiometry

Some solids, especially natural minerals and transition metal compounds, are nonstoichiometric with variable composition.

Chemical classification

The classification of solids into molecular, metallic, covalent (polymeric) and ionic types is useful provided it is recognized that there are no hard boundaries between them.

Related topic

Electronegativity and bond type (B1)

Crystals and glasses

Unlike a molecule or complex ion, which is a finite (often small) assembly of atoms, a solid has no fixed size but can add atoms indefinitely. In a sample of uniform composition the bonding arrangements of atoms are expected to be similar throughout. For example, both crystalline and glassy forms of silica (SiO_2) have structures with each Si surrounded by four oxygen atoms, and each O by two Si. However in **crystalline solids** it is possible to identify a **unit cell** containing a group of atoms that is repeated indefinitely in precise geometric fashion. In practice, all crystals contain defects where this regularity is broken sometimes, but nevertheless crystals are different from **non-crystalline solids** or **glasses**, where there is no regular repetition. The crucial distinction is that of **long-range order**. Local chemical bonding arrangements determine **short-range order**, which may be present even in a glass as in the case of SiO_2. The difference arises from the way these bonds connect together to form an extended network. Glassy forms are metastable, prevented by kinetic factors from achieving the most stable (crystalline) arrangement. Very many, possibly all solids can be made glassy if they are cooled rapidly enough from the gaseous or liquid state. Some substances form glasses especially easily, commonly ones in which atoms are covalently bonded to relatively few (three or four) neighbors.

Macroscopically the distinction can be observed in the definite shapes of crystals, which reflect the regular atomic arrangements: compare, for example, the cubic faces of common salt (sodium chloride) crystals with the irregular and often curved surfaces of fractured window glass. Microscopically the distinction can be made by **X-ray diffraction**, which depends on the fact that the regular atomic spacing in crystals is similar to the wavelength of X-rays. This is the most powerful technique for determining the structures of crystalline substances but cannot be used in the same way for glasses.

A 'complete' specification of the structure of a glass is impossible, but for a crystal it is only necessary to give the details of one unit cell. Substances are said to have the 'same' structure if the arrangement of atoms within a unit cell is essentially similar, although the interatomic distances and the dimensions of the cell are different. **Structure types** are named after a particular example, frequently naturally occurring minerals: thus we talk of the **rocksalt** structure of NaCl or the **rutile** structure of TiO_2. Specifying a definite mineral rather than the compound formula is important, as some compounds show **polymorphism** and can adopt several different crystal forms. TiO_2, for example, is known also as brookite and anatase, in which the arrangement of atoms is different from that in rutile.

Looking at unit cells

To specify the complete structure of a crystalline solid it is only necessary to show one unit cell, but interpreting these pictures requires practice. *Figure 1* shows some views of the cesium chloride structure (CsCl, depicted as MX).

(a) *Figure 1a* is a **perspective view** (more correctly known as a **clinographic projection**), which is the most common way of showing a unit cell.
(b) *Figure 1b* shows a **projection** down one axis of the cell. The position of an atom on the hidden axis is given by a specifying a fractional coordinate (e.g. 0.5 for the central atom showing it is halfway up). No coordinate is given for atoms at the base of the cell.
(c) *Figure 1c* shows the atoms shifted relative to the unit cell, and emphasizes the fact that what is important about a unit cell is its size and shape; its origin is **arbitrary** because of the way in which it is repeated to fill space.
(d) In *Figure. 1d* the drawing has been extended to show some repeated positions of the central atom. This helps in seeing the coordination of the corner atom (see below).

The most important aspect of any structure is its **stoichiometry**, the relative numbers of different types of atoms. The stoichiometry of a unit cell can be determined by counting all the atoms depicted, and then taking account of those that are shared with neighboring cells. Any atom at a corner of a unit cell is shared between eight cells, any at an edge between four, and any on a face between two. Thus the composition MX in *Figure. 1* is arrived at by counting the eight corner M atoms, and then dividing by eight to account for sharing. With some experience, this procedure will seem unnecessary. If one simply imagines the unit cell with a shifted origin as in *Figure. 1c* then it is immediately clear that every cell contains one M and one X atom.

Another feature characteristic of a structure is the **coordination** of each atom. There is usually no difficulty in seeing the coordination of an atom in the middle of a unit cell. (For example, X in *Figure. 1a* can easily be seen to have eight M neighbors forming the corners of cube. In the projection, *Figure. 1b*, one needs to remember that the M atoms at the base of the cell are repeated at the top.) For atoms at corners or edges it is necessary to consider what happens in neighboring cells, and an extended drawing such as *Figure. 1d* may be helpful: this shows each M surrounded by eight X neighbors in the same way as the coordination of X.

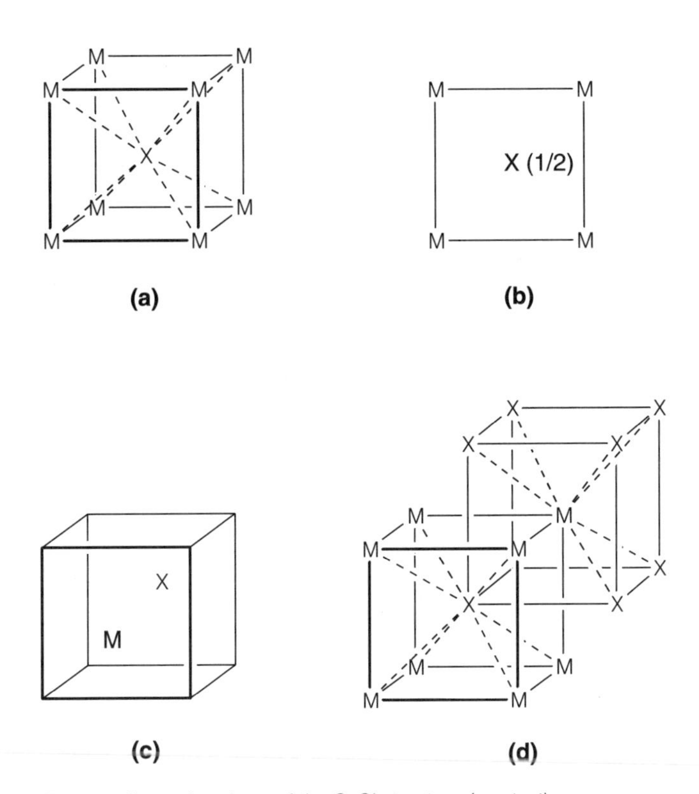

Fig. 1. Alternative views of the CsCl structure (see text).

Nonstoichiometry

Whereas a pure molecular substance has a definite stoichiometry, this is not always true for solids. **Defects** in crystals can include **vacancies** (atoms missing from their expected sites) and **interstitials** (extra atoms in sites normally vacant in the unit cell). An imbalance of defects involving different elements can introduce **nonstoichiometry.** This is common in compounds of transition metals, where variable oxidation states are possible (see Topics D5 and H4). For example, the sodium tungsten bronzes are formulated as Na_xWO_3, where x can have any value in the range 0–0.9.

Another form of nonstoichiometry arises from the partial **replacement** of one element by another in a crystal. It is common in natural minerals, such as the aluminosilicate feldspars $(Na,Ca)(Al,Si)_4O_8$. The notation (Na,Ca) means that Na and Ca can be present in the same crystal sites in varying proportions. Simultaneous (Si,Al) replacement ensures that all elements remain in their normal oxidation states. Even this formulation is approximate, as several other elements may be present in smaller proportions.

Chemical classification

Solids are often classified according to their chemical bonding, structures and properties (see Topic B1):

Molecular solids contain discrete molecular units held by relatively weak intermolecular forces (see Topic C9).

Metallic solids have atoms with high coordination numbers, bound by delocalized electrons that give metallic conduction.

Covalent or *polymeric solids* have atoms bound by directional covalent bonds, giving relatively low coordination numbers in a continuous one-, two- or three-dimensional network.

Ionic solids are bound by electrostatic attraction between anions and cations, with structures where every anion is surrounded by cations and vice versa.

Although these broad distinctions are useful, many solids show a degree of intermediate character, or even several types of bonding simultaneously. Metallic and covalent interactions both arise from overlapping atomic orbitals (see Topics C3–C6) and the distinction in physical properties arises from the energy distribution of electronic levels (see Topic D7). The structures and electronic properties of elements show a gradation in character at the metal-nonmetal borderline (see Topics B2 and D2). A similar gradation is seen between ionic and covalent compounds as the electronegativity difference between two elements changes (see Topics B1 and D4). Furthermore, solids with predominantly ionic bonding between some atoms can also have covalent bonds between others (see Topic D5).

D2 Element structures

Key Notes

Sphere packing

Spheres of equal size may be packed in three dimensions to give hexagonal close-packed (hcp) and cubic close-packed (ccp, also known as face-centered cubic, fcc) structures. The body-centered cubic (bcc) structure is slightly less efficiently close packed.

Metallic elements

Many metallic elements have hcp, fcc or bcc structures. There are some clear group trends in structure, although there are exceptions to these and some metals have less regular structures, especially in the *p* block.

Nonmetallic elements

Most nonmetallic elements have structures that can be understood using simple electron-pair bonding models. C, N and O can form multiple bonds and are exceptional in their groups.

Related topics

Chemical periodicity (B2) Introduction to nonmetals (F1)

Sphere packing

Element structures where chemical bonding is nondirectional are best introduced by considering the packing of equal spheres. **Close-packed structures** are ones that fill space most efficiently. In two dimensions this is achieved in a layer with each sphere surrounded hexagonally by six others. Three-dimensional structures are developed by stacking these layers so that the spheres in one layer fall over the hollows in the one below, as shown in *Figure 1a*. Having placed two layers, labeled A and B, there are alternative positions for the spheres in the third layer. They could be placed directly over spheres in the first layer A to give a sequence denoted ABA. Alternatively, the spheres in the third layer can be placed in positions where there are gaps in layer A; two such spheres labeled C are shown in *Figure. 1a.* A regular packing based on this latter arrangement would then place the fourth layer directly over layer A, giving a sequence denoted ABCA. The simplest three-dimensional close-packed structures are based on these two regular sequences of layer positions:

ABABABAB . . . gives **hexagonal close packing (hcp)**;
ABCABCABC . . . gives **cubic close packing (ccp)**.

These structures are illustrated in *Figure. 1b* and *c*, respectively. In the ccp arrangement, successive close-packed layers are placed along the body diagonal of a cube. The unit cell shown is based on a cube with atoms in the face positions, and the structure is also known commonly as **face-centered cubic (fcc)**.

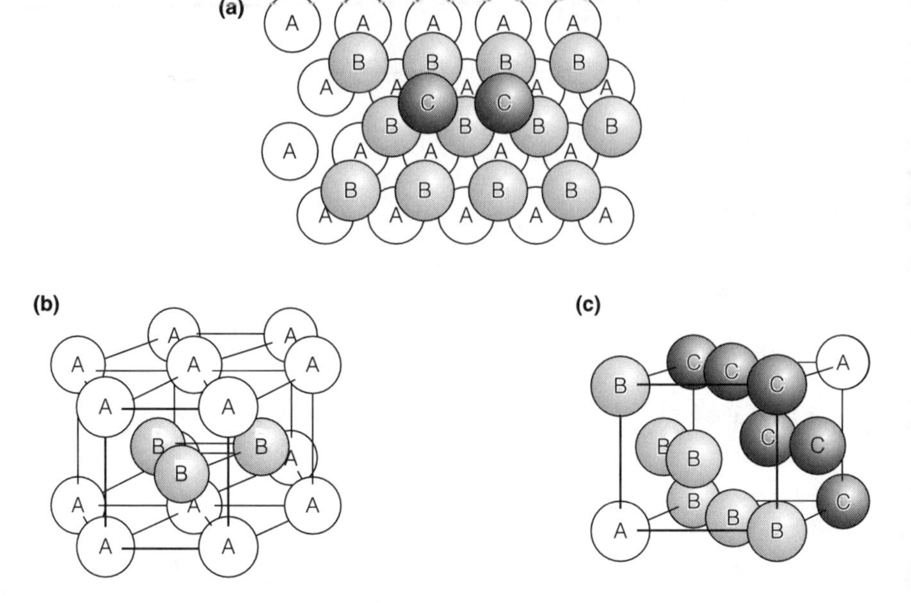

Fig. 1. Close-packed structures. (a) Stacking of layers showing the sequence ABC (see text); (b) the hcp structure; (c) one unit cell of the fcc structure.

In both fcc and hcp structures each sphere is surrounded by 12 others at the same near-neighbor distance. (There are six in the same close-packed layer, and three each in the layers above and below.) If the spheres are in contact both structures give 74% filling of space by the spheres, with the remaining 26% outside them. This is the optimum space filling possible with equal spheres. Similarly close-packed structures can be constructed from more complicated sequences of layers such as ABABCABABC . . . , or even with random sequences. Although these are sometimes found, most close-packed structures are of the simple fcc or bcc types.

Another structure that gives fairly efficient space filling (68% compared with 74% above) is the **body-centered cubic (bcc)** one illustrated in *Fig. 2*. Each atom has eight near-neighbors, but there are six others (also shown in the figure) slightly further away.

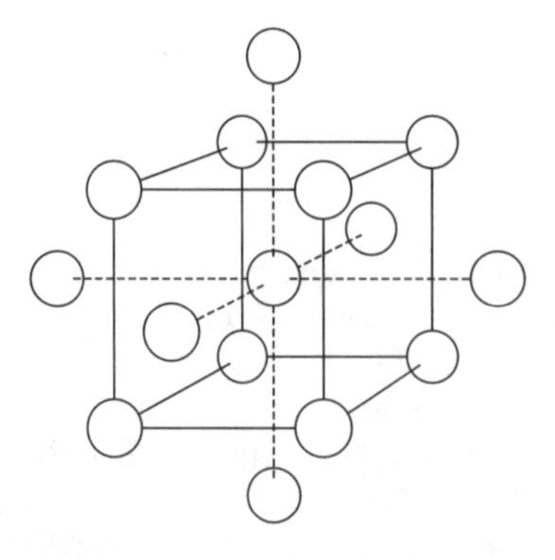

Fig. 2. Bcc structure.

Metallic elements

A high proportion of metallic elements have one of the three structures ccp, hcp or bcc just described. The factors that determine the structure are subtle. In some cases the thermodynamically stable structure depends on temperature and/or pressure, showing that the energy differences between them are small. Nevertheless, some regularities are observed in the periodic table, which suggest that stability depends in a systematic way on the number of valence electrons. The commonest stable structures according to group number are

1: bcc	**2**: varied	**3, 4**: hcp
5, 6: bcc	**7, 8**: hcp	**9–11**: fcc

There are irregularities, however. In the transition metal groups 7, 8 and 9 the *3d* series elements Mn, Fe and Co are exceptions. Some elements also have more complex structures, especially in the *p* block. An understanding of the factors controlling metallic structures requires the band theory of delocalized electrons, not discussed in this book.

Nonmetallic elements

As might be expected from other aspects of its chemistry, boron is exceptional and has elemental structures that cannot be understood in simple bonding terms (see Topic F3). For the remaining nonmetals, the simple concepts of electron-pair bonding and stereochemistry described in Topics C1 and C2 allow the structures to be rationalized although not always predicted. Single-bonded structures where each element achieves an octet lead to the following predictions.

Group 14: four tetrahedral bonds as shown in the diamond structure of C, Si, Ge and Sn, and illustrated in *Fig. 3a*.

Group 15: three bonds in a pyramidal (nonplanar) geometry, which can give rise to P_4 molecules (white phosphorus) or a variety of polymeric structures shown by P and As (see Topic F6). Phosphorus has several allotropes, some with apparently complex structures, but all are based on the same local bonding.

Group 16: two bonds, noncolinear, as found in S_8 rings and in spiral chains with Se and Te (see Topic F8). The different allotropes of sulfur all have this bonding.

Group 17: one bond, giving diatomic molecular structures shown by all the halogens (see Topic F9).

Group 18: no bonds, leading to monatomic structures with atoms held only by van der Waals' forces (see Topics C9 and F10). The normal solid structure of the noble gas elements is fcc.

The structural chemistry of the period 2 elements C, N and O shows a greater tendency to multiple bonding than in lower periods (see Topics C7 and F1). Molecular N_2 (triple bonded) and O_2 (double bonded) are the normal forms of these elements. With carbon, other allotropes in addition to diamond are possible. The thermodynamically stable form at normal pressures is **graphite** (see *Fig. 3b*), where some delocalized π bonding is present along with the three σ bonds formed by each atom. Fullerenes such as C_{60} have similar bonding arrangements (see Topic F4).

Another group trend with *p*-block elements is the increasing tendency towards metallic character in lower periods. As with the chemical trends, the change in structures and properties of the elements appears more of a continuous transition than a sharp borderline (see Topics B2 and D7). The structural distinction between near-neighbor (bonded) atoms and next-near-neighbor (nonbonded) ones

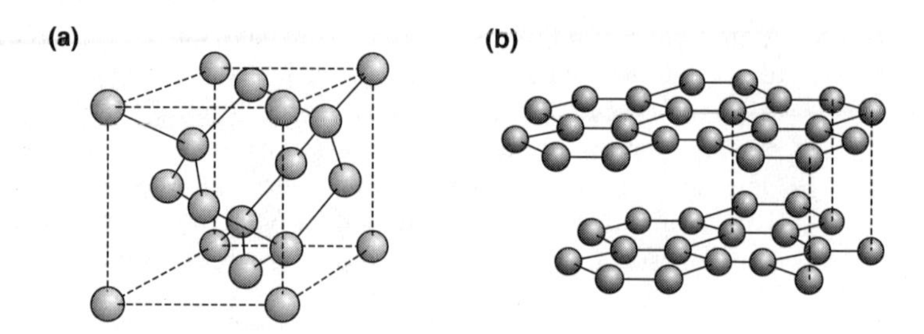

Fig. 3. Structures of (a) diamond and (b) graphite.

becomes less marked down each group. *Table 1* lists the ratio of these distances for some nonmetallic elements of periods 3–5, and shows how the two distances become more nearly equal for heavier elements, especially with Sb and Te, which are close to the metallic borderline. The peculiar structures shown by some *p*-block metals suggests that some influence of directional bonding persists in the metallic state.

Table 1. The ratio of next-near-neighbor to near-neighbor distances in some solid p-block elements

P	1.787	**S**	1.81	**Cl**	1.65
As	1.33	**Se**	1.49	**Br**	1.46
Sb	1.16	**Te**	1.21	**I**	1.33

D3 BINARY COMPOUNDS: SIMPLE STRUCTURES

Key Notes

Coordination number and geometry

The coordination number (CN) and geometry of atoms (or ions) are the most important characteristics of a structure. In regular binary solids the ratio of CN values must reflect the stoichiometry. Both regular and irregular coordination geometries can be found.

Close packing

Many binary structures can be derived from a close-packed array of one element by filling either tetrahedral or octahedral holes between the close-packed layers with atoms of the other kind

Linked polyhedra

An alternative view of binary structures is to consider coordination polyhedra of one element (normally tetrahedra or octahedra), linked together by sharing corners, edges or faces.

Related topics

Element structures (D2)

Binary compounds: factors influencing structure (D4)

Coordination number and geometry

Binary compounds are ones with two elements present. 'Simple' crystal structures may be classed as ones in which each atom (or ion) is surrounded in a regular way by atoms (or ions) of the other kind. Even with this limited scope many structures are possible. *Figure 1* shows a selection of simple ones that exemplify some important principles. Although many are found with ionic compounds, some of these structures are shown by compounds with covalent bonding, and a discussion of the bonding factors involved in favoring one structure rather than another is deferred to Topic D4. *Figure 1* shows the structure name and the stoichiometry (AB, AB_2, etc.). When the two elements A and B are not equivalent A is drawn smaller and with shading. In ionic compounds this is more often the metallic (cationic) element. If the role of anions and cations is reversed we speak of the **anti-structure**: thus Li_2O has the anti-fluorite (CaF_2) structure, and Cs_2O the anti-CdI_2 structure.

From the local point of view of each atom the most important characteristics of a structure are the **coordination number** (CN) and **coordination geometry**. In the examples shown these are the same for all atoms of the same type. Coordination numbers must be compatible with the stoichiometry. In AB both A and B have the same CN, the examples shown being

Zinc blende (4:4); Rocksalt (6:6); NiAs (6:6); CsCl (8:8).

When the stoichiometry is AB_2 the CN of A must be twice that of B:

Rutile (6:3); CdI_2 (6:3); Fluorite (8:4).

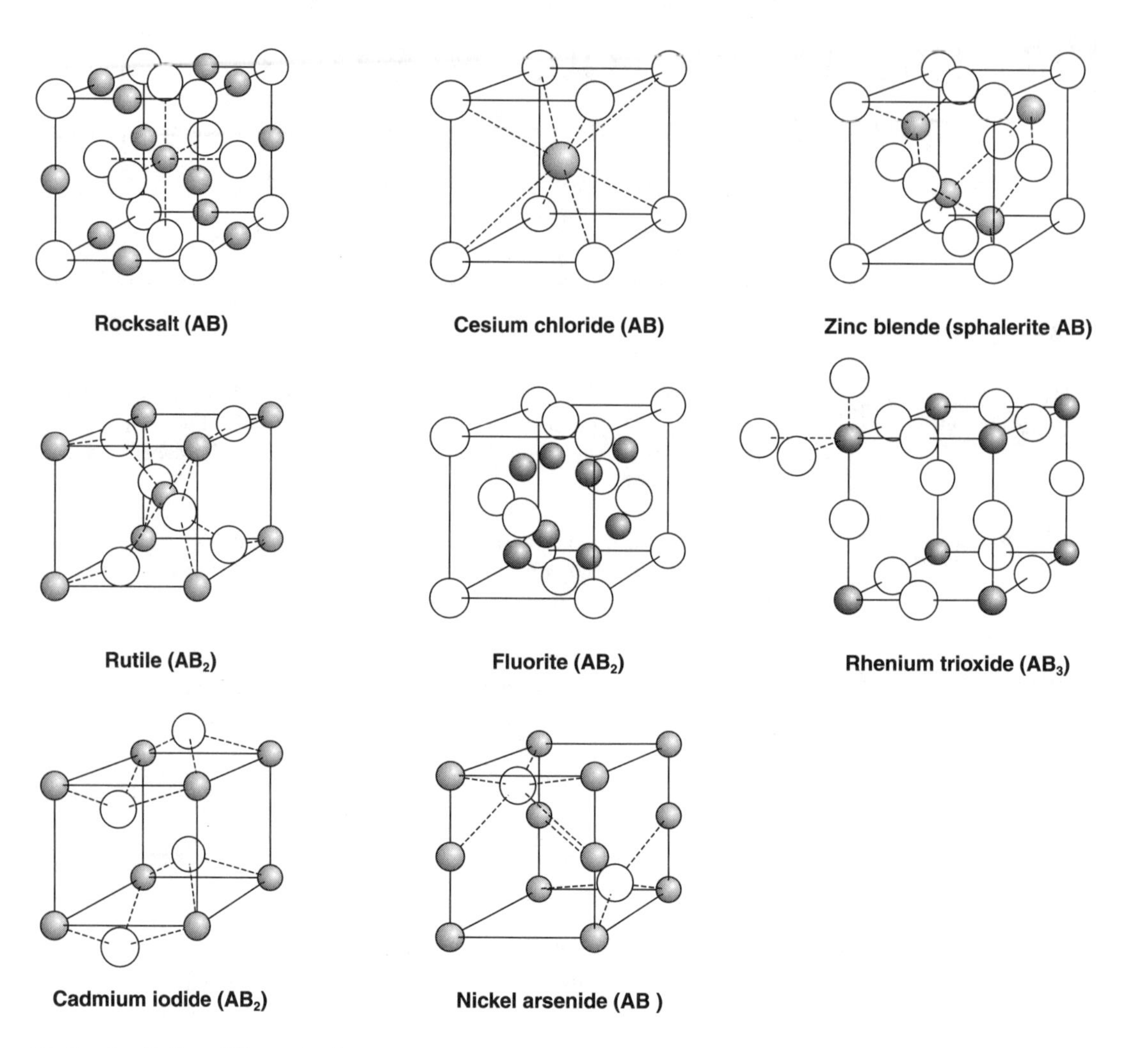

Fig. 1. A selection of binary structures.

In the structures shown many of the atoms have a regular coordination geometry:

CN = 2: linear (B in ReO_3);
CN = 3: planar (B in rutile);
CN = 4: tetrahedral (A and B in zinc blende, B in fluorite);
CN = 6: octahedral (A and B in rocksalt, A in NiAs, rutile and CdI_2);
CN = 8: cubic (A and B in CsCl, A in fluorite).

These geometries are expected in ionic compounds, as they lead to the greatest spacing between ions with the same charge. Other geometries are sometimes found, however, especially for the nonmetal B atom:

CN = 2: bent (SiO_2 structures, not shown);
CN = 3: pyramidal (in CdI_2);
CN = 6: trigonal prismatic (in NiAs).

The explanation of these must involve nonionic factors (see Topic D4).

Close packing

Many binary structures can be derived from close-packed arrays of atoms of one kind (see Topic D2). *Figure 2* shows that between adjacent close-packed layers are **octahedral** and **tetrahedral holes** (labeled O and T) such that atoms of another kind occupying these sites would be octahedrally or tetrahedrally coordinated. For ionic compounds we can imagine the larger ions (usually the anions) forming the close-packed array, and cations occupying some of the holes. In either hexagonal (hcp) or cubic close-packed (ccp or fcc) arrays of B there is **one octahedral** and **two tetrahedral holes** per B atom. *Table 1* shows some binary structures classified in this way. Thus filling all the octahedral holes in a fcc array generates the **rocksalt** structure (in which the original B atoms are also octahedrally coordinated); doing the same in an hcp array gives the **NiAs** structure. Filling all the tetrahedral holes in an fcc anion array gives the **antifluorite** structure, more commonly found with anions and cations reversed as in **fluorite** (CaF_2) itself. A similar arrangement is never found in an hcp array, as the tetrahedral holes occur in pairs that are very close together.

When only a fraction of the holes of a given type are occupied there are several possibilities. The most symmetrical way of filling half the tetrahedral holes gives the **zinc blende** structure with ccp, and the very similar 4:4 **wurtzite** (ZnO) structure with hcp. Both the **rutile** and $\mathbf{CdI_2}$ structures can be derived by filling half the octahedral holes in hcp. The former gives a more regular coordination of the anions (see above) although the resulting structure is no longer hexagonal. The CdI_2 structure arises from alternately occupying every octahedral hole

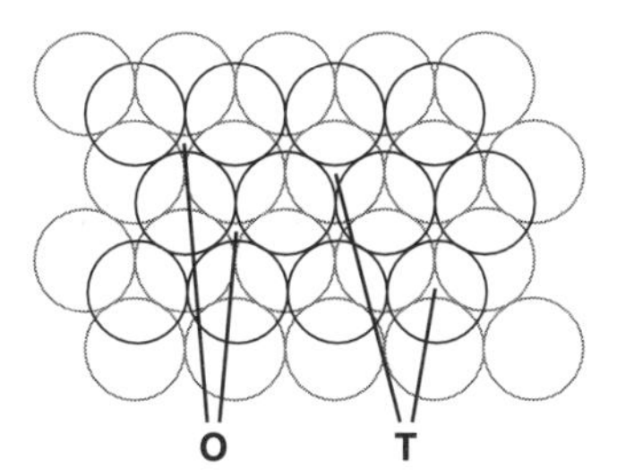

Fig. 2. Octahedral (O) and tetrahedral (T) holes between adjacent close-packed layers.

Table 1. Some binary structures based on close-packed arrays of anions

Array	Holes filled	Structure type	Examples
Fcc	All octahedral	Rocksalt (NaCl)	LiF, MgO
	½ octahedral	Cadmium chloride ($CdCl_2$)[a]	$MgCl_2$
	All tetrahedral	Antifluorite	Li_2O
	½ tetrahedral	Sphalerite (zinc blende)	ZnS, CuCl
	½ tetrahedral	Lead oxide (PbO)[a]	SnO
Hcp	All octahedral	Nickel arsenide (NiAs)	FeS
	½ octahedral	Rutile[b]	MgF_2, TiO_2
	½ octahedral	Cadmium iodide (CdI_2)[a]	TiS_2
	All tetrahedral	*Not found*	–
	½ tetrahedral	Wurtzite	BeO, ZnO

[a] Layer structures.

[b] Filling the holes changes the symmetry; the rutile unit cell is not hexagonal.

between two adjacent close-packed planes, and leaving the next layer of holes empty. It is an example of a **layer structure** based on BAB 'sandwiches' that are stacked with only B–B contacts between them. The $\mathbf{CdCl_2}$ structure is based in a similar way on ccp (rather than hcp) anions, and many other layer structures with formulae such as AB_3 can be formed by only partial filling of the holes between two layers.

Linked polyhedra

An alternative way of analyzing binary structures is to concentrate on the **coordination polyhedra** of one type of atom, and on the way these are linked together. This approach is generally useful in structures with covalent bonding, and/or ones that are more open than those derived from close packing.

If two tetrahedral AB_4 units share one B atom in common (**1**) we talk of **corner sharing**. A corner-shared pair has stoichiometry A_2B_7 and is found in (molecular) Cl_2O_7 and occasionally in silicates. Tetrahedra each sharing corners with two others generate a **chain** or a **ring** (**2**) of stoichiometry AB_3, as found with SO_3 and commonly in silicates (see Topics D5 and G4). These structures are often represented by drawing the tetrahedra without showing the atoms explicitly. Rings and chains with two corners shared are shown in this way (*Fig. 3a* and *b*). Sharing three corners makes a layer or a tetrahedral cluster of stoichiometry A_2O_5; such layers occur in silicates, and the clusters as P_4O_{10} molecules (see Topic F6, Structure 5). Tetrahedra sharing all four corners with others generate a **3D framework** of stoichiometry AB_2, found in the various (crystalline and glassy) structures of SiO_2.

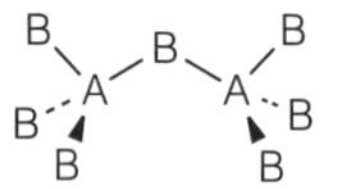

1

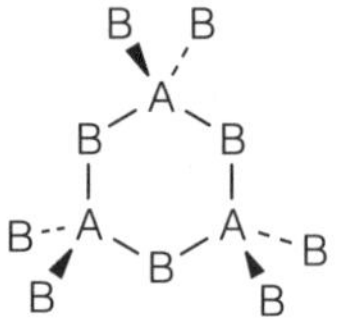

2

Tetrahedra with two B atoms in common are said to be **edge sharing**: examples of isolated edge-sharing pairs are B_2H_6 and Al_2Cl_6 (see Topic C8, Structure 2). A chain of tetrahedra each sharing two edges with others has a stoichiometry AB_2 and is found as the **chain structures** of BeH_2 and SiS_2, shown in *Fig. 3c* and in Topic G3, Structure 3. Face sharing is also possible but is almost never found with tetrahedra as the A atoms would be very close together.

Similar ideas can be used with octahedra. Chains of corner-sharing octahedra are found in $WOBr_4$ and of edge-sharing octahedra in NbI_4. If octahedra share all six corners, the 3D ReO_3 structure results (see *Fig. 3d*; compare *Fig. 1*).

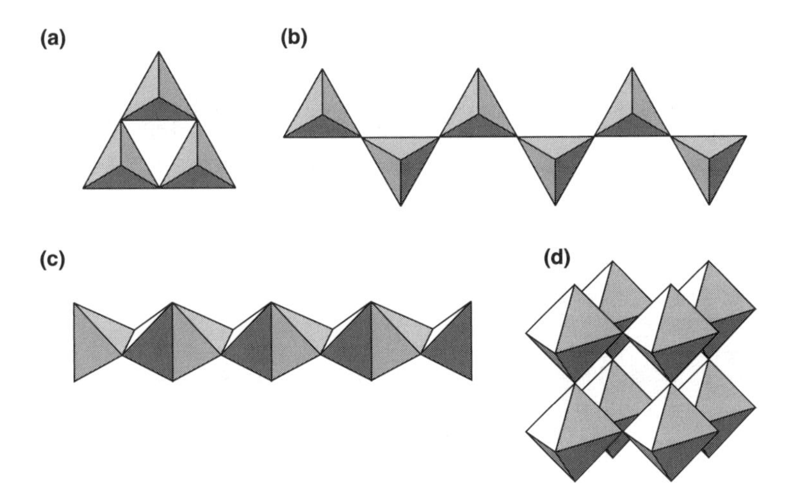

Fig. 3. Structures derived from linking of polyhedra (see text).

D4 BINARY COMPOUNDS: FACTORS INFLUENCING STRUCTURE

Key Notes

Ionic radii

Ionic radii are derived from a somewhat arbitrary division of the observed anion-cation distances. Different assumptions lead to different values, but all sets show similar trends. Ionic radii vary with coordination number.

Radius ratios

Simple geometrical arguments based on hard-sphere ions give predictions of structure according to the ratio of ionic radii. These are qualitatively useful but not quantitatively reliable.

Ion polarizability

The electrostatic polarizability of ions increases with its radius and may be partly responsible for the adoption of structures where coordination geometries are unsymmetrical, and of structures with high coordination numbers.

Covalent bonding

When covalent bonding predominates the coordination numbers and geometries are often those expected by analogy with molecules. Some degree of covalency in 'ionic' compounds can influence the structure, often leading to coordination numbers less than expected.

Related topics

Electronegativity and bond type (B1)

Binary compounds: simple structures (D3)

Ionic radii

The experimentally measured anion-cation distances in highly ionic solids can be interpreted on the assumption that each ion has a nearly fixed radius. For example, the difference in anion-cation distance between the halides NaX and KX is close to 36 pm irrespective of the anion X, and it is natural to attribute this to the difference in radii between Na^+ and K^+. To separate the observed distances into the sum of two **ionic radii** is, however, difficult to do in an entirely satisfactory way. One procedure is to look for the minimum value in the electron density distribution between neighboring ions, but apart from the experimental difficulties involved such measurements do not really support the assumption of constant radius. Sets of ionic radii are therefore all based ultimately on somewhat arbitrary assumptions. Several different sets have been derived, the most widely used being those of Shannon and Prewitt, based on the assumed radius of 140 pm for O^{2-} in six-coordination. Values for a selection of ions are shown in *Table 1*.

Any consistent set of radii should be able to give estimates of the total anion-cation distance and hence the unit cell dimensions if the structure is known. It is essential not to mix values from different sets. Although the values may differ, all sets show the same trends.

Table 1. Ionic radii (pm) for six-coordination, based on a value of 140 pm for O^{2-}

								H^-	146
Li^+	76	Be^{2+}	27			O^{2-}	140	F^-	133
Na^+	102	Mg^{2+}	72	Al^{3+}	53	S^{2-}	182	Cl^-	167
K^+	138	Ca^{2+}	100	Ga^{3+}	62			Br^-	182
Rb^+	149	Sr^{2+}	116					I^-	206
Cs^+	167	Ba^{2+}	149						

(i) For isoelectronic ions, radii decrease with increasing positive charge (e.g. $Na^+ > Mg^{2+} > Al^{3+}$) or decreasing negative charge (e.g. $O^{2-} > F^-$).
(ii) Radii increase down each group (e.g. $Li^+ < Na^+ <$ etc.).
(iii) For elements with variable oxidation state (not shown in *Table 1*) radius decreases with increasing positive charge (e.g. $Fe^{2+} > Fe^{3+}$).
(iv) Most anions are larger than most cations.
(v) Ionic radii **increase with coordination number** (CN). For example, the Shannon and Prewitt radii (in pm) for K^+ with different CN (shown in parenthesis) are: 138 (6); 151 (8); 159 (10); 160 (12).

Trends (i)–(iv) follow the changes expected in the radii of atomic orbitals (see Topic A5). The variation with CN is, however, very important and shows that ions cannot be regarded as hard spheres but have an effective size depending on their environment. This is expected because the equilibrium distance between ions involves a balance of attractive and repulsive forces. Repulsive forces come from the overlap of closed shells and their net importance increases in proportion to the CN. Attractive forces are electrostatic and depend on the long-range summation of the interactions between many ions (see Topic D6). Although they increase with CN the change is much less than for short-range repulsion. Increasing the CN therefore changes the balance in favor of repulsive forces and leads to an increase in distance.

Radius ratios

An ionic solid should achieve maximum electrostatic stability when (i) each ion is surrounded by as many as possible ions of opposite charge, and (ii) the anion-cation distance is as short as possible. There is, however, a play-off between these two factors. Consider an octahedral hole in a close-packed array of anions (see Topic D3). The minimum radius of the hole, obtained when the anions are in contact, is 0.414 times the anion radius. A cation smaller than this will not be able achieve the minimum possible anion-cation distance in octahedral coordination, and a structure with lower coordination (e.g. tetrahedral) may be preferred. These considerations lead to the **radius ratio rules**, which predict the likely CN for the smaller ion (usually the cation) in terms of the ratio $r_</r_>$ where $r_<$ is the smaller and $r_>$ the larger of the two radii. The approximate radius ratios for different CN are:

$r_</r_>$	> 0.7	0.4–0.7	0.2–0.4
CN	8	6	4

The rules provide a useful qualitative guide to the way structures change with the size of ions. For example, the radius ratios and the observed CN of the metal ions M^{2+} in some group 2 fluorides are:

BeF_2:	$r_</r_> = 0.20$	CN = 4
MgF_2:	$r_</r_> = 0.54$	CN = 6
CaF_2:	$r_</r_> = 0.75$	CN = 8

However, radius ratio arguments are not quantitatively reliable, and they even fail to account for the structures of some alkali halides. The predicted coordination number is four in LiI and eight in RbCl, although both compounds have the rocksalt structure (CN = 6) at normal temperature and pressure.

The fact that radius ratio arguments do not always predict the correct structure is sometimes regarded as a serious failure of the ionic model, and an indication that nonionic forces must be involved in bonding. Given the uncertainties in definition of ionic radii, however, and the fact that they are known to vary with CN, it is hardly surprising that predictions based on the assumption of hard spheres are unreliable. It also appears that for some compounds the difference in energy between different structure types is very small, and the observed structure may change with temperature or pressure.

Ion polarizability

The **polarizability** of an ion refers to the ability of an applied electric field to distort the electron cloud and so induce an electric dipole moment. The most polarizable ions are large ones, especially anions from later periods (e.g. S^{2-}, Br^{-}, I^{-}). 'Polarization' is a term often used loosely as meaning 'covalency' but the purely electrostatic polarizability of ions has effects that are entirely separate. In layer and chain structures (see Topic D3) anions are generally in asymmetric environments and experience a strong net electric field from neighboring ions. Polarization lowers the energy of an ion in this situation, giving a stabilizing effect not possible when the coordination is symmetrical. It is notable that layer structures occur frequently with disulfides and dichlorides (and with heavier anions lower in the same groups), but almost never with dioxides and difluorides: compare TiO_2 and FeF_2 (both rutile structure) with TiS_2 and FeI_2 (both CdI_2 structure). Cs_2O is a rare example of the anti-CdI_2 structure, with adjacent layers of Cs^{+}; the high polarizability of the Cs^{+} ion must be a contributing factor.

Another consequence of polarizability is the existence of **van der Waals' forces** between ions (see Topic C9). They are considerably weaker than ionic forces but can have an influence on structures, especially with large ions of high polarizability. Being short-ranged (varying with distance R as R^{-6}) compared with Coulomb energies (R^{-1}) they favor the maximum number of near-neighbors, irrespective of charge. It is very likely that the occurrence of the 8 : 8 CsCl structure with cesium halides (except CsF) is influenced by this effect. Van der Waals' forces between adjacent anions are also responsible for holding together layers and chains, which is another reason why ion polarizability is important for such structures.

Covalent bonding

Purely electrostatic forces between ions are nondirectional, but with increasing covalent character the directional properties of valence orbitals become more important. Compounds between nonmetallic elements have predominantly covalent bonding and the structures can often be rationalized from the expected CN and bonding geometry of the atoms present (see Topics C2 and C5). Thus in SiC both elements have tetrahedral coordination; in SiO_2 silicon also forms four tetrahedral bonds, and oxygen two bonds with a nonlinear geometry.

Compounds of less electropositive metals also show structural effects that can be attributed to partial covalent bonding. CuCl and ZnO have structures with tetrahedral coordination although from radii the (octahedral) rocksalt structure would seem more likely. Partial covalent bonding involves some transfer of electrons back from the anions, into the empty 4*s* and 4*p* orbitals on Cu^{+} and Zn^{2+}. Tetrahedral coordination is the normal bonding geometry when a complete set of *s* and *p* orbitals is used in this way (see Topics C2 and C5). Mercury forms an extreme example of the lower coordination numbers often found with post-transition metals: Hg^{II} compounds are of generally low ionic character, and two-coordination is common (see Topic G4).

Covalent bonding effects sometimes dictate less regular structures than those shown by Cu^{+} and Zn^{2+}. Specific *d* electron effects operate in compounds such as CuO and PdO (see Topics H4 and H5), and some post-transition metal compounds such as SnO and PbO apparently show the structural influence of nonbonding electron pairs on the cation (see Topic G6).

Covalent bonding interactions can also occur between atoms of the same element. Section D5 describes some structures that can arise in this way. Here it is worthwhile noting that the NiAs structure (see Topic D3), never expected for purely ionic compounds because cations are closer together than in the rocksalt structure, is often found with transition metals in combination with less electronegative nonmetals such as S, P and As. The compounds formed are of low ionic character and frequently show metallic conduction. The close contacts between metal atoms facilitate direct bonding interaction.

D5 MORE COMPLEX SOLIDS

Key Notes

Homoelement bonding

Binary solids with bonds between atoms of the same type include compounds with ions such as O_2^{2-}, Zintl compounds formed between electropositive metals and *p*-block elements of period 3 and below, and compounds with metal-metal bonding often formed by 4*d* and 5*d* transition metals.

Ternary structures

Some ternary oxides and halides may have discrete complex ions such as CO_3^{2-}, others have structures with no such discrete ions. Silicates show a range of intermediate possibilities. The compound formula alone does not indicate the structure type.

Microporous solids

Zeolites are solids with aluminosilicate frameworks having pores and channels. When these are occupied by hydrated ions the compounds are used as ion exchangers; when the pores are empty they have useful catalytic properties.

Intercalation and insertion compounds

Intercalation compounds are formed from layered structures with additional atoms or molecules between the layers, insertion compounds when atoms enter a three-dimensional framework. Many of these compounds are nonstoichiometric.

Related topics

Binary compounds: simple structures (D3)

Oxygen (F7)

Homoelement bonding

Bonding between atoms of the same kind may often be present when a binary compound shows an apparently anomalous stoichiometry. For example, the solids with empirical formulae NaO, KO_2, LiS, CaC_2 and NaN_3 contain the ions O_2^{2-}, O_2^-, S_2^{2-}, C_2^{2-} and N_3^-, respectively. Combination of an electropositive metal with a *p*-block element of intermediate electronegativity gives so-called **Zintl compounds**. Some contain discrete polyatomic units such as Ge_4 tetrahedra in KGe; in others there are continuous bonded networks such as Si chains in CaSi, or layers in $CaSi_2$. Often these structures can be understood by isoelectronic analogy with the non-metallic elements (see Topic D2): thus Ge_4^{4-} (in KGe) has the same valence electron count as P_4; Si^{2-} (in CaSi) is similarly isoelectronic to S, and Si^- (in $CaSi_2$) to P. Although this analogy is useful the ionic formulation may be misleading, as the solids are often metallic in appearance and are semiconductors.

The term **metal-metal bonding** is used when such homoelement bonding involves the more electropositive element of a binary pair. Again, it may sometimes be present when an unusual oxidation state is found. For example, HgCl contains molecular Hg_2Cl_2 units with Hg–Hg bonds, and GaS also has Ga–Ga bonds (see Topics G4 and G5). **Metal-rich compounds** are formed by early transition metals, with formulae such as Sc_2Cl_3 and ZrCl, and structures showing extensive metal-metal bonding. They are especially common with elements of the 4*d* and 5*d* series and sometimes may not be suspected from the stoichiometry. An example is $MoCl_2$, which contains the cluster $[Mo_6Cl_8]^{4+}$ with a metal-metal bonded Mo octahedron (see Topic H5). Metal-metal bonding often gives rise to anomalous magnetic or other properties, but the surest criterion is a structural one, with metal-metal distances comparable with or shorter than those found in the metallic element.

Ternary structures

Ternary structures are ones with three elements present, examples being $CaCO_3$ and $CaTiO_3$. Oxides are the commonest examples of such structures and exemplify some of the important principles (see Topic F7). Two fundamentally different structural features are possible, as follows.

- **Complex oxides** are compounds containing complex ions, which appear as discrete structural units. For example, calcium carbonate has a structure based on rocksalt with the different sites occupied by Ca^{2+} and CO_3^{2-} ions.
- **Mixed oxides** are exemplified by $CaTiO_3$, which, although often called 'calcium titanate', does **not** have discrete titanate ions. The **perovskite** structure (*Fig. 1*) shows a corner-sharing network of TiO_6 octahedra (essentially the ReO_3 structure; see Topic D3, *Figs 1* and *3*) with Ca^{2+} occupying the large central site coordinated by 12 oxygen ions.

This division is not absolute, however, and the varied structures of **silicates** provide examples of intermediate cases. $ZrSiO_4$ (zircon) has discrete ions SiO_4^{4-}, but silicates such as $CaSiO_3$ do **not** contain individual SiO_3^{2-} units but are formed from tetrahedral SiO_4 groups sharing corners to make rings or infinite chains (see Topic D3, *Fig. 3*). Further sharing of corners can make two- and three-dimensional networks. The different structures of carbonates and silicates reflect some typical and very important differences in bonding preference between periods 2 and 3 in the *p* block (see Topics F1 and F4).

Complex oxides are normally found when a nonmetal is present, with oxoanions such as nitrate NO_3^-, carbonate CO_3^{2-}, phosphate PO_4^{3-} or sulfate SO_4^{2-}, but are also sometimes formed by metals in high oxidation states (e.g. permanganate MnO_4^- in $KMnO_4$). When a compound contains two metallic elements the mixed oxide form is more normal, but it is important to note that the compound formula itself provides very little guide to the structure (compare $CaCO_3$ and $CaSiO_3$ above). A similar structural variety is found with complex halides. For example, the K_2NiF_4 structure is based on layers of corner-sharing NiF_6 octahedra with no discrete complex ions, whereas K_2PtCl_4 contains individual square planar ions $[PtCl_4]^{2-}$. These differences reflect the bonding preferences of Ni^{II} and Pt^{II} (see Topics H4 and H5).

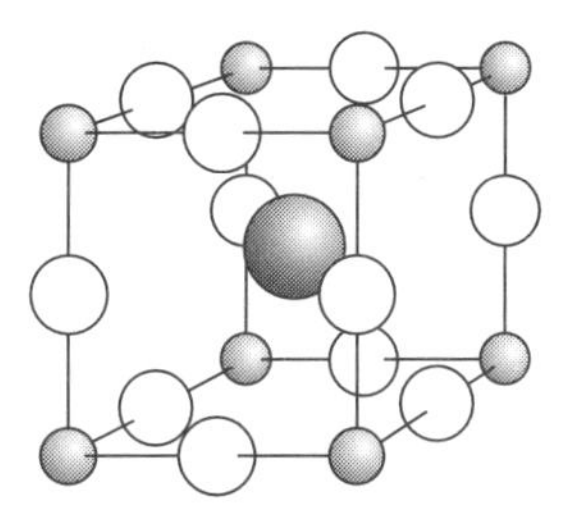

Fig. 1. Unit cell of the perovskite structure of $CaTiO_3$.

Microporous solids

Zeolites are **aluminosilicate** solids based on a framework of corner-sharing SiO_4 and AlO_4 tetrahedra. These frameworks contain pores and channels of molecular dimensions, which in natural minerals (or after laboratory synthesis) contain species such as water and hydrated ions. Removal of these species (e.g. by careful heating under vacuum) leads to **microporous** materials with empty channels and pores. It is possible to make synthetic zeolites of composition SiO_2 with no aluminum, but when Al^{III} is present the framework formula is $[Al_xSi_{1-x}O_2]^{x-}$ and the charge must be compensated by extra-framework cations. In as-prepared zeolites these may be alkali cations, NH_4^+ or organic amines, but when the pore materials are removed they are replaced by H^+, which forms strong **Brønsted acid sites** within the pores.

The structure of the zeolite **faujasite** is shown in *Figure 2*. In this conventional representation the framework structure is shown without depicting atoms directly. Each line represents an Si—O—Si or Si—O—Al connection. Four lines meet at tetrahedral vertices representing the positions of the four-coordinate Si or Al atoms. Space-filling models of this zeolite show that the pores can accommodate molecules up to about 750 pm in diameter.

In their hydrated forms zeolites are used for **ion exchange** purposes, for example, water softening by replacement of Ca^{2+} with Na^+ or another ion (see Topic J4). When dehydrated they have important catalytic applications, promoted by the Brønsted acid sites, and by the large area of 'internal surface'. They are used for the cracking of petroleum and for the isomerization of hydrocarbons, where limited pore size exerts a 'shape selectivity', which allows one desirable product to be formed in high yield (see Topic J5).

Intercalation and insertion compounds

Alkali metals and bromine react with graphite to form solids known as **intercalation compounds**, where the foreign atoms are inserted between the intact graphite layers (see Topic D2). Many other layered solids, for example dichalcogenides such as TaS_2, which have structures similar to CdI_2 (Topic D3), will also form intercalation compounds. The inserted species may be alkali metals, or electron donor molecules such as amines or organometallic compounds. Sometimes compounds of definite composition may be formed, such as KC_6 or C_8Br, but in other cases intercalated phases may be nonstoichiometric, such as Li_xTiS_2 ($0 < x < 1$). Most intercalation reactions involve electron transfer between the guest and the host, and modify the electronic properties.

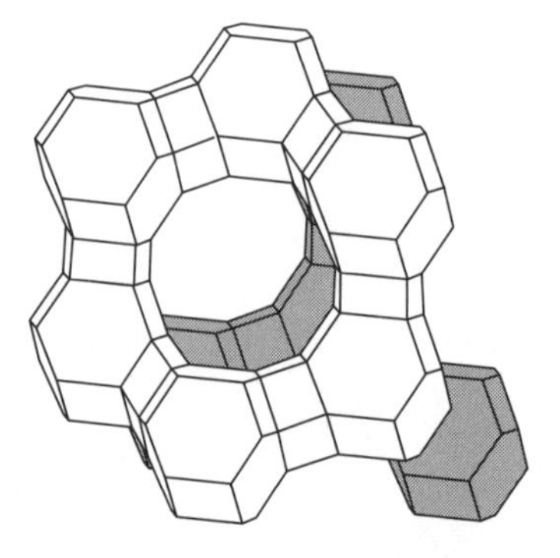

Fig. 2. Representation of the structure of faujasite (see text).

The term **insertion compound** is used for solids where atoms or ions enter a three-dimensional framework without disrupting its essential structure. Many **oxide bronzes** are of this type, based on transition metal oxides with inserted alkali or other electropositive metals. For example, the **sodium tungsten bronzes** are of composition Na_xWO_3, where x can range from zero up to about 0.9. Their structures are based on the ReO_3 framework (see Topic D3) with Na occupying the large vacant site. The structure therefore resembles that of perovskite (*Fig. 1*) except that the site occupied by Ca in $CaTiO_3$ is only partially occupied in Na_xWO_3. As with intercalation, electron transfer is also involved, and Na_xWO_3 has a metallic appearance and good electronic conductivity whereas pure WO_3 is a pale yellow insulator (see Topic D7).

D6 LATTICE ENERGIES

Key Notes

The Born-Haber cycle

The lattice energy of an ionic compound is the energy required to separate the solid into gas-phase ions. It may be estimated using Hess' Law from a sequence of steps known as a Born-Haber cycle.

Theoretical estimates

Theoretical estimates of lattice energies using the Born-Landé or Born-Mayer equations agree well with Born-Haber values for many compounds. The Kapustinskii equation gives a useful approximate estimate. Both experimental and theoretical lattice energies increase as ions become smaller or more highly charged.

Applications

Lattice energies may be used to understand many important chemical trends, including the characteristic oxidation states of metallic elements, the stabilization of high oxidation states by oxide and fluoride, and trends in the thermal stability of oxoanion salts such as carbonates.

Related topics

Stability and reactivity (B3)   Solubility of ionic substances (E4)

The Born-Haber cycle

The lattice energy U_L of a solid compound is defined as the energy required to transform it into gas-phase ions, for example,

$$NaCl(s) \rightarrow Na^+(g) + Cl^-(g)$$

(**Note**: sometimes the reverse process is used as a definition, which makes U_L a negative quantity rather than positive as here.) It is generally assumed that the compound concerned is ionic, but a lattice energy can be defined without that assumption, provided the ions formed in the gas phase are clearly specified.

Lattice energies may be estimated from a thermodynamic cycle known as a **Born-Haber cycle**, which makes use of Hess' Law (see Topic B3). Strictly speaking, the quantities involved are enthalpy rather than energy changes and one should write H_L for the **lattice enthalpy**. From *Fig. 1*, which shows a cycle for NaCl, we see that

$$H_L(NaCl) = -\Delta H_f(NaCl) + \Delta H_{at}(Na) + \tfrac{1}{2}B(Cl_2) + I(Na) - A(Cl)$$

where the terms on the right-hand side are, in order: the enthalpy of formation of NaCl, the enthalpy of atomization of Na solid, the bond enthalpy of Cl_2, the ionization energy of Na and the electron affinity of Cl (see Topics A5 and C7). When multiply charged ions are involved the cycle can be adapted by summing higher ionization energies or electron affinities as appropriate.

$I(Na)$ is greater than $A(Cl)$ in the equation above. This shows that in the gas phase, Na and Cl atoms are more stable than the ions Na^+ and Cl^-, and it is the lattice energy that stabilizes the ionic charge distribution in solid NaCl. A similar result is found for all ionic solids.

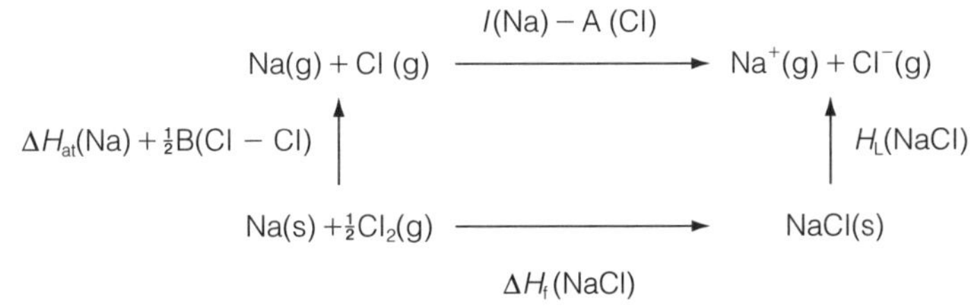

$$
\begin{array}{ccc}
 & I(\mathrm{Na}) - A(\mathrm{Cl}) & \\
\mathrm{Na(g)} + \mathrm{Cl(g)} & \longrightarrow & \mathrm{Na^+(g)} + \mathrm{Cl^-(g)} \\
\Delta H_{at}(\mathrm{Na}) + \tfrac{1}{2}B(\mathrm{Cl-Cl}) \uparrow & & \uparrow H_L(\mathrm{NaCl}) \\
\mathrm{Na(s)} + \tfrac{1}{2}\mathrm{Cl_2(g)} & \longrightarrow & \mathrm{NaCl(s)} \\
 & \Delta H_f(\mathrm{NaCl}) &
\end{array}
$$

Fig. 1. Born-Haber cycle for determining the lattice enthalpy of NaCl.

Theoretical estimates

Theoretical lattice energies can be calculated if some interaction potential between ions is specified. The most important term in the ionic model is the long-range **Coulomb interaction** between charges. A complex summation is necessary over the different pairs of unlike and like charges appearing at different distances in the crystal structure, and gives the Coulomb energy per mole of lattice as

$$U_C = \frac{N_0 A z_+ z_- e^2}{4\pi\varepsilon_0 r_0}$$

where N_0 is Avogadro's constant, z_+e and z_-e are the charges on the ions and r_0 is the distance between them, and A is the **Madelung constant** coming from the long-range summation of ionic interactions. A depends on the structure, and increases slowly with the coordination number. (For example, values for the simple AB structures discussed in Topics D3 and D4 are: zinc blende (CN = 4) 1.638; rocksalt (CN = 6) 1.748; CsCl (CN = 8) 1.763.)

The attractive Coulomb energy needs to be balanced against the contribution from the short-range repulsive forces that occur between ions when their closed shells overlap. There is no accurate simple expression for this repulsion. In the **Born-Landé model** it is assumed proportional to $1/r^n$, where n is a constant that varies in the range 7–12 depending on the ions. The resulting expression for the lattice energy is

$$U_L = \frac{N_0 A z_+ z_- e^2}{4\pi\varepsilon_0 r_0}\left(1 - \frac{1}{n}\right)$$

The **Born-Mayer** equation is an alternative (and possibly more accurate) form based on the assumption of an exponential form for the repulsive energy. Both equations predict lattice energies for compounds such as alkali halides that are in reasonably close agreement with the 'experimental' values from the Born-Haber cycle. Some examples are shown in *Table 1*. A strict comparison requires some corrections. Born-Haber values are generally enthalpies, not total energies, and are estimated from data normally measured at 298 K not absolute zero; further corrections can be made, for example, including van der Waals' forces between ions.

Table 1. Comparison of lattice energies (all kJ mol^{-1}) determined by different methods

Compound	Born-Haber cycle	Born-Landé	Extended	Kapustinskii
NaCl	772	757	770	765
CsCl	652	623	636	617
CaF_2	2597	2594	2610	2779
AgCl	902	–	833	732

When these extended calculations are compared with experiment many compounds agree well (see *Table 1*). Significant deviations do occur, however; for example, in compounds of metals in later groups where bonding is certainly less ionic (e.g. AgCl).

One of the disadvantages of the fully theoretical approach is that it is necessary to know the crystal structure and the interionic distances to estimate the lattice energy. The **Kapustinskii equations** overcome this limitation by making some assumptions. The Madelung constant A and the repulsive parameter n are put equal to average values, and it is also assumed that the interionic distance can be estimated as the sum of anion and cation radii r_+ and r_- (see Topic D4). The simpler of the Kapustinskii equations for a binary solid is

$$U_L = \frac{C \nu z_+ z_-}{r_+ + r_-} \quad (1)$$

ν is the number of ions in the formula unit (e.g. two for NaCl, three for MgF_2 and five for Al_2O_3) and C is a constant equal to 1.079×10^5 when U_L is in kJ mol^{-1} and the radii are in pm. The Kapustinskii equation is useful for rough calculations or where the crystal structure is unknown. It emphasizes two essential features of lattice energies, which are true even when the bonding is not fully ionic:

- lattice energies increase strongly with increasing charge on the ions;
- lattice energies are always larger for smaller ions.

Calculations can be extended to complex ions such as carbonate and sulfate by the use of **thermochemical radii**, chosen to give the best match between experimental lattice energies and those estimated by the Kapustinskii equation.

Applications

Even though ionic model calculations do not always give accurate predictions of lattice energies (and especially when the approximate Kapustinskii equation is used) the trends predicted are usually reliable and can be used to rationalize many observations in inorganic chemistry.

(i) Group oxidation states

The occurrence of ions such as Na^+, Mg^{2+} and Al^{3+} depends on the balance between the energies required to form them in the gas phase and the lattice energies that stabilize them in solids. Consider magnesium. The gas-phase ionization energy (IE) required to form Mg^{2+} is considerably greater than for Mg^+. However, the lattice energy stabilizing the ionic structure MgF_2 is much larger than that of MgF, and amply compensates for the extra IE. It is possible to estimate the lattice energy of MgF, and (depending on what assumptions are used about the ionic radius of Mg^+) its formation from the elements may be exothermic. However, the enthalpy of formation of MgF_2 is predicted to be much more negative, and the reason why MgF(s) is unknown is that it spontaneously disproportionates:

$$2MgF(s) \rightarrow Mg(s) + MgF_2(s)$$

Ionization beyond the closed-shell configuration Mg^{2+} involves the removal of a much more tightly bound 2*p* electron (see Topics A4 and A5). The third IE is therefore very large and can never be compensated by the extra lattice energy of a Mg^{3+} compound.

(ii) Stabilization of high and low oxidation states

When an element has variable oxidation states, it is often found that the highest value is obtained with oxide and/or fluoride (see, e.g. Topic H4). The ionic model again suggests that a balance between IE and lattice energy is important. Small and/or highly charged ions provide the highest lattice energies according to Equation 1, and the increase in lattice energy with higher oxidation state is more likely to compensate for the high IE.

By contrast, a large ion with low charge such as I^- is more likely to stabilize a low oxidation state, as the smaller lattice energy may no longer compensate for high IE input. Thus CuF is not known but the other halides CuX are. Presumably the lattice energy increase from CuF to CuF_2 is sufficient to force a disproportionation like that of MgF but this is not so with larger halide ions. By contrast, CuX_2 is stable with X =F, Cl and Br, but not I.

(iii) Stabilization of large anions or cations

It is a useful rule that **large cations stabilize large anions**. Oxoanion salts such as carbonates are harder to decompose thermally when combined with large cations. It is also found that solids where both ions are large are generally less soluble in water than ones with a large ion and a small one. These trends are sometimes erroneously ascribed to 'lattice packing' effects, with the implication that two large ions together have a larger lattice energy than a large and a small ion. Theoretical (and experimental) estimates of lattice energies contradict this view, and a satisfactory explanation depends on a balance of energies (see also Topic E4).

Consider the decomposition of a group 2 metal carbonate MCO_3:

$$MCO_3(s) \rightarrow MO(s) + CO_2(g)$$

Figure 2 shows a thermochemical cycle, which predicts that the enthalpy change in this reaction is

$$\Delta H = X + H_L(MCO_3) - H_L(MO) \quad (2)$$

where X is enthalpy input required for the gas-phase decomposition of CO_3^{2-}, and H_L are the lattice enthalpies. X is positive, but according to Equation 1 the lattice energy of MO will always be larger than that of MCO_3 because the oxide ion is smaller. The difference of lattice energies in Equation 2 therefore gives a negative contribution to the overall ΔH. If we have a larger M^{2+} ion, **both** lattice energies become smaller, but the important thing is that their **difference** becomes smaller. Thus larger M^{2+} gives a more endothermic decomposition reaction, which therefore requires a higher temperature to accomplish.

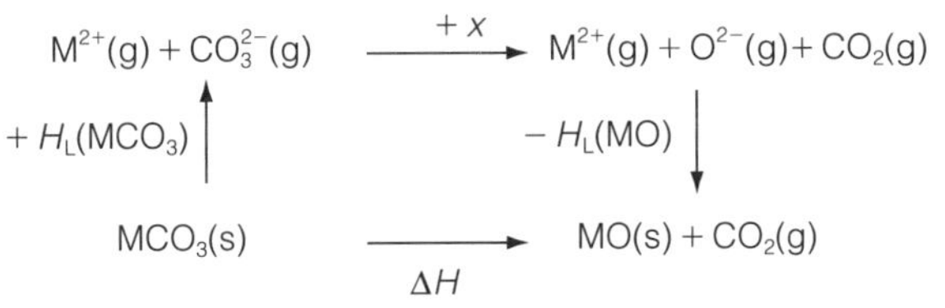

$M^{2+}(g) + CO_3^{2-}(g)$ —— $+X$ ——→ $M^{2+}(g) + O^{2-}(g) + CO_2(g)$

$+H_L(MCO_3)$ ↑ $-H_L(MO)$ ↓

$MCO_3(s)$ —— ΔH ——→ $MO(s) + CO_2(g)$

Fig. 2. Thermochemical cycle for the decomposition of MCO_3.

D7 ELECTRICAL AND OPTICAL PROPERTIES OF SOLIDS

Key Notes

The band model

Metallic solids have a continuous band of electronic energy levels with the top filled level, the Fermi level, within it. In nonmetallic solids there is a bandgap separating the filled valence band from the empty conduction band.

Bandgaps

Bandgaps determine the optical absorption of a nonmetallic solid and the possibility of semiconduction. Bandgaps in binary solids decrease with decreasing electronegativity difference between the elements. In both ionic and covalent solids bandgaps are smaller with elements in lower periods.

Dielectric properties

The static dielectric constant of a solid arises from the displacement of ions in an electric field and may be particularly large for some ionic solids. The high-frequency dielectric constant depends on electronic polarizability and determines the optical refractive index.

Influence of defects

Defects including impurities have a major influence on the electrical properties of nonmetallic solids. They can provide extra electrons or holes, which enhance semiconduction, and they can also facilitate conduction by ions.

Related topic

Element structures (D2)

The band model

The **band model** of solids is an extension of the molecular orbital (MO) method (see Topics C3–C6). The overlap of atomic orbitals in an extended solid gives rise to continuous bands of electronic energy levels associated with different degrees of bonding. In a simple monatomic solid the bottom of the band is made up of orbitals bonding between all neighboring atoms; orbitals at the top of the band are antibonding, and levels in the middle have an intermediate bonding character. Different atomic orbitals can, in principle, give rise to different bands, although they may overlap in energy.

The fundamental distinction between **metallic** and **nonmetallic** solids arises from the way in which orbitals are filled (see *Fig. 1*). Metallic behavior results from a band partially occupied by electrons, so that there is no energy gap between the top filled level (known as the **Fermi level**) and the lowest empty one. On the other hand, a nonmetallic solid has a **bandgap** between a completely filled band (the **valence band** VB) and a completely empty one (the **conduction band** CB). In a filled band the motion of any electron is matched by another one moving in the opposite direction, so that there is no net motion of electric charge. For conduction to occur in a nonmetallic solid, therefore, some electrons must be excited from the VB to the CB. This gives rise to an **activation energy**, and conductivity increases

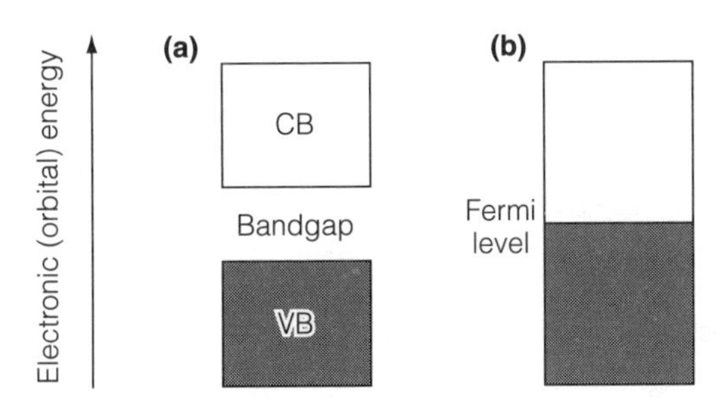

Fig. 1. Band picture for (a) nonmetallic and (b) metallic solid; occupied electronic levels are shown shaded.

with rise in temperature approximately in accordance with the **Arrhenius equation** used in chemical kinetics (see Topic B3).

Nonmetallic solids include ionic and covalent compounds. In the former case, the VB is made up of the top filled anion levels (e.g. the 3*p* orbitals of Cl^-, which are filled in making the ion) and the CB of the lowest empty cation levels (e.g. in Na^+ the 3*s* level from which an electron has been removed to make the cation). In covalent solids such as diamond the VB consists of bonding orbitals (e.g. C–C) and the CB of antibonding orbitals.

Simple metallic solids are elements or alloys with close-packed structures where the large number of interatomic overlaps gives rise to wide bands with no gaps between levels from different atomic orbitals. Metallic properties can arise, however, in other contexts. In transition metal compounds a partially occupied *d* shell can give rise to a partly filled band. Thus rhenium in ReO_3 has the formal electron configuration $5d^1$ (see Topic H1) and is metallic. WO_3 (formally $5d^0$) is not metallic but $Na_{0.7}WO_3$ is, as electrons from sodium occupy the band made up of W 5*d* orbitals (see Topic D5).

Bandgaps

The bandgap in a nonmetallic solid is important for electrical and optical properties. A solid with a small bandgap is a **semiconductor** with a conductivity that (unlike the case with a metal) increases as temperature is raised. The bandgap also determines the minimum photon energy required to excite an electron from the VB to the CB, and hence the threshold for **optical absorption** by a solid.

In a covalent solid the bandgap is related to the energy splitting between bonding and antibonding orbitals (see Topic C3) and thus to the strength of bonding. In an ionic solid the bandgap is determined by the energy required to transfer an electron back from the anion to cation, which is related to the lattice energy (see Topic D6). Bandgaps for elements and binary compounds follow some systematic trends.

- In a series of isoelectronic solids such as CuBr – ZnSe – GaAs – Ge the bandgap decreases with decreasing electronegativity difference between the two elements. This trend reflects the decreasing energy difference between 'anion' and 'cation' orbitals.
- In series such as C – Si – Ge or LiF – NaF – KF the bandgap decreases as the group is descended and atoms or ions become larger. This trend reflects the decline in bond or lattice energies with larger atoms or ions (see Topics C7 and D6).

A comparison between compounds of pre-transition metals (e.g. Ca) and corresponding post-transition metals (e.g. Cd) provides a good example of the influence of the electronegativity differences (see Topic G1). Bandgaps are smaller in compounds of the less electropositive post-transition metals. The colors of CdS and CdSe (used as yellow and red pigments) come from strong absorption of blue light, as the bandgaps correspond to photon energies in the visible spectrum. Analogous calcium compounds are not colored as the larger bandgaps correspond to UV radiation.

Dielectric properties

The **dielectric constant** of a medium is a measure of the electrostatic polarization, which reduces the forces between charges (see Topics C9 and E1 for liquids). Two different mechanisms contribute to the dielectric properties of a solid according to the time-scale involved. The **static dielectric constant** depends on the displacement of ions from their regular positions in an applied electric field. It is applicable for static fields, or frequencies of electromagnetic radiation up into the microwave range. The **high-frequency dielectric constant** is measured at frequencies faster than the vibrational motion of ions. It is applicable in the visible region of the spectrum, and determines the **refractive index**, which governs the transmission of light in transparent media.

As expected, ionic substances have higher static dielectric constants than nonionic ones. Especially large values arise when ions can be easily displaced from their positions in the regular structure. For example, barium titanate $BaTiO_3$ has a very high dielectric constant that varies with temperature. In the perovksite structure (see Topic D5) the large Ba^{2+} ion imposes a relatively large O–O distance so that Ti^{4+} can move easily out of the center of its octahedral site. Below 120°C a permanent distortion sets in, which gives each unit cell a dipole moment. This type of behavior is called **ferroelectric** and has important applications, for example, in capacitors for electronic circuits.

Large high-frequency dielectric constants (and hence refractive indices) depend not on ionic motion but on electronic polarizability. Large ions contribute to this, and glasses containing Pb^{2+} are traditionally used for lenses where a high refractive index is necessary. Electronic polarizability can also be large in compounds with small bandgaps. A gap outside the visible spectrum is necessary for a colorless material in optical applications. TiO_2 is used as a white pigment because it has the right optical properties combined with cheapness, chemical stability and nontoxicity. The bandgap is only just in the UV, and the refractive index in the visible spectrum is high. Each grain is highly reflective, and a powdered sample appears white because light is reflected in random directions.

Influence of defects

All solids contain defects where the regularity of the ideal periodic lattice is broken. **Line** and **plane defects** (dislocations, grain boundaries, etc.) are important for mechanical properties but it is **point defects** that are most significant for electrical properties. They include

- **vacancies** or atoms missing from regular lattice positions;
- **interstitials** or atoms in positions not normally occupied;
- **impurities** either accidentally present or introduced as deliberate **doping**.

Defects that introduce extra electrons, or that give missing electrons or 'holes', have a large influence on electronic conduction in nonmetallic solids. Most semiconductor devices use doped or **extrinsic semiconductors** rather than the intrinsic semiconduction of the pure material. Doping Si with P replaces some tetrahedrally bonded Si atoms in the diamond lattice (see Topic D2) with P. Each replacement provides one extra valence electron, which requires only a small energy to escape into the CB of silicon. This is an **n-type** semiconductor. On the other hand, replacing an Si atom with Al gives a missing electron or 'hole', which may move in the VB giving a **p-type** semiconductor. Some other types of non-metallic solid can be doped, especially compounds of transition metals, which have variable oxidation states. Thus slight reduction of TiO_2 introduces electrons and gives n-type behavior. Similarly, oxidation of NiO removes some electrons and it becomes a p-type semiconductor.

Instead of providing electrons, atoms in defect sites may themselves be mobile and thus provide **ionic conduction** in a solid. Conductivity of this type can often be correlated with the number of defects, and may be especially large in disordered solids. For example, above a transition temperature of 70°C, AgI adopts a structure with a bcc array of I^-. The Ag^+ ions move freely between a variety of sites where they have almost equal energy. One cannot think strictly of defects in a case like this, rather it is the absence of a unique ordered structure that gives rise to high ionic conductivity. Anions are mobile in some compounds with the fluorite structure, for example, ZrO_2 doped with compounds such as CaO that increase the number of oxide vacancies. This material is used as a 'solid electrolyte' in electrochemical sensors for measuring oxygen partial pressure (see Topic J5).

E1 SOLVENT TYPES AND PROPERTIES

Key Notes

Polarity and solvation

Strongly polar molecules form solvents with high dielectric constants that are good at solvating charged species. At a molecular level solvation involves specific donor–acceptor interactions and other types of intermolecular force.

Donor and acceptor properties

Most good solvents have donor (Lewis base) and acceptor (Lewis acid) properties, responsible for solvation and other chemical reactions.

Ion-transfer solvents

The solvent-system acid–base concept depends on the possibility of ion transfer from one solvent molecule to another. Protic solvents act as H^+ donors and can support Brønsted acid–base reactions. Oxide and halide ions may be transferred in other solvents.

Related topics

Lewis acids and bases (C8)

Molecules in condensed phases (C9)

Polarity and solvation

A **solvent** is a liquid medium in which dissolved substances are known as **solutes**. Solvents are useful for storing substances that would otherwise be in inconvenient states (e.g. gases) and for facilitating reactions that would otherwise be hard to carry out (e.g. ones involving solids). The physical and chemical characteristics of a solvent are important in controlling what substances dissolve easily, and what types of reactions can be performed. The chemical as well as the physical state of solutes may be altered by interaction with the solvent. A list of useful solvents is given in *Table 1*.

The most important physical property of a solvent is its **polarity**. Molecules with large dipole moments such as water and ammonia form **polar solvents**. The macroscopic manifestation is the **dielectric constant** (ε_r), the factor by which electrostatic forces are weakened in comparison with those in a vacuum (see Topic C9). For example, in water $\varepsilon_r = 82$ at 25°C, and so attractive forces between anions and cations will be weaker by this factor.

At a microscopic level, solutes in polar solvents undergo strong **solvation**. For example, the **Born model** predicts that the Gibbs free energy of an ion with charge q (in Coulombs) and radius r will be changed in the solvent compared with the gas phase by an amount

$$\Delta G_{solv} = -\frac{q^2}{8\pi\varepsilon_0 r}\left(1 - \frac{1}{\varepsilon_r}\right) \qquad (1)$$

This estimate of the **solvation energy** is highly approximate, as it assumes that the solvent can be treated as a continuous dielectric medium on a microscopic scale. Nevertheless, it gives a rough guide that is useful in interpreting solubility trends (see Topic E4).

In reality, solvation involves **donor–acceptor interactions**, which may not be

Table 1. Properties of some solvents, showing normal melting and boiling points (MP and BP, respectively), dielectric constant (ε_r, at 25°C or at the boiling point if that is lower), and donor and acceptor numbers (DN and AN, respectively)

Solvent	MP/°C	BP/°C	ε_r	DN	AN
Acetonitrile CH_3CN (MeCN)	−41	82	14	19	36
Ammonia NH_3	−78	−33	22	–	–
Benzene C_6H_6	6	80	2	0	8
Bromine trifluoride BF_3	9	126	–	–	–
Dimethylsulfoxide $(CH_3)_2SO$ (DMSO)	18	189	45	30	19
Hydrogen fluoride HF	−83	19	84	–	–
n-Hexane C_6H_{14}	−95	69	2	0	0
Propanone $(CH_3)_2CO$	−95	57	21	13	21
Pyridine C_5H_6N	−42	115	12	14	12
Sulfur dioxide SO_2	−75	−10	15	–	–
Sulfuric acid H_2SO_4	10	(300)[a]	100	–	–
Tetrahydrofuran C_4H_8O			7	20	8
Water H_2O	0	100	82	18	55

[a]Decomposes.

purely electrostatic in nature (see below), so that neutral molecules may also be strongly solvated. Solvent molecules are ordered round the solute, not only in the **primary solvation sphere** but (especially with ions) affecting more distant molecules. Solvation therefore produces a decrease in entropy, which can be substantial with small highly charged ions, and contributes to acid–base strength, complex formation and solubility trends (see Topics E2–E4).

Nonpolar solvents such as hexane have molecules with little or no dipole moment and low dielectric constants. They are generally better at dissolving nonpolar molecules and for carrying out reactions where no ions are involved. The molecules interact primarily through van der Waals' forces (see Topic C9). Nonpolar media are generally poor solvents for polar molecules because the weak intermolecular forces cannot compete with the stronger ones in the pure solute. Similarly, nonpolar solutes cannot compete with the strong intermolecular forces in a polar solvent and so may not be very soluble. These generalizations have many limitations. Ionic substances can dissolve in solvents of lower polarity if the ions are efficiently solvated by appropriate donor and acceptor interactions (see Topic E4). As the electrostatic forces between solvated ions remain relatively strong, however, they tend to form ion pairs. Although liquid ammonia ($\varepsilon_r = 22$) is a good solvent for some ionic compounds, ion pairing is much commoner than in water ($\varepsilon_r = 82$).

Donor and acceptor properties

Most polar solvents have **donor** or **Lewis base** properties resulting from lone-pair electrons (see Topic C8). Good donor solvents include water, ammonia and pyridine, and are efficient at solvating cations and other Lewis acids. **Acceptor** or **Lewis acid** behavior is important for solvating anions, and results from empty orbitals or from hydrogen bonding. **Donor** and **acceptor numbers** have been defined by measuring the strength of interaction between solvent molecules and the 'standard' acceptor ($SbCl_5$) and donor ($OPCl_3$) molecules, respectively. Values are shown in *Table 1*, and can provide a useful guide although they ignore many specific details of the interaction, and in particular make no distinction between 'hard' and 'soft' character. As an example of this limitation, benzene is listed as having no appreciable donor strength, yet will dissolve silver perchlorate $AgClO_4$

Table 2. Some ion-transfer solvents, with the characteristic solvent-system acid and base species, and other examples of acids and bases

Ion transfered	Solvent-system species			Other examples of	
	Solvent	Acid	Base	Acids	Bases
H^+	H_2O	H_3O^+	OH^-	HNO_3	NH_3
	NH_3	NH_4^+	NH_2^-	H_2O	Na_2O
	H_2SO_4	$H_3SO_4^+$	HSO_4^-	HSO_3F	HNO_3
F^-	BrF_3	BrF_2^+	BrF_4^-	SnF_4	KF
O^{2-}	$CaSiO_3$	SiO_2	O^{2-}	P_2O_5	$CaCO_3$

because of a strong 'soft' donor–acceptor interaction between Ag^+ and a benzene molecule.

In many cases a donor–acceptor interaction may be only the first step in a more substantial **solvolysis** reaction. These reactions are common with nonmetal halides and oxides in water and ammonia; for example,

$$OPCl_3 + 6H_2O \rightarrow OP(OH)_3 + 3H_3O^+ + 3Cl^-$$

$$OPCl_3 + 6NH_3 \rightarrow OP(NH_2)_3 + 3NH_4^+ + 3Cl^-$$

An example of the variety of products formed in different donor solvents is provided by the reactions of $FeCl_3$, where S represents a coordinated solvent molecule:

$$FeCl3 \rightarrow [FeCl_3.S] \quad \text{in pyridine;}$$

$$\rightarrow [FeCl_2.S_4]^+ + Cl^- \quad \text{in DMSO;}$$

$$\rightarrow [FeCl_2.S_4]^+ + [FeCl_4]^- \quad \text{in MeCN.}$$

These differences are thought to result from the lower polarity of pyridine compared with the other two solvents, and the better solvation of small ions such as Cl^- in DMSO compared with MeCN.

Ion-transfer solvents

Water, ammonia and other **protic solvents** undergo a reaction known as **autoprotolysis**:

$$2NH_3 \rightleftharpoons NH_4^+ + NH_2^-$$

Although the equilibrium constants may be small (around 10^{-30} for ammonia) the possibility of such reactions leads to a definition of acids and bases based on a **solvent system** (see *Table* 2). An acid is the positive species formed (NH_4^+ in the above example) or any solute that gives rise to it; similarly, a base is the negative species (NH_2^-) or anything producing it in solution. With protic solvents this corresponds to the Brønsted definition of acids and bases (see Topic E2). The examples in *Table* 2 show that something acting as an acid in one solvent can be a base in another.

Aprotic solvents do not have transferable H^+ but some other ion such as halide or oxide can be involved. *Table* 2 shows the example of BrF_3, which undergoes some **autoionization** with F^- transfer. Substances dissolving to produce F^- ions act as bases, and Lewis acids that can react with F^- act as acids:

$$2BrF_3 + SnF_4 \rightarrow 2BrF_2^+ + SnF_6^-$$

In oxide melts the solvent system corresponds to the **Lux-Flood acid/base**

definition: an oxide ion donor is a base, and an oxide acceptor an acid. In the reaction

$CaO + SiO_2 \rightarrow CaSiO_3$

the calcium oxide is basic, and the silica acidic.

E2 BRØNSTED ACIDS AND BASES

Key Notes

Definitions

A Brønsted acid is a proton donor and undergoes protolysis when a base is present. Acids and bases form conjugate pairs. Water and some other substances are both acidic and basic.

pH

Water undergoes autoprotolysis (self-ionization) giving H_3O^+ and OH^-. The pH scale is related to these concentrations. Water at pH 7 is neutral, that with pH < 7 is acidic (H_3O^+ dominating) and with pH > 7 alkaline or basic (OH^- dominating).

Strong and weak behavior

The acidity constant and the related pK_a value give the equilibrium constant for protolysis. This reaction goes nearly to completion with strong acids, which are leveled to H_3O^+, the strongest acid possible in water. Weak acids have incomplete protolysis. Strong and weak bases show analogous behavior, the former being leveled to OH^-.

Trends in pK values

The acid strength of nonmetal hydrides increases towards the right and to the bottom of the periodic table. Acid strengths of oxoacids can be predicted approximately from their formulae by Pauling's rules. Metal cations with polarizing character are acidic in water, and some form amphoteric oxides or hydroxides.

Related topics

Lewis acids and bases (C8)
Solvent types and properties (E1)
Hydrogen (F2)

Definitions

A **Brønsted acid** is a proton donor, and a **Brønsted base** a proton acceptor. In this definition an acid–base reaction involves **protolysis**:

$$HA + B \rightarrow A^- + HB^+$$

HA is called the **conjugate acid** to A^-, and A^- the **conjugate base** to HA; HB^+ and B form another conjugate acid–base pair. Examples of some conjugate pairs (with the acid given first) are:

HCl/Cl^- H_2O/OH^- H_3O^+/H_2O NH_4^+/NH_3

Water is both an acid and a base, and this also happens with **polyprotic** (or **polybasic**) acids such as H_2SO_4, which can undergo successive protolysis steps to give HSO_4^- and SO_4^{2-}; thus HSO_4^- is the conjugate base of H_2SO_4 but the conjugate acid of SO_4^{2-}.

This definition of acids and bases should not be confused with the Lewis definition (Topic C8) although there is a connection: H^+ is a Lewis acid, and Brønsted bases are also Lewis bases, but in general Lewis acids such as BF_3 are **not** Brønsted acids, and Brønsted acids such as HCl are **not** Lewis acids.

Brønsted acidity is solvent dependent. Substances such as HCl are covalent molecules that undergo protolysis only in solvents polar enough to solvate ions, and when a base is present (which may be a solvent molecule). The following discussion concentrates on water, the commonest solvent in which protolytic reactions are studied. (See Topics E1, F5 and F8 for some other protic solvents.)

pH

Being simultaneously acidic and basic, water undergoes **autoprotolysis**, also called **self-ionization**:

$$2H_2O \rightleftharpoons H_3O^+ + OH^-$$

The equilibrium constant is

$$K_W = [H_3O^+][OH^-] = 1.0 \times 10^{-14} \text{ at } 298 \text{ K}$$

H_3O^+ in these equations is often simply written H^+. In pure water and in solutions that do not provide any additional source of H^+ or OH^- both ions have molar concentrations equal to 10^{-7}. Addition of an acid increases $[H_3O^+]$ and hence decreases $[OH^-]$; addition of a base has the reverse effect.

The **pH scale** is defined by

$$pH = -\log_{10}[H_3O^+]$$

Neutral water has a pH of 7, **acidic solutions** have lower values (typically 0–7), and **alkaline** or **basic solutions** higher values (7–14). In alkaline solutions $[OH^-]$ is thus greater than $[H^+]$.

Strong and weak behavior

The equilibrium constant of the protolysis reaction

$$HA + H_2O \rightleftharpoons H_3O^+ + A^-$$

is known as the **acidity constant** or the **acid dissociation constant** (K_a) of HA:

$$K_a = \frac{[H_3O^+][A^-]}{[HA]}$$

It is often expressed as a **pK_a value**, defined as

$$pK_a = -\log_{10}K_a$$

(Note that a larger K_a value corresponds to a smaller pK_a.) A selection of K_a and pK_a values is given in *Table 1*. If $pK_a < 0$ (i.e. $K_a > 1$) the equilibrium lies strongly to the right, and HA is called a **strong acid**. Acids with $pK_a > 0$ (i.e. $K_a < 1$) are **weak acids** and undergo only partial protolysis. Strong acids in water include HCl and H_2SO_4, whereas HF and HSO_4^- are weak acids.

In a similar way it is possible to define the **basicity constant** K_b and the corresponding pK_b from the equilibrium

$$B + H_2O \rightleftharpoons BH^+ + OH^-$$

We can distinguish **strong bases** with $pK_b < 0$ and the equilibrium lying to the right-hand side (examples being O^{2-} and NH_2^-) and weak bases with $pK_b > 0$ (e.g. NH_3 and F^-, with pK_b equal to 4.75 and 10.55, respectively). However, the use of pK_b is unnecessary, as the base reaction above may be combined with autoprotolysis to show that

$$pK_b = 14 - pK_a$$

Table 1. Some acidity constants in water at 25°C

Acid	HA	A^-	K_a	pK_a
Hydrochloric	HCl	Cl^-	10^7	−7
Sulfuric	H_2SO_4	HSO_4^-	10^2	−2
Nitric	HNO_3	NO_3^-	25	−1.4
Hydronium ion	H_3O^+	H_2O	1	0
Hydrogensulfate ion	HSO_4^-	SO_4^{2-}	1.2×10^{-2}	1.92
Phosphoric	H_3PO_4	$H_2PO_4^-$	7.5×10^{-3}	2.12
Iron (III) ion	$[Fe(H_2O)_6]^{3+}$	$[Fe(H_2O)_5OH]^{2+}$	10^{-2}	2
Hydrofluoric	HF	F^-	3.5×10^{-4}	3.45
Aluminum (III) ion	$[Al(H_2O)_6]^{3+}$	$[Al(H_2O)_5OH]^{2+}$	10^{-5}	5
Carbonic[a]	H_2CO_3	HCO_3^-	4.3×10^{-7}	6.37
Ammonium ion	NH_4^+	NH_3	5.6×10^{-10}	9.25
Water	H_2O	OH^-	10^{-14}	14
Ammonia	NH_3	NH_2^-	10^{-38}	38

[a]See Topic F4 for the anomalous behavior of carbonic acid.

where pK_a refers to the conjugate acid BH^+. Thus the pK_a values in *Table 1* can be used to calculate the pK_b values for the conjugate bases A^-.

As a strong acid such as HCl is fully protolyzed it is impossible to study this species itself in water. H_3O^+ is effectively the strongest acid possible there, and any stronger acid is said to be **leveled**. In a similar way, strong bases such as NH_2^- are leveled to the strongest base possible in water, OH^-. **Solvent leveling** limits the range of acid–base behavior that can be observed in a given solvent, and is one reason for using other solvents with different leveling ranges. For example, liquid ammonia is very basic compared with water, and H_2SO_4 is very acidic (see Topics F5 and F8).

Trends in pK values

A complete thermodynamic analysis of protolysis requires a cycle that includes the solvation of both HA and A^-. Entropy is important because of the ordering of water molecules around small ions and species with strongly localized charges (see Topic E1). Entropy changes will therefore tend to reduce the acid strength of any species giving a conjugate base with strongly localized negative charge. For positive ions protolysis reduces the charge and entropy contributions will increase the acid strength. Although solvation effects make a rigorous analysis difficult, some straightforward trends can be observed.

- **AH_n compounds**: acid strength increases from left to right in the periodic table, for example,

$$CH_4 \ll NH_3 \ll H_2O \ll HF$$

This trend is most simply related to the electronegativity increase of the element attached to hydrogen, which gives more bond polarity in the direction $A^{\delta-}-H^{\delta+}$ (see Topic B1). Acid strength also increases down the group, for example,

$$HF < HCl < HBr$$

which is not the order expected from electronegativity. Changes of solvation are important, but one simple contribution to the trend is the decreasing H–X bond strength down the group (see Topic C7).

The **oxides** of nonmetallic elements are generally acidic and give **oxoacids** in water (e.g. HNO_3 and H_2SO_4). Oxides and hydroxides of metals tend to be **basic** and form **aqua cations**. However, metals in high oxidation states can also form oxoacids (see Topics B1 and F7).

- The strengths of **oxoacids** can be predicted roughly by **Pauling's rules**.
  (i) Writing the formula as $XO_p(OH)_q$ the pK_a depends largely on the value of p, being roughly equal to $8 - 5p$ irrespective of q. Examples with their pK_a values are: $p = 0$ (HOCl, 7.2), $p = 1$ (H_3PO_4, 2.1), $p = 2$ (H_2SO_4, −2) and $p = 3$ ($HClO_4$, −10).
  (ii) For polyprotic acids, pK_a increases by about five units for each subsequent protolysis step (e.g. $H_2PO_4^-$, 7.4; HPO_4^{2-}, 12.7).
  Although solvation plays a role in these trends, the simplest explanation of rule (i) is that larger values of p give more scope for the negative charge to be delocalized over the anion. For example, in ClO^- (**1**) the formal charge is confined to one oxygen atom, whereas in ClO_4^- (**2**; only two of the four equivalent resonance structures are shown) it is spread equally over four.

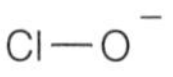

$Cl{-}O^-$ (**1**) $O{=}Cl({=}O)_2{-}O^- \longleftrightarrow O{=}Cl(=O)_2{-}O^-$ (**2**)

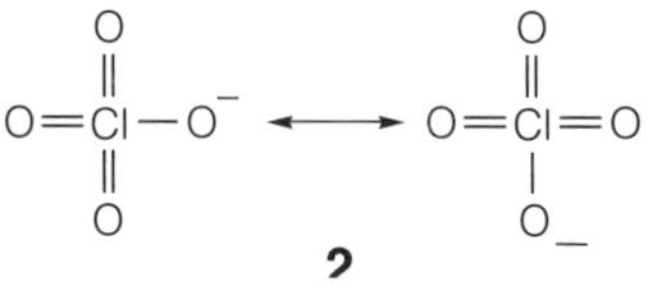

- **Aqua cations**: many metal cations are acidic in water. *Table 1* shows that aqueous Fe^{3+} is a stronger acid than HF. Acidity may be correlated with the 'polarizing' power of a cation associated with deviations from the ionic model (see Topic B1). Strongly acidic cations have either a high charge/size ratio (e.g. Be^{2+}, Al^{3+}, Fe^{3+}) or are derived from metals with low electropositive character (e.g. Hg^{2+}). Salts containing these ions form rather acidic solutions, and if the pH is increased successive protolysis may lead to polymerization and precipitation of an insoluble oxide or hydroxide such as $Al(OH)_3$. Some of these compounds show **amphoteric behavior** and dissolve in alkaline solution to give oxoanions. Thus $Al(OH)_3$, which is very insoluble in a neutral pH range, dissolves at pH > 10 to form $[Al(OH)_4]^-$ (see Topic G5).

E3 Complex formation

Key Notes

Equilibrium constants

Complexes are formed in aqueous solution when a ligand molecule or ion replaces solvating water molecules. Successive ligands may be attached, giving a series of stepwise formation (equilibrium) constants.

Hard and soft behavior

Class a (hard) cations complex more strongly with small electronegative ligands whereas class b (soft) cations have more affinity for less electronegative and more polarizable ligands. The difference involves entropic and enthalpic solvation terms.

Chelates and macrocycles

Polydentate or chelating ligands have more than one atom available for coordination to the metal, and form stronger complexes than monodentate ligands. The effect is enhanced in macrocyclic ligands, which have more rigid structures.

Effect of pH

Basic ligands become protonated at low pH and complex formation is suppressed.

Related topics

Lewis acids and bases (C8)
Group 12: zinc, cadmium and mercury (G4)
3*d* series: aqueous ions (H3)
Complexes: structure and isomerism (H6)

Equilibrium constants

A **complex** in general is any species formed by specific association of molecules or ions by donor–acceptor interactions (see Topic C8). In aqueous solution the most important complexes are those formed between a metal cation and **ligands**, which may be ions (e.g. halides, cyanide, oxalate) or neutral molecules (e.g. ammonia, pyridine). The ligand acts as a donor and replaces one or more water molecules from the primary solvation sphere, and thus a complex is distinct from an **ion pair**, which forms through purely electrostatic interactions in solvents of low polarity (see Topic E1). Although complex formation is especially characteristic of transition metal ions it is by no means confined to them.

Several steps of complex formation may be possible, and the successive equilibrium constants for the reactions

$$M + L \rightleftharpoons ML$$

$$ML + L \rightleftharpoons ML_2$$

and so on are known as the **stepwise formation constants** K_1, K_2 . . . . The overall equilibrium constant for the reaction

$$M + nL \rightleftharpoons ML_n$$

is given by

$$\beta_n = K_1 K_2 \ldots K_n$$

Successive stepwise formation constants often decrease regularly $K_1 > K_2 > \ldots$ of the maximum value being determined by the number of ligands that can be accommodated: this is often six except for chelating ligands (see below). The decrease can be understood on entropic (statistical) grounds, as each successive ligand has one less place available to attach. Exceptional effects may result from the charge and size of ligands, and a reversal of the normal sequence can sometimes be attributed to specific electronic or structural effects. It is important to remember that each ligand replaces one or more solvating water molecules. For example, in the Cd^{2+}/Br^- system $K_4 > K_3$ as the octahedral species $[Cd(H_2O)_3Br_3]^-$ is converted to tetrahedral $[CdBr_4]^{2-}$ with an entropy gain resulting from the increased freedom of three water molecules.

Hard and soft behavior

For cations formed from metals in early groups in the periodic table the complexing strength with halide ions follows the sequence

$$F^- \gg Cl^- > Br^- > I^-$$

whereas with some later transition metals and many post-transition metals the reverse sequence is found (e.g. Pt^{2+}, Hg^{2+}, Pb^{2+}; see Topics G4, G6 and H5). The former behavior is known as **class a** and the latter as **class b** behavior, and the difference is an example of **hard** and **soft** properties, respectively (see Topic C8). Class b ions form strong complexes with ligands such as ammonia, which are softer than water, whereas class a ions do not complex with such ligands appreciably in water.

Solvation plays an essential part in understanding the factors behind class a and b behavior. Trends in bond strengths show that almost every ion would follow the class a sequence in the gas phase, and the behavior in water is a partly a consequence of the weaker solvation of larger anions. With a class b ion such as Hg^{2+} the bond strengths decrease more slowly in the sequence Hg–F > Hg–Cl > . . . than do the solvation energies of the halide ions. With a class a ion such as Al^{3+}, on the other hand, the change in bond strengths is more marked than that in the solvation energies.

In solution the difference between the two classes is often manifested in different thermodynamic behavior. Class b complex formation is **enthalpy dominated** (i.e. driven by a negative ΔH) whereas class a formation is often **entropy dominated** (driven by positive ΔS). The strongest class a ions are small and highly charged (e.g. Be^{2+}, Al^{3+}) and have very negative entropies of solvation (see Topic E1). Complexing with small highly charged ions such as F^- reduces the overall charge and hence frees up water molecules, which are otherwise ordered by solvation. Hard cations with low charge/size ratio, such as alkali ions, form very weak complexes with all ligands except macrocycles (see below).

Some polyatomic ions such as NO_3^-, ClO_4^- and PF_6^- have very low complexing power to either class a or b metals. They are useful as counterions for studying the thermodynamic properties of metal ions (e.g. electrode potentials; see Topic E5) unaffected by complex formation.

Chelates and macrocycles

Chelating ligands are ones with two or more donor atoms capable of attaching simultaneously to a cation: they are described as **bidentate, tridentate**, . . . according to the number of atoms capable of binding. Chelating ligands include bidentate ethylenediamine (**1**) and ethylenediamine tetraacetate (EDTA, **2**), which is hexadentate, having two nitrogen donors and four oxygens (one from each carboxylate). Chelating ligands generally form stronger complexes than unidentate ones with similar donor properties. They are useful for analysis of metal ions by **complexometric titration** and for removing toxic metals in cases of poisoning (see Topic J3).

1

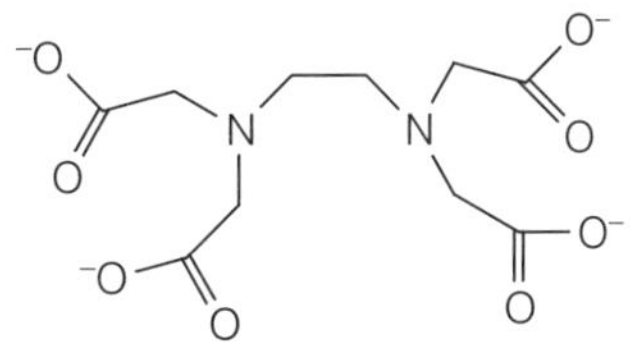

2

The origin of the **chelate effect** is entropic. Each ligand molecule can replace more than one solvating water molecule, thus giving a favorable entropy increase. Structural requirements occasionally subvert the effect: for example, Ag^+ does not show the expected increase of K_1 with ethylenediamine compared with ammonia, because it has a strong bonding preference for two ligand atoms in a linear configuration, which is structurally impossible with the bidentate ligand.

The length of the chain formed between ligand atoms is important in chelate formation, the most stable complexes generally being formed with four atoms (including the donors) so that with the metal ion a five-membered ring is formed. Smaller ring sizes are less favorable because of the bond angles involved, larger ones because of the increased configurational entropy of the molecule (coming from free rotation about bonds), which is lost in forming the complex. Limiting the possibility of bond rotation increases the complexing power even with optimum ring sizes, so that phenanthroline (**3**) forms stronger complexes than bipyridyl (**4**).

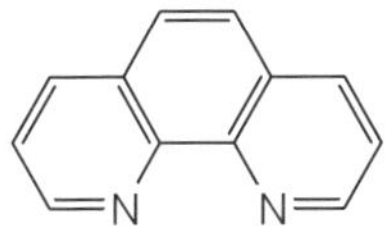

3

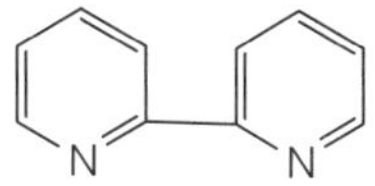

4

Reducing the configurational entropy is important in **macrocyclic ligands**, where several donor atoms are already 'tied' by a molecular framework into the optimal positions for complex formation. Examples are the cyclic **crown ethers** (e.g. 18-crown-6, **5**) and bicyclic **cryptands** (e.g. [2.2.1]-crypt, **6**). As expected, complexing strength is enhanced, and the resulting **macrocyclic effect** allows complexes to be formed with ions such as those of group 1, which otherwise have very low complexing power (see Topic G2). Another feature of macrocyclic ligands is the **size selectivity** corresponding to different cavity sizes. Thus with ligand **6** complex stability follows the order $Li^+ < Na^+ > K^+ > Rb^+$ and the selectivity can be altered by varying the ring size.

Chelating and macrocyclic effects are important in biological chemistry (see Topic J3). Metal binding sites in **metalloproteins** contain several ligand atoms, with appropriate electronegativities, and arranged in a suitable geometrical arrangement, to optimize the binding of a specific metal ion.

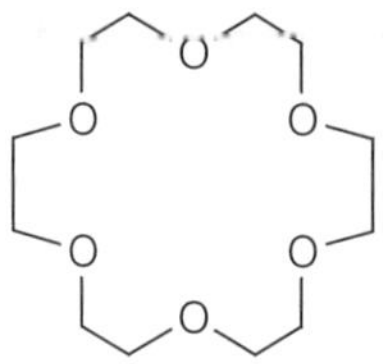

5

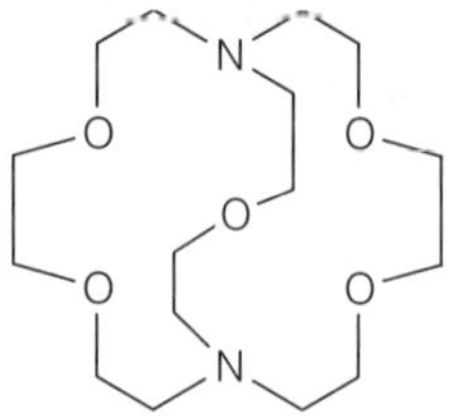

6

Effect of pH

pH changes will affect complex formation whenever any of the species involved has Brønsted acidity or basicity (see Topic E2). Most good ligands (except Cl^-, Br^- and I^-) are basic, and protonation at low pH will compete with complex formation. This is important in analytical applications. For example, in titrations with EDTA, Fe^{3+} (for which K_1 is around 10^{25}) can be titrated at a pH down to two, but with Ca^{2+} (where K_1 is about 10^{10}) a pH of at least seven is required because at lower pH values complex formation is incomplete.

E4 SOLUBILITY OF IONIC SUBSTANCES

Key Notes

Thermodynamics The equilibrium constant for dissolving an ionic substance is known as the solubility product. It is related to a Gibbs free energy change that depends on a balance of lattice energy and solvation energies, together with an entropy contribution.

Major trends in water Solids tend to be less soluble when ions are of similar size or when both are multiply charged. Covalent contributions to the lattice energy reduce solubility.

Effect of pH and complexing Solubility increases in acid conditions when the anion is derived from a weak acid, for example hydroxide, sulfide or carbonate. Amphoteric substances may dissolve again at high pH. Complexing agents also increase solubility.

Other solvents Highly polar solvents show parallels with water. Compounds with multiply charged ions are often insoluble in less polar ones, but different donor properties and polarizability play a role.

Related topic Lattice energies (D6)

Thermodynamics

Consider an ionic solid that dissolves in water according to the equation:

$$M_nX_m(s) \rightleftharpoons nM^{m+}(aq) + mX^{n-}(aq) \qquad (1)$$

The equilibrium constant for this reaction,

$$K_{sp} = [M^{m+}]^n[X^{n-}]^m$$

is known as the **solubility product** of M_nX_m. The form of this equilibrium is important in understanding effects such as the influence of pH and complexing (see below) and the **common ion effect**: it can be seen that adding one of the ions M^{m+} or X^{n-} will shift Reaction 1 to the left and so reduce the solubility of the salt. Thus AgCl(s) is much less soluble in a solution containing 1 M Ag^+ (e.g. from soluble $AgNO_3$) than otherwise.

Equilibrium constants in solution should correctly be written using **activities** not concentrations. The difference between these quantities is large in concentrated ionic solutions, and K_{sp} is quantitatively reliable as a guide to solubilities (measured in concentration units) only for very dilute solutions. Nevertheless, a thermodynamic analysis of the factors determining K_{sp} is useful for understanding general solubility trends. According to

$$\Delta G^{\ominus} = -RT \ln K$$

(see Topic B3), K_{sp} is related to the standard Gibbs free energy change of solution. *Figure 1* shows a thermodynamic cycle that relates the overall ΔG to two separate steps: (i) the formation of gas-phase ions; (ii) their subsequent solvation. The enthalpy contributions involve a balance between the lattice energy of the solid and the solvation enthalpies of the ions (see Topics D6 and E1). In a solvent such as water with a very high dielectric constant these contributions almost cancel. Nevertheless, some of the solubility trends summarized below can be understood in terms of the changing balance between lattice energies (proportional to $1/(r_+ + r_-)$, where r_+ and r_- are radii of individual ions) and the sum of the individual solvation enthalpies (each roughly proportional to $1/r$). For example, a small ion has a large (negative) solvation energy, but when partnered by a large counterion cannot achieve an especially large lattice energy. With ions of very different size, therefore, solvation is relatively favored and solubility tends to be larger than with ions of similar size. Entropy terms are, however, also important. The first step in *Fig. 1* involves an entropy increase, but solvation produces an ordering of solvent molecules and a negative ΔS contribution. Overall ΔS values for dissolving ions with multiple charges are usually negative, an effect that tends to lower the solubility as mentioned below.

Major trends in water

The aqueous solubility of ionic compounds is important in synthetic and analytical chemistry, and in the formation of minerals by geochemical processes (see Topic J2). The most significant trends are as follows.

(i) Soluble salts are more often found when ions are of very different size rather than similar size. Thus in comparing salts with different alkali metal cations, lithium compounds are the least soluble of the series with OH^- and F^-, but the most soluble with larger cations such as Cl^- or NO_3^-. This principle is often useful in preparative reactions and separations. If it is desired to precipitate a large complex anion, a large cation such as tetrabutyl ammonium $[(C_4H_9)_4N]^+$ can be helpful.

(ii) Salts where both ions have multiple charges are less likely to be soluble than ones with single charges. Thus carbonates (CO_3^{2-}) and sulfates (SO_4^{2-}) of the larger group 2 cations are insoluble. An important factor is the negative solvation entropies of the ions.

(iii) With ions of different charges, especially insoluble compounds result when the lower charged one is smaller (as this gives a very large lattice energy). Thus with M^{3+} ions, fluorides and hydroxides are generally very insoluble, whereas heavier halides and nitrates are very soluble.

(iv) Lower solubility results from covalent contributions to the lattice energy (see Topic D6). This happens especially with ions of less electropositive metals in combination with more polarizable cations. Late transition and post-transition elements often have insoluble sulfides (see Topic J2); insoluble halides (but not generally fluoride) also occur, for example with Ag^+ and Pb^{2+}.

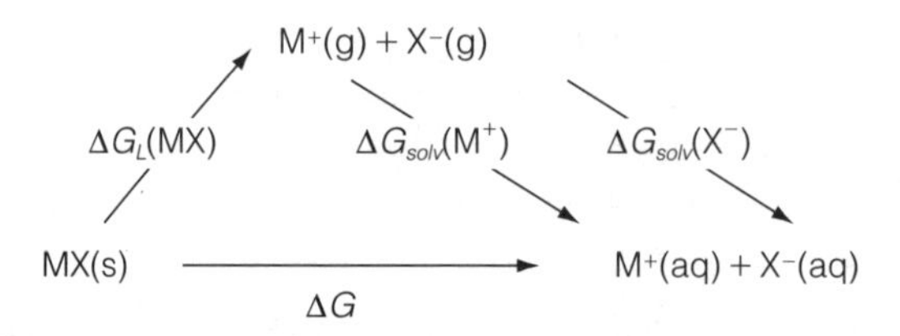

Fig. 1. Thermodynamic cycle for the solution of an ionic solid MX.

Influence of pH and complexing

Any substance in solution that reacts with one of the ions formed in Reaction 1 will shift the equilibrium to the right and hence increase the solubility of the solid. pH will therefore influence the solubility in a range where one of the ions has significant Brønsted acid or base properties (see Topic E2). The solubility of NaCl, for example, should not be affected by pH, but when the anion is the conjugate base of a weak acid solubility will increase at low pH. Metal oxides and hydroxides dissolve in acid solution, and conversely such solids may be precipitated from a solution containing a metal ion as the pH is increased. The solubility range depends on the K_{sp} value: for example, $Fe(OH)_3$ is precipitated at much lower pH than the more soluble $Fe(OH)_2$. At high pH the acidity of the hydrated metal ion may come into play and **amphoteric** substances such as $Al(OH)_3$ will dissolve in alkaline solution to give $[Al(OH)_4]^-$.

Sometimes the conjugate acid of the anion is volatile, or decomposes to form a gas. Thus action of an acid on a sulfide will liberate H_2S, and on a carbonate CO_2 from the decomposition of carbonic acid.

Any ligand that complexes with the metal ion will also increase solubility. AgCl dissolves in aqueous ammonia by the formation of $[Ag(NH_3)_2]^+$. Addition of Cl^-, which initially decreases the solubility of AgCl through the common ion effect (see above), will at high concentrations increase the solubility by forming $[AgCl_2]^-$. The solubility of some amphoteric oxides and hydroxides at high pH can be interpreted as a similar complexing effect, with OH^- acting as the ligand.

Other solvents

With its combination of high dielectric constant, good donor ability and hydrogen bonding capability (which contributes to the solvation of anions) water is one of the best solvents for ionic substances. Two other solvents of comparable polarity are HF and H_2SO_4 (see Topic E1, *Table 1*). Solubility trends for metal fluorides in HF show close parallels with those for hydroxides and oxides in water. Thus they follow the sequence $MF > MF_2 > MF_3$, and for a given charge the solubility tends to increase with the cation size.

As the dielectric constant decreases, the solvation energies become less able to compensate for the lattice energy. This is especially true for solids with multiply charged ions, and such compounds are much less soluble in liquid ammonia than in water. Detailed comparisons are complicated, however, by the occurrence of ion pairing, and by the increased importance of other interactions. Ammonia is a better donor than water for soft class b cations such as Ag^+ (see Topic E3) and compounds such as AgCl are much more soluble. The solubility trend $AgI > AgBr > AgCl$ in ammonia is also the reverse of that found in water, reflecting another difference: ammonia has a larger polarizability than water and so van der Waals' forces are more important. They contribute significantly to the solvation of heavier anions such as I^-. Some iodides such as LiI are soluble in solvents of low polarity, a fact that is sometimes wrongly used to suggest that the solids have appreciable covalent character. In fact, LiI often dissolves as an ion pair, the donor solvent coordinating Li^+ and the van der Waals' forces solvating I^-.

E5 ELECTRODE POTENTIALS

Key Notes

Standard potentials — An electrode potential is a measure of the thermodynamics of a redox reaction. It may be expressed as the difference between two half-cell potentials, which by convention are measured against a hydrogen electrode. Tabulated values refer to standard conditions (ions at unit activity).

Direction of reaction — Comparison of two electrode potentials allows prediction of the favorable direction of a redox reaction, and of its equilibrium constant. Only the differences between electrode potentials are significant; individual potentials have no meaning.

Nonstandard conditions — The Nernst equation shows how electrode potentials vary with activity (approximately equal to concentration). Potentials may be influenced by pH and complexing.

Diagrammatic representations — Latimer and Frost diagrams are different ways of representing the electrode potentials for different oxidation states of an element. Frost diagrams are useful for visual comparisons between elements, and for showing which species are likely to disproportionate.

Kinetic limitations — Electrode potentials give no information about the rate of a redox reaction. Reactions where covalent bonds are involved may be very slow.

Related topics — Oxidation and reduction (B4); 3*d* series: aqueous ions (H3)

Standard potentials

In an **electrochemical cell** a redox reaction occurs in two halves (see Topic B4). Electrons are liberated by the oxidation half reaction at one **electrode** and pass through an electrical circuit to another electrode where they are used for the reduction. The **cell potential *E*** is the potential difference between the two electrodes required to balance the thermodynamic tendency for reaction, so that the cell is in equilibrium and no electrical current flows. *E* is related to the molar Gibbs free energy change in the overall reaction (see Topic B3) according to

$$\Delta G = -\,n\,F\,E \tag{1}$$

where *F* is the Faraday constant (9.6485×10^4 C mol^{-1}) and *n* the number of moles of electrons passed per mole of reaction.

It is useful to think of the cell potential as the difference between the potentials associated with the two **half-cell reactions**, although these are not separately measurable. **Standard electrode potentials** are the half-cell potentials measured against a **hydrogen electrode**, where the half-cell reaction is

Table 1. Some standard electrode potentials in aqueous solution at pH = 0 and 25°C

Couple	Half-cell reaction	$E^{\ominus}$(V)
MnO_4^-/Mn^{2+}	$MnO_4^- + 8H^+ + 5e^- \rightarrow Mn^{2+} + 4H_2O$	+1.51
O_2/H_2O	$O_2 + 4H^+ + 4e^- \rightarrow 2H_2O$	+1.23
Br_2/Br^-	$Br_2 + 2e^- \rightarrow 2Br^-$	+1.09
Fe^{3+}/Fe^{2+}	$Fe^{3+} + e^- \rightarrow Fe^{2+}$	+0.77
H^+/H_2	$2H^+ + 2e^- \rightarrow H_2$	0
Zn^{2+}/Zn	$Zn^{2+} + 2e^- \rightarrow Zn$	−0.72
Na^+/Na	$Na^+ + e^- \rightarrow Na$	−2.71

$$H^+ + e^- \rightarrow \tfrac{1}{2} H_2$$

all reagents being under standard conditions (unit activity and pressure). Some values are shown in *Table 1* for species in aqueous solution. By convention, tabulated potentials refer to **reduction** reactions, with electrons on the left as in the above equation. Only **differences** in electrode potential are significant, the absolute values having no meaning except in comparison with the H^+/H_2 potential (zero by definition).

Direction of reaction

Comparing two couples Ox-Red, a more positive potential means that the corresponding species Ox is a stronger oxidizing agent. Thus from *Table 1* Br_2 is a stronger oxidizing agent than Fe^{3+} and will so oxidize Fe^{2+}, the products being Br^- and Fe^{3+}. Conversely, a lower (more negative) potential means that the corresponding Red is a stronger reducing agent. Thus zinc metal is a stronger reducing agent than dihydrogen, and will reduce H^+:

$$2H^+ (aq) + Zn(s) \rightarrow H_2 (g) + Zn^{2+} (aq) \quad (2)$$

The **equilibrium constant** K for such a reaction can be calculated from

$$RT \ln K = nF \Delta E^{\ominus}$$

where R is the gas constant, T the temperature in kelvin, $\Delta E^{\ominus}$ the difference in the two electrode potentials and n the number of electrons in each half reaction: this must be the same for both half reactions in a balanced equation (see example in Topic B4). In Equation 2 $n = 2$, which gives K around 10^{24} at 298 K for this reaction. As the potential of a single half-cell is not measurable, so an equilibrium constant based on a single potential has no meaning.

Nonstandard conditions

Under nonstandard conditions the electrode potential of a couple can be calculated from the **Nernst equation**:

$$E = E^{\ominus} + \frac{RT}{nF} \ln \frac{[\mathrm{Ox}]}{[\mathrm{Red}]}$$

where [Ox] and [Red] are the **activities** of the species involved; it is a common approximation, especially in dilute solutions, to assume that these are the same as the molar concentrations. With $n = 1$ at 298 K, a factor of 10 difference in the activity changes E by 0.059 V.

When a reaction involves H^+ or OH^- ions, these must be included in the Nernst equation to predict the **pH dependence** of the couple. Thus for the MnO_4^-/Mn^{2+} half-cell reaction shown in *Table 1*, a factor of $[H^+]^8$ should be included in the [Ox] term, leading to a reduction in potential of $(8/5) \times 0.059 = 0.094$ V per unit increase in pH. pH changes may also have a more subtle influence by altering the species involved. For example, in alkaline solution the ion Mn^{2+} precipitates as $Mn(OH)_2$. Standard potentials at pH = 14 refer to reactions written with OH^- rather than H^+ (see Topic B4).

Potentials may also be strongly influenced by **complex formation**. In general, any ligand that complexes more strongly with the higher oxidation state will reduce the potential. For example, cyanide (CN^-) complexes much more strongly with Mn^{3+} than with Mn^{2+}, and at unit activity reduces the Mn^{3+}/Mn^{2+} potential from its standard value of +1.5 V to +0.22 V. Conversely, the potential increases if the lower oxidation state is more strongly complexed.

Diagrammatic representations

A **Latimer diagram** shows the standard electrode potentials associated with the different oxidation states of an element, as illustrated in *Fig. 1* for manganese. Potentials not given explicitly can be calculated using Equation 1 and taking careful account of the number of electrons involved. Thus the free energy change for the Mn^{3+}/Mn reduction is the sum of those for Mn^{3+}/Mn^{2+} and Mn^{2+}/Mn. From Equation 1 therefore

$$3\,E^{\ominus}(Mn^{3+}/Mn) = E^{\ominus}(Mn^{3+}/Mn^{2+}) + 2\,E^{\ominus}(Mn^{2+}/Mn)$$

from which $E^{\ominus}(Mn^{3+}/Mn) = -0.29$ V.

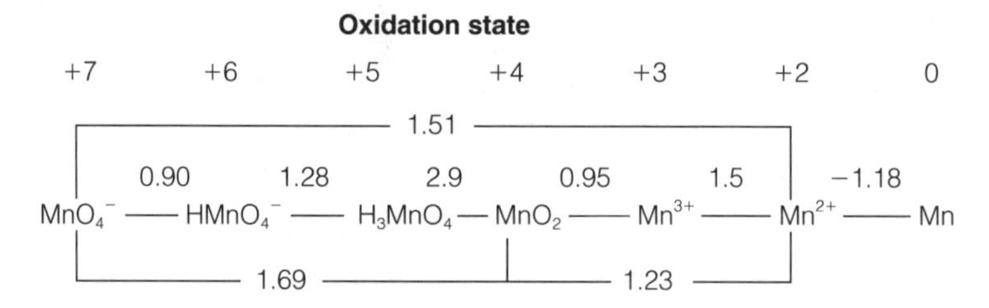

Fig. 1. Latimer diagram for Mn at pH 5 0.

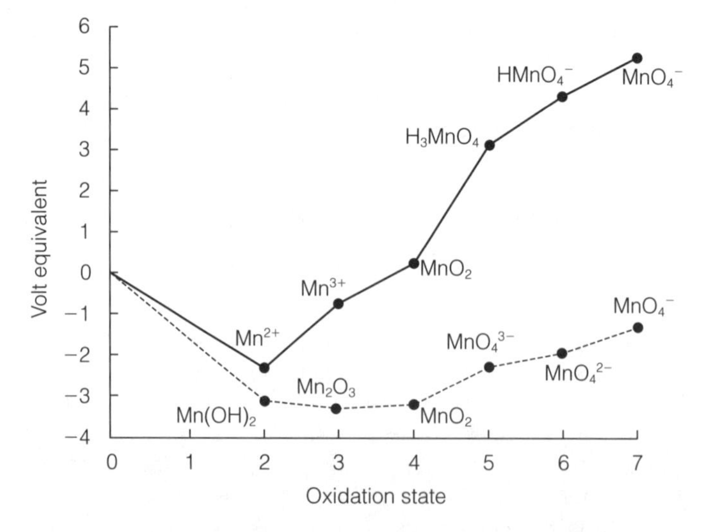

Fig. 2. Frost diagram for Mn at pH = 0 (solid line) and pH = 14 (dashed line).

In a **Frost** or **oxidation state diagram** (see *Fig. 2*) each oxidation state (n) is assigned a **volt equivalent** equal to n times its $E^{\ominus}$ value with respect to the element. The potential $E^{\ominus}$ in volts between any two oxidation states is equal to the **slope** of the line between the points in this diagram. Steep positive slopes show strong oxidizing agents, steep negative slopes strong reducing agents. Frost diagrams are convenient for displaying the comparative redox properties of elements (see Topics F9 and H3).

Frost diagrams also provide a visual guide to when **disproportionation** of a species is expected. For example, in *Fig. 2* the Mn^{3+} state at pH = 0 is found above the line formed by joining Mn^{2+} with MnO_2. It follows that the Mn^{3+}/Mn^{2+} potential is more positive than MnO_2/Mn^{3+}, and disproportionation is predicted:

$$2Mn^{3+}(aq) + 2H_2O \rightarrow MnO_2(s) + Mn^{2+}(aq) + 4H^{+}(aq)$$

The equilibrium constant of this reaction can be calculated by noting that it is made up from the half reactions for MnO_2/Mn^{3+} and Mn^{3+}/Mn^{2+} each with $n = 1$, and has $E^{\ominus} = 1.5 - 0.95 = 0.55$ V from *Fig. 1*, giving $K = 2 \times 10^{9}$. The states Mn^{V} and Mn^{VI} are similarly unstable to disproportionation at pH = 0, whereas at pH = 14, also shown in *Fig. 2*, only Mn^{V} will disproportionate.

Latimer and Frost diagrams display the same information but in a different way. When interpreting electrode potential data, either in numerical or graphical form, it is important to remember that a single potential in isolation has no meaning,

Kinetic limitations

Electrode potentials are **thermodynamic quantities** and show nothing about how fast a redox reaction can take place (see Topic B3). Simple electron transfer reactions (as in Mn^{3+}/Mn^{2+}) are expected to be rapid, but redox reactions where covalent bonds are made or broken may be much slower (see Topics F9 and H7). For example, the MnO_4^{-}/MnO_2 potential is well above that for the oxidation of water (see O_2/H_2O in *Table 1*), but the predicted reaction happens very slowly and aqueous permanganate is commonly used as an oxidizing agent (although it should always be standardized before use in volumetric analysis).

Kinetic problems can also affect redox reactions at electrodes when covalent substances are involved. For example, a practical hydrogen electrode uses specially prepared platinum with a high surface area to act as a catalyst for the dissociation of dihydrogen into atoms (see Topic J5). On other metals a high **overpotential** may be experienced, as a cell potential considerably larger than the equilibrium value is necessary for a reaction to occur at an appreciable rate.

F1 INTRODUCTION TO NONMETALS

Key Notes

Covalent chemistry

Hydrogen and boron stand out in their chemistry. In the other elements, valence states depend on the electron configuration and on the possibility of octet expansion which occurs in period 3 onwards. Multiple bonds are common in period 2, but are often replaced by polymerized structures with heavier elements.

Ionic chemistry

Simple anionic chemistry is limited to oxygen and the halogens, although polyanions and polycations can be formed by many elements.

Acid–base chemistry

Many halides and oxides are Lewis acids; compounds with lone-pairs are Lewis bases. Brønsted acidity is possible in hydrides and oxoacids. Halide complexes can also be formed by ion transfer.

Redox chemistry

The oxidizing power of elements and their oxides increases with group number. Vertical trends show an alternation in the stability of the highest oxidation state.

Related topics

Electronegativity and bond type (B1)
Chemical periodicity (B2)
Electron pair bonds (C1)

Covalent chemistry

Nonmetallic elements include hydrogen and the upper right-hand portion of the *p* block (see Topic B2, *Fig. 1*). Covalent bonding is characteristic of the elements, and of the compounds they form with other nonmetals. The bonding possibilities depend on the **electron configurations** of the atoms (see Topics A4 and C1). **Hydrogen** (Topic F2) is unique and normally can form only one covalent bond. **Boron** (Topic F3) is also unusual as compounds such as BF_3 have an incomplete octet. **Electron deficiency** leads to the formation of many unusual compounds, especially hydrides (see also Topic C6).

The increasing number of valence electrons between groups 14 and 18 has two possible consequences. In simple molecules obeying the **octet rule** the valency falls with group number (e.g. in CH_4, NH_3, H_2O and HF, and in related compounds where H is replaced by a halogen or an organic radical). On the other hand, if the number of valence electrons involved in bonding is not limited, then a wider range of valencies becomes possible from group 15 onwards. This is most easily achieved in combination with the highly electronegative elements O and F, and the resulting compounds are best classified by the **oxidation state** of the atom concerned (see Topic B4). Thus the maximum possible oxidation state increases from +5 in group 15 to +8 in group 18. The +5 state is found in all periods (e.g. NO_3^-, PF_5) but higher oxidation states in later groups require **octet expansion** and occur only from period 3 onwards (e.g. SF_6 and ClO_4^-; in group 18 only xenon can do this, e.g. XeO_4).

Octet expansion or **hypervalence** is often attributed to the involvement of *d* orbitals in the same principal quantum shell (e.g. 3*d* in period 3; see Topics A3 and A4). Thus six octahedrally directed bonds as in SF_6 could be formed with sp^3d^2 hybrid orbitals (see Topic C5). In a similar way the multiple bonding normally drawn in species such as PO_4^{3-} (**1**) is often described as ***d*π-*p*π bonding**. These models certainly overestimate the contribution of *d* orbitals. It is always possible to draw valence structures with no octet expansion provided that nonzero formal charges are allowed. For example, the orthonitrate ion NO_4^{3-} is drawn without double bonds (**2**), and PO_4^{3-} could be similarly represented. One of many equivalent valence structures for SF_6 where sulfur has only eight valence-shell electrons is shown in **3**. Three-center four-electron bonding models express similar ideas (see Topic C5) . Such models are also oversimplified. It is generally believed that *d* orbitals do play some role in octet expansion, but that two other factors are at least as important: the larger size of elements in lower periods, which allows higher coordination numbers, and their lower electronegativity, which accommodates positive formal charge more easily.

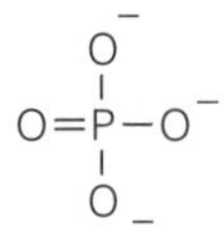

1 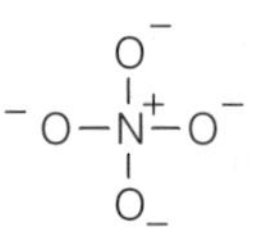 **2** 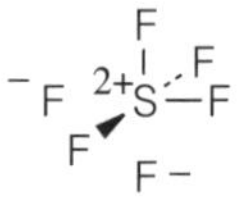 **3**

Another very important distinction between period 2 elements and others is the ready formation of **multiple bonds** by C, N and O (see Topic C7). Many of the compounds of these elements have stoichiometries and structures not repeated in lower periods (e.g. oxides of nitrogen; see Topic F5).

Some of these trends are exemplified by the selection of molecules and complex ions in *Table 1*. They have been classified by (i) the total number of valence electrons (VE), and (ii) the **steric number** of the central atom (SN), which is calculated by adding the number of lone-pairs to the number of bonded atoms and used for interpreting molecular geometries in the VSEPR model (see Topic C2). The species listed in *Table 1* illustrate the wide variety of **isoelectronic relationships** that exist between the compounds formed by elements in different groups and periods. Species with SN = 4 are found throughout the *p* block, but ones with lower steric numbers and/or multiple bonding are common only in period 2. In analogous compounds with heavier elements the coordination and steric numbers are often increased by **polymerization** (compare CO_2 and SiO_2, CO_3^{2-} and $[SiO_3^{2-}]_n$) or by a change of stoichiometry (e.g. SiO_4^{4-}). Species with steric numbers higher than four require octet expansion and are not found in period 2. Many of the species listed in *Table 1* are referred to in Topics dealing with the appropriate elements.

Ionic chemistry

Simple **monatomic anions** are formed by only the most **electronegative elements**, in groups 16 and 17 (e.g. O^{2-}, Cl^-). Although C and N form some compounds that could be formulated in this way (e.g. Li_3N and Al_4C_3), the ionic model is not very appropriate for these. There are often structural differences between oxides or fluorides and the corresponding compounds from later periods. These are partly due to the larger size and polarizability of ions, but compounds of S, Se and Te are also much less ionic than oxides (see Topics D4, F7, F8 and F9).

Table 1. A selection of molecules and ions (including polymeric forms) classified according to the valence electron count (VE) and the steric number (SN) of the central atom shown in bold type

VE	SN	Molecule or ion	Structure
8	4	$H_2\mathbf{O}$, $\mathbf{N}H_2^-$	Bent
		$\mathbf{N}H_3$, $H_3\mathbf{O}^+$, $\mathbf{Si}H_3^-$	Pyramidal
		$\mathbf{Si}H_4$, $\mathbf{B}H_4^-$, $\mathbf{N}H_4^+$	Tetrahedral
10	2	N_2, CO, CN^-, C_2^{2-}, NO^+	Triple bond
14	4	C_2H_6, P_2H_4, Cl_2, O_2^{2-}, XeF^+	Single bond
16	2	$\mathbf{C}O_2$, $N_2\mathbf{O}$...	Linear, double bonds
	4	$[\mathbf{Si}O_2]_\infty$	Polymeric 3D network
18	3	$\mathbf{N}OF$, $\mathbf{O}_3$, $\mathbf{S}O_2$, $\mathbf{N}O_2^-$	Bent
	4	$[\mathbf{Se}O_2]_\infty$, $[\mathbf{As}O_2^-]_\infty$	Polymeric chain
24	3	$\mathbf{B}F_3$, $(CH_3)_2\mathbf{C}O$, $\mathbf{S}O_3$, $\mathbf{N}O_3^-$, $\mathbf{C}O_3^{2-}$	Planar
	4	$[(CH_3)_2\mathbf{Si}O]_n$, $[\mathbf{P}(Cl)_2N]_n$, $[\mathbf{S}O_3]_n$, $[\mathbf{P}O_3^-]_n$, $[\mathbf{Si}O_3^{2-}]_n$	Polymeric ring or chain
26	4	$(CH_3)_3\mathbf{N}$, $\mathbf{P}Cl_3$, $\mathbf{S}OCl_2$, $\mathbf{S}F_3^+$, $\mathbf{Cl}O_3^-$	Pyramidal
32	4	$\mathbf{C}F_4$, $\mathbf{Si}Cl_4$, $\mathbf{Xe}O_4$, $\mathbf{N}OF_3$, $\mathbf{P}Cl_4^+$, $\mathbf{Cl}O_4^-$, $\mathbf{P}O_4^{3-}$, $\mathbf{N}O_4^{3-}$, $\mathbf{Si}O_4^{4-}$	Tetrahedral
48	6	$\mathbf{Si}F_6^{2-}$, $\mathbf{P}Cl_6^-$, $\mathbf{S}F_6$, $\mathbf{Br}F_6^+$, $\mathbf{Te}(OH)_6$, $\mathbf{Xe}O_6^{4-}$	Octahedral

Many **polyanions** are known. Those with multiple bonding are characteristic of period 2 (e.g. C_2^{2-} and N_3^-); ones with single bonding are often more stable for heavier elements (e.g. S_n^{2-}), and some form polymerized structures (see Topic D5). Simple cations are not a feature of nonmetal chemistry but some **polycations** such as O_2^+ and S_4^{2+} can be formed under strongly oxidizing conditions. **Complex cations and anions** are discussed below.

Acid–base chemistry

Many nonmetal oxides and halides are **Lewis acids** (see Topic C8). This is not so when an element has its maximum possible steric number (e.g. CF_4, NF_3 or SF_6) but otherwise acidity generally increases with oxidation state. Such compounds react with water to give **oxoacids**, which together with the salts derived from them are common compounds of many nonmetals (see Topics D5 and F7). Compounds with lone-pairs are potential **Lewis bases**, base strength declining with group number (15 > 16 > 17). In combination with 'hard' acceptors the donor strength decreases down a group (e.g. N ≫ P > As) but with 'soft' acceptors the trend may be reversed.

Ion-transfer reactions give a wide variety of complex ions, including ones formed from proton transfer (e.g. NH_4^+, H_3O^+, NH_2^- and OH^-), halide complexes (e.g. $[PCl_4]^+$, $[SF_5]^-$), and oxoanions and cations (e.g. SO_4^{2-}, NO_2^+). Such ions are formed in appropriate polar solvents (see Topic E1) and are also known in solid compounds. The trends in **Brønsted acidity** of hydrides and oxoacids in water are described in Topic E2. pK_a values of oxoacids may change markedly down a group as the structure changes (e.g. HNO_3 is a strong acid, H_3PO_4 a weak acid; the elements Sb, Te and I in period 5 form octahedral species such as $[Sb(OH)_6]^-$, which are much weaker acids). **Brønsted basicity** of compounds with lone pairs follows the 'hard' sequence discussed above (e.g. $NH_3 > H_2O > HF$, and $NH_3 \gg PH_3 > AsH_3$).

Redox chemistry

The elements O, F, Cl and Br are good oxidizing agents. Compounds in high oxidation states (e.g. oxides and halides) are potentially oxidizing, those in low oxidation states (e.g. hydrides) reducing. Oxidizing power increases with group number, and reducing power correspondingly declines. The trends down each group are dominated by bond strength changes (see Topic C7). Between periods 2 and 3 bonds to hydrogen become weaker (and so hydrides become more reducing and the elements less oxidizing) whereas bonds to oxygen and halogens become stronger (and so oxides and halides become less oxidizing). Compounds of As^{V}, Se^{VI} and Br^{VII} in period 4 are more strongly oxidizing than corresponding ones in periods 3 or 5. This **alternation effect** can be related to irregular trends in ionization energies, associated with the way that electron shells are filled in the periodic table (see Topics A4 and A5).

F2 HYDROGEN

Key Notes

The element — Hydrogen occurs on Earth principally in water, and is a constituent of life. The dihydrogen molecule has a strong covalent bond, which limits its reactivity. It is an important industrial chemical.

Hydrides of nonmetals — Nonmetallic elements form molecular hydrides. Bond strengths and stabilities decline down each group. Some have Brønsted acidic and basic properties.

Hydrides of metals — Solid hydrides with some ionic character are formed by many metals, although those of *d*- and *f*-block elements are often nonstoichiometric and metallic in character. Hydride can form complexes such as AlH_4^- and many examples with transition metals.

The hydrogen bond — Hydrogen bound to a very electronegative element can interact with a similar element to form a hydrogen bond. Hydrogen bonding is important in biology, and influences the physical properties of some simple hydrides.

Deuterium and tritium — Deuterium is a stable isotope occurring naturally; tritium is radioactive. These isotopes are used in research and in thermonuclear weapons.

Related topics — Chemical periodicity (B2) Brønsted acids and bases (E2)

The element

Hydrogen is the commonest element in the Universe and is a major constituent of stars. It is relatively much less common on Earth but nevertheless forms nearly 1% by mass of the crust and oceans, principally as **water** and in hydrates and hydroxide minerals of the crust. It is ubiquitous in biology (see Topics J1–J3).

The **dihydrogen molecule H_2** is the stable form of the element under normal conditions, although atomic hydrogen can be made in the gas phase at high temperatures, and hydrogen may become a metallic solid or liquid at extremely high pressures. At 1 bar pressure, dihydrogen condenses to a liquid at 20 K and solidifies at 14 K, these being the lowest boiling and melting points for any substance except helium. The H–H bond has a length of 74 pm and a dissociation enthalpy of 436 kJ mol^{-1}. This is the shortest bond known, and one of the strongest single covalent bonds. Although it is thermodynamically capable of reacting with many elements and compounds, these reactions often have a large kinetic barrier and require elevated temperatures and/or the use of catalysts (see Topic J5).

Dihydrogen is an important industrial chemical, mostly made from the **steam re-forming** of hydrocarbons from petroleum and natural gas. The simplest of these reactions,

$$CH_4 + H_2O \rightarrow CO + 3H_2$$

is endothermic, and temperatures around 1400 K are needed to shift the equilibrium to the right. Major uses of hydrogen are in the synthesis of ammonia, the hydrogenation of vegetable fats to make margarine, and the production of organic chemicals and hydrogen chloride (see Topic J4).

Hydrides of nonmetals

Hydrogen forms molecular compounds with nonmetallic elements. *Table 1* shows a selection. With the exception of the boranes (see Topic F3) hydrogen always forms a single covalent bond. Complexities of formula or structure arise from the possibility of **catenation**, direct element–element bonds as in hydrogen peroxide, H–O–O–H, and in many organic compounds. The International Union of Pure and Applied Chemistry (IUPAC) has suggested systematic names ending in -ane, but for many hydrides 'trivial' names are still generally used (see Topic B5). In addition to binary compounds, there are many others with several elements present. These include nearly all organic compounds, and inorganic examples such as hydroxylamine, H_2NOH. The **substitutive system** of naming inorganic compounds derived from hydrides is similar to the nomenclature used in organic chemistry (e.g. chlorosilane, SiH_3Cl; see Topic B5).

Table 1 shows the bond strengths and the standard free energies of formation of hydrides. Bond strengths and thermodynamic stabilities decrease down each group. Compounds such as boranes and silanes are strong reducing agents and may inflame spontaneously in air. Reactivity generally increases with catenation.

Table 1. A selection of nonmetal hydrides (E indicates nonmetal)

Hydride formulae and names		Normal boiling point (°C)	E–H bond enthalpy ($kJ\ mol^{-1}$)	$\Delta G_f^\ominus$ ($kJ\ mol^{-1}$ at 298 K)
B	B_2H_6 diborane	–93	–	+87
C	CH_4 methane	–162	413	–51
	C_2H_6 ethane	–89	–	–33
	C_2H_4 ethene	–104	–	+68
Si	SiH_4 silane	–112	318	+57
	Si_2H_6 disilane	–14	–	+127
Ge	GeH_4 germane	–88	285	+113
N	NH_3 ammonia (*azane*)[a]	–33	391	–17
	N_2H_4 hydrazine (*diazane*)[a]	113	–	+149
	HN_3 hydrogen azide	*36*[b]	–	+327
P	PH_3 phosphine (*phosphane*)[a]	–88	321	+13
	P_2H_4 diphosphane	*64*[b]	–	–
As	AsH_3 arsine (*arsane*)[a]	–55	296	69
O	H_2O water (*oxidane*)[a]	100	464	–237
	H_2O_2 hydrogen peroxide (*dioxidane*)[a]	152	–	–120
S	H_2S hydrogen sulfide (*sulfane*)[a]	–60	364	–34
	H_2S_2 disulfane	*70*[b]	–	–
Se	H_2Se hydrogen selenide (*selane*)[a]	–42	313	+16
F	HF hydrogen fluoride	19	568	–297
Cl	HCl hydrogen chloride	–85	432	–95
Br	HBr hydrogen bromine	–68	366	–53
I	HI hydrogen iodide	–35	289	+2

[a]IUPAC recommended systematic names that are rarely used.

[b]Extrapolated values for compounds decomposing before boiling at atmospheric pressure.

General routes to the preparation of hydrides include:

(i) direct combination of elements:

$$N_2 + 3H_2 \rightarrow 2NH_3$$

(ii) reaction of a metal compound of the element with a protonic acid such as water:

$$Ca_3P_2 + 6H_2O \rightarrow 2PH_3 + 3Ca(OH)_2$$

(iii) reduction of a halide or oxide with $LiAlH_4$ or $NaBH_4$:

$$3NaBH_4 + 4BF_3 \rightarrow 2B_2H_6 + 3NaBF_4$$

Route (ii) or (iii) is required when direct combination is thermodynamically unfavorable. Catenated hydrides can often be formed by controlled pyrolysis of the mononuclear compound.

Brønsted acidity arises from the possibility of transferring a proton to a base, which may sometimes be the same compound (see Topic E2 for discussion of trends). **Basicity** is possible when nonbonding electron pairs are present (see Topics C1 and C8). Basicity towards protons decreases towards the right and down each group in the periodic table, so that ammonia is the strongest base among simple hydrides.

Compounds with metals

Not all metallic elements form hydrides. Those that do may be classified as follows.

- Highly electropositive metals have solid hydrides often regarded as containing the H^- ion. They have structures similar to halides, although the ionic character of hydrides is undoubtedly much lower. Examples include LiH (rocksalt structure) and MgH_2 (rutile structure; see Topic D3).
- Some *d*- and *f*-block elements form hydrides that are often metallic in nature, and of variable (nonstoichiometric) composition. Examples include TiH_2 and CeH_{2+x}.
- Some heavier *p*-block metals form molecular hydrides similar to those of nonmetals in the same group, examples being digallane (Ga_2H_6) and stannane (SnH_4), both of very low stability.

Hydrides of more electropositive elements can be made by direct reaction between elements. They are very strong reducing agents and react with water to give dihydrogen:

$$CaH_2 + 2H_2O \rightarrow Ca(OH)_2 + 2H_2$$

The hydride ion can act as a ligand and form **hydride complexes** similar in some ways to those of halides, although their stability is often limited by the reducing properties of the H^- ion. The most important complexes are the tetrahedral ions BH_4^- and AlH_4^- normally found as the salts $NaBH_4$ and $LiAlH_4$. They may be made by the action of NaH or LiH on a halide or similar compound of B or Al, and are used as reducing agents and for the preparation of hydrides of other elements.

Many transition metal complexes containing hydrogen are known, including the unusual nine-coordinate ion $[ReH_9]^{2-}$ (see Topic H5). Hydride is a very strong σ-donor ligand and is often found in conjunction with π-acid ligands and in organometallic compounds (see Topics H9 and H10).

The hydrogen bond

A hydrogen atom bound to an electronegative atom such as N, O or F may interact in a noncovalent way with another electronegative atom. The resulting **hydrogen bond** has an energy in the range 10–60 kJ mol^{-1}, weak by standards of covalent bonds but strong compared with other intermolecular forces (see Topic C9). The strongest hydrogen bonds are formed when a fluoride ion is involved, for example in the symmetrical $[F–H–F]^-$ ion. Symmetrical bonds are occasionally formed with oxygen but in most cases the hydrogen is not symmetrically disposed, a typical example being in liquid water where the normal O–H bond has a length of 96 pm and the hydrogen bond a length around 250 pm. Hydrogen bonding arises from a combination of electrostatic (ion–dipole and dipole–dipole) forces and orbital overlap; the latter effect may be treated by a three-center molecular orbital approach (see Topic C5).

Hydrogen bonding is crucial for the secondary structure of biological molecules such as proteins and nucleic acids, and for the operation of the genetic code. Its influence can be seen in the boiling points of simple hydrides (see *Table 1* and Topic C9, *Fig. 1*). The exceptional values for NH_3, H_2O and HF result from strong hydrogen bonding in the liquid.

Deuterium and tritium

Deuterium (^2D) and tritium (^3T) are heavier isotopes of hydrogen (see Topic A1). The former is stable and makes up about 0.015% of all normal hydrogen. Its physical and chemical properties are slightly different from those of the light isotope ^1H. For example, in the electrolysis of water H is evolved faster and this allows fairly pure D_2 to be prepared. Tritium is a radioactive β-emitter with a half-life of 12.35 years, and is made when some elements are bombarded with neutrons. Both isotopes are used for research purposes. They also undergo very exothermic **nuclear fusion reactions**, which form the basis for thermonuclear weapons ('hydrogen bombs') and could possibly be used as a future energy source.

F3 Boron

Key Notes

The element	Boron has an unusual chemistry characterized by electron deficiency. It occurs in nature as borates. Elemental structures are very complex.
Hydrides	There is a vast range of neutral compounds and anions. Except in the BH_4^- ion, the compounds show complex structures, which cannot be interpreted using simple electron pair bonding models.
Halides	BX_3 compounds are Lewis acids, with acceptor strength in the order $BI_3 > BBr_3 > BCl_3 > BF_3$.
Oxygen compounds	B_2O_3 and the very weak acid $B(OH)_3$ give rise to a wide range of metal borates with complex structures containing both three- and four-coordinate boron.
Other compounds	Some boron–nitrogen compounds have similar structures to those of carbon. Structurally complex borides are formed with many metals.
Related topic	Rings and clusters (C6)

The element

The only nonmetallic element in group 13 (see Topic B2), boron has a strong tendency to covalent bonding. Its uniquely complex structural chemistry arises from the $(2s)^2(2p)^1$ configuration, which gives it one less valence electron than the number of orbitals in the valence shell. Simple compounds such as BCl_3 have an incomplete octet and are strong Lewis acids (see Topics C1 and C8), but boron often accommodates its **electron deficiency** by forming clusters with multicenter bonding.

Boron is an uncommon element on the Earth overall (about 9 p.p.m. in the crust) but occurs in concentrated deposits of borate minerals such as borax $Na_2[B_4O_5(OH)_4].8H_2O$, often associated with former volcanic activity or hot springs. It is used widely, mostly as borates in glasses, enamels, detergents and cosmetics, and in lesser amounts in metallurgy.

Boron is not often required in its elemental form, but it can be obtained by electrolysis of fused salts, or by reduction either of B_2O_3 with electropositive metals or of a halide with dihydrogen, the last method giving the purest boron. The element has many allotropic structures of great complexity; their dominant theme is the presence of icosahedral B_{12} units connected in different ways. Multicenter bonding models are required to interpret these structures.

Hydrides

The simplest hydrogen compounds are salts of the **tetrahydroborate ion** BH_4^-, which is tetrahedral and isoelectronic with methane (see Topic C1). $LiBH_4$ is prepared by reducing BF_3 with LiH. It is more widely used as the sodium salt, which is a powerful reducing agent with sufficient kinetic stability to be used in aqueous solution. Reaction of $NaBH_4$ with either I_2 or BF_3 in diglyme $(CH_3OCH_2)_2O$ gives **diborane** B_2H_6, the simplest molecular hydride. Its structure with bridging hydrogen atoms requires three-center two-electron bonds (see Topics C1 and C5):

$$2NaBH_4 + 2I_2 \rightarrow B_2H_6 + 2NaI + 2HI$$

Heating B_2H_6 above 100ºC leads to pyrolysis and generates a variety of more complex boranes of which tetraborane(10) B_4H_{10} and decaborane(14) $B_{10}H_{14}$ are the most stable. Other reactions can lead to anionic species, such as the icosahedral dodecahydrododecaborate(2–) $[B_{12}H_{12}]^{2-}$, prepared at 180ºC:

$$5B_2H_6 + 2NaBH_4 \rightarrow Na_2[B_{12}H_{12}] + 13H_2$$

The structural classification and bonding in boranes is described in Topic C6; especially striking are the anions $[B_nH_n]^{2-}$ with closed polyhedral structures. Boranes with heteroatoms can also be prepared, such as $B_{10}C_2H_{12}$, which is isoelectronic with $[B_{12}H_{12}]^{2-}$.

Boranes are strong reducing agents and the neutral molecules inflame spontaneously in air, although the anions $[B_nH_n]^{2-}$ have remarkable kinetic stability. Diborane itself reacts with Lewis bases (see Topic C8). The simplest products can be regarded as donor–acceptor complexes with BH_3, which is a 'soft' Lewis acid and forms adducts with soft bases such as CO (**1**). More complex products often result from unsymmetrical cleavage of B_2H_6, for example,

$$B_2H_6 + 2NH_3 \rightarrow [NH_3BH_2NH_3]^+[BH_4]^-$$

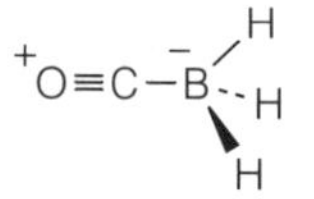

$^+O{\equiv}C{-}B^-H_3$

1

Halides

Molecular BX_3 compounds are formed with all halogens. They have the trigonal planar structure predicted by VSEPR (see Topic C2), although there appears to be a certain degree of π bonding (strongest in BF_3) involving halogen lone-pairs and the empty boron $2p$ orbital (see **2** for one of the possible resonance forms). The halides are strong Lewis acids, BF_3 and BCl_3 being used as catalysts (e.g. in organic Friedel–Crafts acylations). Interaction with a donor gives a tetrahedral geometry around boron as with the analogous BH_3 complex **1**. The π bonding in the parent molecule is lost and for this reason BF_3, where such bonding is strongest, is more resistant to adopting the tetrahedral geometry than are the heavier halides. Thus the acceptor strengths follow the order

$$BF_3 < BCl_3 < BBr_3 < BI_3$$

which is the reverse of that found with halides of most other elements (see Topic

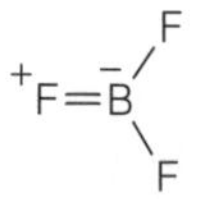

$^+F{=}B^-F_2$

2

F9). Strongest interaction occurs with hard donors such as F^- (forming the stable tetrafluoroborate ion $[BF_4]^-$) and with oxygen donors such as water. Except with BF_3 (where the B—F bonds are very strong) complex formation often leads to solvolysis, forming $B(OH)_3$ in water.

Pyrolysis of BX_3 compounds leads to halides with B—B bonds, for example, B_2X_4 (**3** with X = F or Cl) and polyhedral B_nCl_n molecules (n = 4, 8, 9).

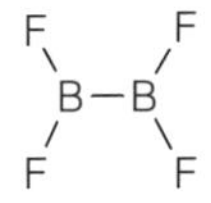

F₂B—BF₂

3

Oxygen compounds

Boric oxide B_2O_3 is very hard to crystallize; the glass has a linked covalent network in which both bridging B—O—B and terminal B=O bonds may be present. The hydroxide **boric acid** $B(OH)_3$ is formed by the hydrolysis of many boron compounds. It has a layer structure made up of planar molecules linked by hydrogen bonding. It is a Lewis acid that acts as a Brønsted acid in protic solvents. In water the equilibrium

$$B(OH)_3 + 2H_2O \rightleftharpoons B(OH)_4^- + H_3O^+$$

gives a $pK_a = 9.25$ but complexing can increase the acidity; for example, in anhydrous H_2SO_4 it forms $[B(HSO_4)_4]^-$ and is one of the few species that can act as a strong acid in that solvent (see Topic F8).

Borates can be formed with all metals, although those of groups 1 and 2 are best known. The structural features are complex and rival those of silicates (see Topic D5). Boron can occur as planar BO_3 or tetrahedral BO_4 groups, often linked by B—O—B bonds as in silicates. For example, **4** shows the ion found in borax $Na_2[B_4O_5(OH)_4].8H_2O$, where both three- and four-coordinate boron is present. **Borosilicate glasses** (such as 'Pyrex') have lower coefficients of thermal expansion than pure silicate glasses and so are more resistant to thermal shock.

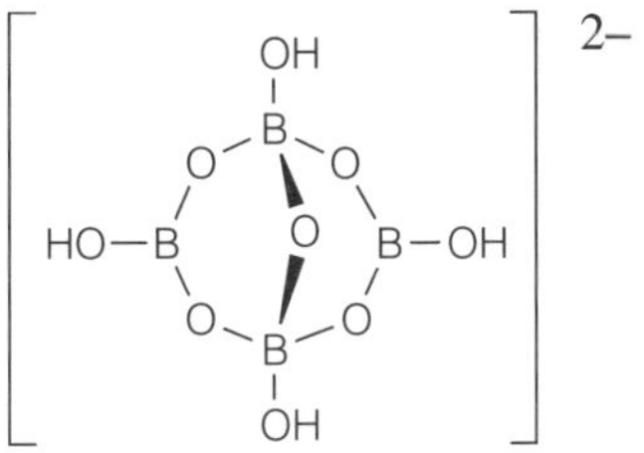

$[B_4O_5(OH)_4]^{2-}$

4

Other compounds

Boron forms many compounds with nitrogen. Some of these are structurally analogous to carbon compounds, the pair of atoms BN being isoelectronic with CC. (For example, the ion $[NH_3BH_2NH_3]^+$ is analogous to propane, $CH_3CH_2CH_3$.) **Boron nitride** BN can form two solid structures, one containing hexagonal BN layers similar to graphite, and the other with tetrahedral sp^3 bonding like diamond (see Topic D2). **Borazine** $B_3N_3H_6$ has a 6-π-electron ring like benzene (5 shows one resonance form; see Topic C6). Although BN is very hard and resistant to chemical attack, borazine is much more reactive than benzene and does not undergo comparable electrophilic substitution reactions. The difference is a result of the polar B–N bond, and the more reactive B–H bonds.

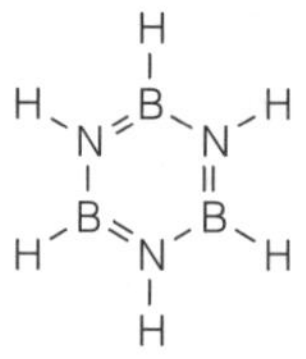

5

Boron forms a binary carbide, often written B_4C but actually nonstoichiometric, and compounds with most metals. The stoichiometries and structures of these solids mostly defy simple interpretation. Many types of chains, layers and polyhedra of boron atoms are found. Simple examples are CaB_6 and UB_{12}, containing linked octahedra and icosahedra, respectively.

F4 Carbon, silicon and germanium

Key Notes

The elements

Carbonates and reduced forms of carbon are common on Earth, and silicates make up the major part of the crust; germanium is much less common. All elements can form the diamond structure; graphite and other allotropes are unique to carbon.

Hydrides and organic compounds

Silanes and germanes are less stable than hydrocarbons. Double bonds involving Si and Ge are very much weaker than with C.

Halides

Halides of all the elements have similar formulae and structures. Those of Si and Ge (but not of C) are Lewis acids and are rapidly hydrolyzed by water.

Oxygen compounds

Carbon oxides are molecular with multiple bonds, those of Si and Ge polymeric in structure. Carbonates contain simple CO_3^{2-} ions, but silicates and germanates have very varied and often polymeric structures.

Other compounds

Compounds with S and N also show pronounced differences between carbon and the other elements. Many compounds with metals are known but these are not highly ionic. Metal–carbon bonds occur in organometallic compounds.

Related topics

Introduction to nonmetals (F1)
Organometallic compounds (H10)
Geochemistry (J2)

The elements

With the valence electron configuration s^2p^2 the nonmetallic elements of group 14 can form compounds with four tetrahedrally directed covalent bonds. Only carbon forms strong multiple bonds, and its compounds show many differences in structure and properties from those of Si and Ge. Like the metallic elements of the group (Sn and Pb), germanium has some stable divalent compounds.

The abundances of the elements by mass in the crust are: C about 480 p.p.m., Si 27% (second only to oxygen), and Ge 2 p.p.m. Carbon is present as carbonate minerals and in smaller amounts as the element and in hydrocarbon deposits. It is important in the atmosphere (as the greenhouse gas CO_2; see Topic J6) and is the major element of life. Silicate minerals are the dominant chemical compounds of the crust and of the underlying mantle (see Topic J2). Germanium is widely but thinly distributed in silicate and sulfide minerals.

All three elements can crystallize in the tetrahedrally bonded **diamond structure** (see Topic D2). Si and Ge are semiconductors (see Topic D7). Carbon has other allotropes. **Graphite** is the thermodynamically stable form at ordinary pressures, diamond at high pressures. More recently discovered forms include **buckminsterfullerene** C_{60}, higher fullerenes such as C_{70}, and **nanotubes** composed of graphite sheets rolled into cylinders. In these structures carbon forms three σ bonds, the remaining valence electron being in delocalized π orbitals analogous to those in benzene (see Topic C6).

The elements can be produced by reduction of oxides or halides. Highly divided carbon black is used as a catalyst and black pigment, and impure carbon (coke) for reducing some metal oxides (e.g. in the manufacture of iron; see Topic B4). Pure silicon prepared by reduction of $SiCl_4$ with Mg is used in electronics ('silicon chips') although much larger quantities of impure Si are used in steels.

Hydrides and organic compounds

Compounds of carbon with hydrogen and other elements form the vast area of organic chemistry. **Silanes** and **germanes** are Si and Ge analogs of methane and short-chain saturated hydrocarbons, and can be prepared by various methods, such as reduction of halides with $LiAlH_4$:

$$ECl_4 + LiAlH_4 \rightarrow EH_4 + LiAlCl_4 \text{ (E = Si or Ge)}$$

They are much more reactive than corresponding carbon compounds and will inflame spontaneously in air. Stability decreases with chain length in series such as

$$EH_4 > E_2H_6 > E_3H_8 > \ldots$$

Many derivatives can be made where H is replaced by monofunctional groups such as halide, alkyl, $-NH_2$. Many Si and Ge compounds are similar in structure to those of carbon, but trisilylamine $(SiH_3)_3N$ and its germanium analog differ from $(CH_3)_3N$ in being nonbasic and having a geometry that is planar rather than pyramidal about N. This suggests the involvement of the N lone-pair electrons in partial multiple bonding through the valence expansion of Si or Ge (see Topic C2, Structure 8).

Si and Ge analogs of compounds where carbon forms double bonds are much harder to make. $(CH_3)_2SiO$ is not like propanone $(CH_3)_2C{=}O$, but forms **silicone polymers** with rings or chains having single Si–O bonds (**1**). Attempts to make alkene analogs $R_2Si{=}SiR_2$ (where R is an organic group) generally result in single-bonded oligomers, except with very bulky R– groups such as mesityl $(2,4,6(CH_3)_3C_6H_2-)$, which prevent polymerization.

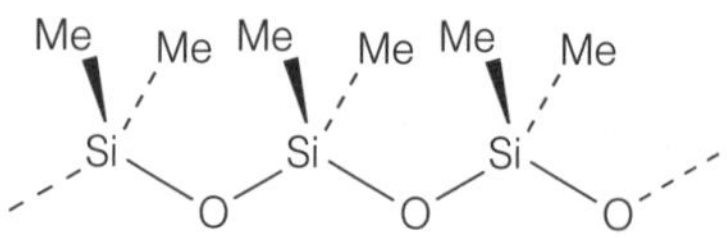

1

Halides

All halides EX_4 form tetrahedral molecules. Mixed halides are known, as well as fully or partially halogen-substituted catenated alkanes, silanes and germanes (e.g. Ge_2Cl_6). Unlike the carbon compounds, halides of Si and Ge are Lewis acids and readily form complexes such as $[SiF_6]^{2-}$. Attack by Lewis bases often leads to decomposition, and thus rapid hydrolysis in water, unlike carbon halides, which are kinetically more inert.

Divalent halides EX_2 can be made as reactive gas-phase species, but only for Ge are stable noncatenated Ge^{II} compounds formed. They have polymeric structures with pyramidal coordination as with Sn^{II} (see Topic G6). The compound CF formed by reaction of fluorine and graphite has one F atom bonded to every C, thus disrupting the π bonding in the graphite layer but retaining the σ bonds and giving tetrahedral geometry about carbon. (Bromine forms intercalation compounds with graphite; see Topic D5.)

Oxygen compounds

Whereas carbon forms the molecular oxides CO and CO_2 with multiple bonding (see Topics C1 and C4), stable oxides of Si and Ge are polymeric. **Silica** SiO_2 has many structural forms based on networks of corner-sharing SiO_4 tetrahedra (see Topic D3). GeO_2 can crystallize in silica-like structures as well as the rutile structure with six-coordinate Ge. This structure is stable for SiO_2 only at very high pressures, the difference being attributable to the greater size of Ge. Thermodynamically unstable solids SiO and GeO can be made but readily disproportionate to the ioxide.

CO_2 is fairly soluble in water but true **carbonic acid** is present in only low concentration:

$$CO_2(aq) + H_2O = H_2CO_3 \qquad K = 1.6 \times 10^{-3}$$
$$H_2CO_3 + H_2O = HCO_3^- + H_3O^+ \qquad K = 2.5 \times 10^{-4}$$

The apparent K_a given by the product of these two equilibria is 4.5×10^{-7} ($pK_a = 6.3$), much smaller than the true value for carbonic acid, which is more nearly in accordance with Pauling's rules ($pK_a = 3.6$; see Topic E2). The hydration of CO_2 and the reverse reaction are slow, and in biological systems are catalyzed by the zinc-containing enzyme carbonic anhydrase (see Topic J3).

SiO_2 and especially GeO_2 are less soluble in water than is CO_2, although solubility of SiO_2 increases at high temperatures and pressures. **Silicic acid** is a complex mixture of polymeric forms and only under very dilute conditions is the monomer $Si(OH)_4$ formed. SiO_2 reacts with aqueous HF to give $[SiF_6]^{2-}$.

The structural chemistry of carbonates, silicates and germanates shows parallels with the different oxide structures. All **carbonates** (e.g. $CaCO_3$) have discrete planar CO_3^{2-} anions (see Topic C1, Structure 11). **Silicate** structures are based on tetrahedral SiO_4 groups, which can be isolated units as in Mg_2SiO_4, but often polymerize by Si—O—Si corner-sharing links to give rings, chains, sheets and 3D frameworks (see Topics D3, D5 and J2). Many germanates are structurally similar to silicates, but germanium more readily adopts six-coordinate structures.

Other compounds

Carbon disulfide CS_2 has similar bonding to CO_2, but SiS_2 differs from silica in having a chain structure based on edge-sharing tetrahedra, and GeS_2 adopts the CdI_2 layer structure with octahedral Ge (see Topic D3).

Nitrogen compounds include the toxic species cyanogen $(CN)_2$ (**2**) and the **cyanide ion** CN^-, which forms strong complexes with many transition metals (see Topics H2 and H6). Si_3N_4 and Si_2N_2O are polymeric compounds with single Si—N bonds, both forming refractory, hard and chemically resistant solids of interest in engineering applications.

N≡C—C≡N

2

Compounds with metals show a great diversity. A few carbides and silicides of electropositive metals, such as Al_3C_4 and Ca_2Si, could be formulated with C^{4-} and Si^{4-} ions although the bonding is certainly not very ionic. Compounds with transition metals are metallic in character, those of Si and Ge being normally regarded as **intermetallic compounds**, those of carbon as **interstitial compounds** with small carbon atoms occupying holes in the metal lattice. Some such as TaC and WC are remarkably hard, high melting and chemically unreactive, and are used in cutting tools. Fe_3C occurs in steel and contributes to the mechanical hardness.

Many compounds with E–E bonding are known (see Topic D5). CaC_2 has C_2^{2-} ions (isoelectronic with N_2) and reacts with water to give ethyne C_2H_2. On the other hand, KSi and $CaSi_2$ are **Zintl compounds** with single-bonded structures. Ge (like Sn and Pb) forms some polyanions such as $[Ge_9]^{4-}$ (see Topics C6 and G6).

Organometallic compounds containing metal–carbon bonds are formed by nearly all metals, and are discussed under the relevant elements (see especially transition metals, Topic H10). Some analogous Si and Ge compounds are known.

F5 NITROGEN

Key Notes

The element

Nitrogen has a strong tendency to form multiple bonds. Dinitrogen is a major constituent of the atmosphere. The great strength of the triple bond limits its reactivity.

Ammonia and derivatives

Ammonia is basic in water and a good ligand. It is an important industrial and laboratory chemical. Related compounds include hydrazine and organic derivatives of ammonia (amines).

Oxygen compounds

The many known nitrogen oxides have unusual structures, all with some degree of multiple bonding. Oxocations and oxoacids can be formed, of which nitric acid is the most important. All compounds with oxygen are potentially strong oxidizing agents, but reactivity is often limited by kinetic factors.

Other compounds

Fluorides are the most stable halides. Many metals form nitrides but these are not highly ionic.

Related topics

Introduction to nonmetals (F1)

Phosphorus, arsenic and antimony (F6)

The element

Nitrogen is a moderately electronegative element but the great strength of the triple bond makes N_2 kinetically and thermodynamically stable. The atom can form three single bonds, generally with a pyramidal geometry (see Topics C1 and C2), but also has a notable tendency to multiple bonding. Its unusually rich redox chemistry is illustrated in the Frost diagram in *Fig. 1* (see below).

Dinitrogen makes up 79 mol % of dry air. The element is essential for life and is one of the elements often in short supply, as fixation of atmospheric nitrogen to form chemically usable compounds is a difficult process (see Topics J3 and J6).

Nitrogen is obtained from the atmosphere by liquefaction and fractional distillation. Its normal boiling point (77 K or –196ºC) and its ready availability make it a useful coolant. It reacts directly with rather few elements and is often used as an inert filling or 'blanket' for metallurgical processes. The majority of industrial nitrogen, however, is used to make ammonia and further compounds (see Topic J4).

Ammonia and derivatives

Ammonia NH_3 is manufactured industrially in larger molar quantities than any other substance. The **Haber process** involves direct synthesis from the elements at around 600 K at high pressure and in the presence of a potassium-promoted iron catalyst. Ammonia is used to make nitric acid and other chemicals including many plastics and pharmaceuticals.

Ammonia is a good Lewis base and an important ligand in transition metal

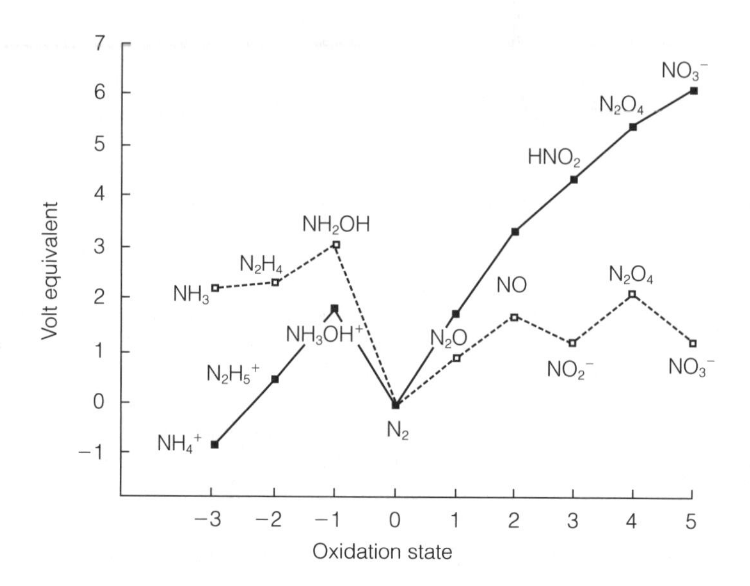

Fig. 1. Frost diagram showing the redox states of nitrogen in water at pH = 0 (continuous line) and pH = 14 (dashed line).

complexes (see Topics C8, E3 and H3). In water it acts as a Brønsted base through the equilibrium

$$NH_3 + H_2O \rightleftharpoons NH_4^+ + OH^- \quad K_b = 1.8 \times 10^{-5}$$

The **ammonium ion** forms salts and has a similar radius to K^+, although the structures are sometimes different because NH_4^+ can undergo hydrogen bonding. For example, NH_4F has the tetrahedral wurtzite structure rather than the rocksalt structure of KF; the tetrahedral coordination is ideal for formation of hydrogen bonds between NH_4^+ and F^- ions. Ammonium salts often dissociate reversibly on heating:

$$NH_4Cl(s) \rightleftharpoons NH_3(g) + HCl(g)$$

Ammonia has a normal boiling point of –33ºC. As with water, this value is much higher than expected from the normal group trend, a manifestation of strong hydrogen bonding. Liquid ammonia also undergoes autoprotolysis although to a lesser extent than water (see Topics E1 and E2). It is a good solvent for many ionic substances, and is much more basic than water. Ammonium salts act as acids and amides as bases. Ammonia is kinetically inert under strongly reducing conditions, and will dissolve alkali metals to give solutions with free solvated electrons present (see Topic G2).

Hydrazine N_2H_4 (**1**) can be made by the Rauschig synthesis:

H–N–N–H (with H on each N)

1

$$2NH_3 + NaOCl \rightarrow N_2H_4 + NaCl + H_2O$$

Its combustion to give N_2 and H_2O is extremely exothermic ($\Delta H = -620$ kJ mol^{-1})

and it has been used as a rocket fuel. The explosive hydrogen azide HN_3 is the conjugate acid of the **azide ion** N_3^- (**2**). Another hydrogen compound is hydroxylamine NH_2OH.

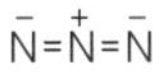

$\bar{N}=\overset{+}{N}=\bar{N}$

2

Nitrogen forms an enormous variety of organic compounds. **Amines** such as methylamine CH_3NH_2 and trimethylamine $(CH_3)_3N$ can be regarded as derived from ammonia by replacing one or more H atoms with alkyl or aryl groups. Like ammonia, amines are basic and form complexes with transition metals. **Tetraalkyl ammonium ions** such as $[(C_4H_9)_4N]^+$ are useful when large anions are required in inorganic synthesis (see Topic D6). Nitrogen also forms **heterocyclic compounds** such as pyridine C_5H_5N.

Oxygen compounds

The most commonly encountered oxides, oxocations and oxoanions, are shown in *Fig. 2*. All these species have some multiple bonding, the single N—N and N—O bonds being comparatively weak. **Nitrous oxide** N_2O can be made by heating ammonium nitrate. It is isoelectronic with CO_2 and somewhat unreactive, and is used as an anaesthetic ('laughing gas') and as a propellant for aerosols. **Nitric oxide** NO and **nitrogen dioxide** NO_2 are the normal products of reaction of oxygen and nitrogen at high temperatures, or of the oxidation of ammonia. They are both odd-electron molecules. NO_2 dimerizes reversibly at low temperatures to make N_2O_4, but NO has very little tendency to dimerize in the gas phase, probably because the odd electron is delocalized in a π antibonding orbital (see Topic C4; the molecular orbital diagram is like that for CO but with one more electron). NO reacts with oxygen to give NO_2. It can act as a ligand in transition metal complexes. The other oxides of nitrogen are less stable: N_2O_3 is shown in *Fig. 2*; N_2O_5 is normally found as $[NO_2]^+[NO_3]^-$; and NO_3 is an unstable radical that (like NO and NO_2) plays a role in atmospheric chemistry.

NO^+ and NO_2^+ (isoelectronic with CO and CO_2, respectively) can be formed by the action of strong oxidizing agents on NO or NO_2 in acid solvents such as H_2SO_4, and are known as solid salts (e.g. $NO^+[AsF_6]^-$). The **nitrite** and **nitrate** ions NO_2^- and NO_3^- are formed respectively from **nitrous acid** HNO_2 and **nitric acid** HNO_3. As expected from Pauling's rules, HNO_2 is a weak acid in water and HNO_3 a strong acid (see Topic E2). Metal nitrates and nitrites are strong oxidizing agents, generally very soluble in water. Other less stable oxoacids are known, mostly containing N—N bonds. Although the free acid corresponding to phosphoric acid

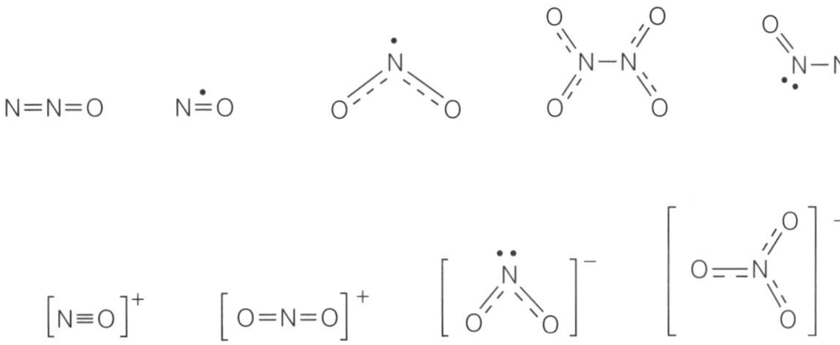

N=N=O   N=O $[N\equiv O]^+$ $[O=N=O]^+$

Fig. 2. Structures of some oxides, oxocations and oxoanions of nitrogen.

H_3PO_4 is unknown, it is possible to make **orthonitrates** containing the tetrahedral NO_4^{3-} ion (see Topic F1, Structure 2). Nitric acid is a major industrial chemical made from ammonia by catalytic oxidation to NO_2, followed by reaction with water and more oxygen:

$$2NO_2 + H_2O + 1/2\ O_2 \rightarrow 2HNO_3$$

It is used to make NH_4NO_3 fertilizer, and in many industrial processes (see Topic J4).

The redox chemistry of nitrogen compounds in aqueous solution is illustrated in the Frost diagram in *Fig. 1* (see Topic E5 for construction and use). All oxides and oxoacids are strong oxidizing agents, and all oxidation states except –3, 0 and +5 are susceptible to disproportionation. The detailed reactions are, however, mostly controlled by kinetic rather than thermodynamic considerations. In conjunction with oxidizable groups, as in ammonium nitrate NH_4NO_3 or in organic nitro compounds, N—O compounds can be powerful explosives.

Other compounds

Compounds with sulfur are described in Topic F8. Apart from its fluorides, nitrogen **halides** are thermodynamically unstable and very explosive. The trifluoride NF_3 can be prepared by direct reaction of NH_3 and F_2. It is kinetically inert and nontoxic. Further fluorination gives the N^V species NF_4^+:

$$NF_3 + 2F_2 + SbF_3 \rightarrow [NF_4]^+[SbF_6]^-$$

The oxofluoride ONF_3 is also known. Like NF_4^+ it is isoelectronic with NO_4^{3-} and must be described by a similar valence structure (**3**). N_2F_4 is interesting in that like N_2O_4 it readily dissociates into NF_2 radicals. Double-bonded N_2F_2 exists in *cis* (**4**) and *trans* (**5**) forms, the former being thermodynamically more stable.

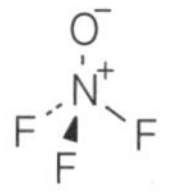

3 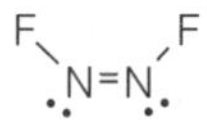 **4** 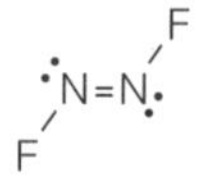 **5**

Nitrogen reacts directly with some electropositive metals to form **nitrides** such as Li_3N and Ca_3N_2. Although these can be formulated with nitride ion N^{3-} the bonding may be partially covalent. Other compounds with metals are **amides** and **imides** (containing NH_2^- and NH^{2-}, respectively) and **azides** containing N_3^-. Metal azides are thermodynamically unstable and often explosive.

F6 PHOSPHORUS, ARSENIC AND ANTIMONY

Key Notes

The elements

Elemental structures are based on E_4 molecules or three-coordinate polymeric structures. Phosphates are widespread minerals, As and Sb being found as sulfides.

Hydrides and organic derivatives

Hydrides are less stable than ammonia and less basic. Many organic derivatives can be made.

Halides

Compounds in the +3 and +5 oxidation state are known, although As^{V} is strongly oxidizing. Some halides are good Lewis acids, and halide transfer reactions are common.

Oxides and oxoacids

Oxides in the +3 and +5 oxidation state are increasingly polymeric with heavier elements. They form oxoacids, of which phosphoric acid is the most important.

Other compounds

These include many sulfides, phosphonitrilic compounds with ring and chain structures, and compounds with metals, which are generally of low ionic character.

Related topics

Introduction to nonmetals (F1) Nitrogen (F5)

The elements

The heavier elements in the same group (15) as nitrogen are occasionally known as 'pnictogens' and their compounds with metals as 'pnictides'. Although the elements form some compounds similar to those of nitrogen, there are very pronounced differences, as is found in other nonmetal groups (see Topics F1 and F5).

Phosphorus is moderately abundant in the Earth's crust as the phosphate ion; the major mineral source is **apatite** $Ca_5(PO_4)_3(F,Cl,OH)$, the notation (F,Cl,OH) being used to show that F^-, Cl^- and OH^- can be present in varying proportions. Arsenic and antimony are much rarer. They occur in minerals such as realgar As_4S_4 and stibnite Sb_2S_3, but are mostly obtained as byproducts from the processing of sulfide ores of other elements. Elemental P is obtained by reduction of calcium phosphate. The complex reaction approximates to:

$$2Ca_3(PO_4)_2 + 6SiO_2 + 10C \rightarrow 6CaSiO_3 + 10CO + P_4$$

Most phosphates are used more directly without conversion to the element.

Phosphorus has many allotropes. It is most commonly encountered as **white**

phosphorus, which contains tetrahedral P_4 molecules (**1**). Other forms, which are more stable thermodynamically but kinetically harder to make, contain polymeric networks with three-coordinate P. White phosphorus is highly reactive and toxic. It will combine directly with most elements, glows in air at room temperature as a result of slow oxidation, and combusts spontaneously at a temperature above 35ºC. Arsenic can also form As_4 molecules, but the common solid forms of this element and Sb are polymeric with three-coordination. They are markedly less reactive than phosphorus.

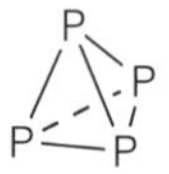

P P P P

1

Enormous quantities of phosphates are used, in fertilizers, food products, detergents and other household products. For fertilizer applications apatite is converted by the action of acid to the much more soluble compound $Ca(H_2PO_4)_2$, known as 'superphosphate' (see Topic J4).

Hydrides and organic derivatives

The hydrides **phosphine** PH_3, **arsine** AsH_3 and **stibine** SbH_3 can be prepared by hydrolysis of metal phosphides, or by reduction of molecular compounds such as PCl_3. They are very toxic gases, with decreasing thermal stability $P > As > Sb$. Unlike ammonia they are not basic in water. The hydrazine analog diphosphane P_2H_4 and a few other catenated compounds with P–P bonds can be made, although their stability is low.

Organic derivatives include alkyl and aryl phosphines such as triphenyl phosphine $(C_6H_5)_3P$. As with the hydrides these compounds are much less basic than the corresponding nitrogen compounds towards acceptors such as H^+, but are good ligands for transition metals in low oxidation states, as they have π-acceptor properties (see Topic H9). Cyclic polyarsanes such as $(AsPh)_6$ (where Ph is a phenyl group, C_2H_5) with As—As bonds are readily made, and with very bulky organic groups it is possible to prepare compounds with E=E double bonds, for example,

$$2((CH_3)_3Si)_3CAsCl_2 + 2Li(C_4H_9) \rightarrow ((CH_3)_3Si)_3CAs{=}AsC((CH_3)_3Si)_3 + 2LiCl + 2C_4H_9Cl$$

(compare C, Si and Ge; Topic F4). Unlike with nitrogen, the five-coordinate compounds Ph_5E are known. The P and As compounds have the normal trigonal bipyramidal geometry (Topic C2) but Ph_5Sb is unexpectedly square pyramidal (**2**).

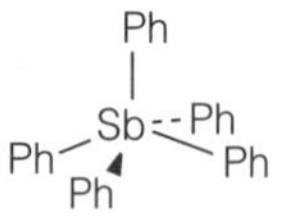

Ph Sb Ph Ph Ph Ph

2

Halides

Phosphorus forms the binary compounds P_2X_4 (with a P—P bond), PX_3 and PX_5 with all halogens. With As and Sb a complete set of EX_3 compounds is known, but the only E^V halides stable under normal conditions are AsF_5, SbF_5 and $SbCl_5$. $AsCl_5$ has been identified from the UV irradiation of PCl_3 in liquid Cl_2 but decomposes above –50ºC. Most known halides can be obtained by direct reaction of the elements in appropriate proportions, but P and F together form only PF_5 and the trihalide can be prepared by reacting PCl_3 with ZnF_2 or HgF_2. The molecular substances have the expected structures, pyramidal for EX_3 and trigonal bipyramidal for EX_5 (see Topic C2). However, some have a marked tendency to undergo halide transfer, and in the solid state PCl_5 and PBr_5 form the ionic structures $[PCl_4]^+[PCl_6]^-$ and $[PBr_4]^+Br^-$, respectively. Presumably it is the lattice energy associated with an ionic solid that stabilizes these forms. Many halide complexes are known. AsF_5 and SbF_5 are Lewis acids with a very strong affinity for F^-, giving $[AsF_6]^-$ or fluoride bridged species such as $[Sb_2F_{11}]^-$ (**3**).

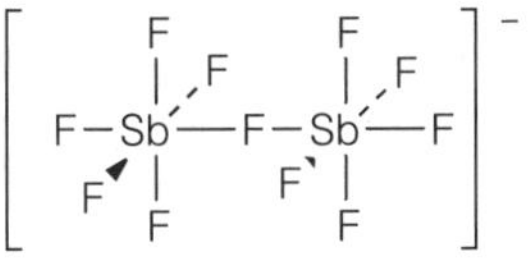

3

Oxohalides EOX_3 form tetrahedral molecules with E=P, but polymeric structures with As and Sb. $POCl_3$ is an important intermediate in the manufacture of organophosphorus compounds, used, for example, as insecticides.

Oxides and oxoacids

P_4O_6 (**4**) and P_4O_{10} (**5**) can be obtained by direct reaction of the elements, the P^V compound 'phosphorus pentoxide' being the normal product when phosphorus burns in air. Under carefully controlled conditions intermediate oxides P_4O_n ($n = 7, 8, 9$) can be made. The oxides of As and Sb have polymeric structures, and include a mixed valency compound Sb_2O_4 with Sb^{III} in pyramidal coordination and octahedral Sb^V.

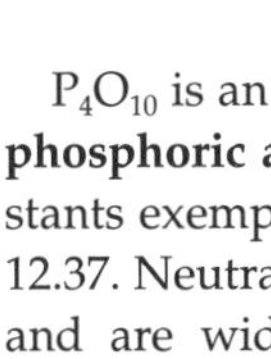

4

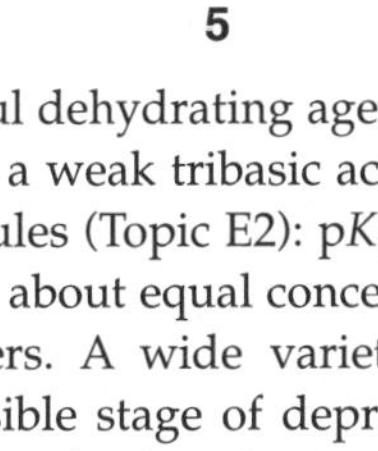

5

P_4O_{10} is an extremely powerful dehydrating agent, reacting with water to form **phosphoric acid** H_3PO_4. This is a weak tribasic acid with successive acidity constants exemplifying Pauling's rules (Topic E2): $pK_1 = 2.15$, $pK_2 = 7.20$ and $pK_3 = 12.37$. Neutral solutions contain about equal concentrations of $H_2PO_4^-$ and HPO_4^{2-} and are widely used as buffers. A wide variety of metal **orthophosphates**, containing ions with each possible stage of deprotonation, are known. Further addition of P_4O_{10} to concentrated phosphoric acid results in the formation **polyphosphates** with P–O–P linkages as in silicates. These linkages are kinetically stable in aqueous solution and are important in biology (see Topic J3). **Metaphosphates** such as KPO_3 have infinite chains of corner-sharing octahedra as in the isoelectronic metasilicates such as $CaSiO_3$ (see Topic D5).

The P^{III} oxoacid **phosphorous acid** H_3PO_3 does not have the structure $P(OH)_3$ that its formula suggests, but is tetrahedral with a PH bond: $HPO(OH)_2$. It is thus diprotic with a similar pK_1 to phosphoric acid. The trend is continued with hypophosphorous acid $H_2PO(OH)$. Both acids are strong reducing agents.

Arsenic acid H_3AsO_4 is similar to phosphoric acid but is a relatively strong oxidizing agent. Sb^V oxo compounds have different structures and are based on the octahedral $[Sb(OH)_6]^-$ ion. Aqueous As^{III} and Sb^{III} species are hard to characterize; they are much more weakly acidic than phosphorous acid and are probably derived from $As(OH)_3$ and $Sb(OH)_3$. The corresponding salts tend to have polymeric structures, for example, $NaAsO_2$ with oxygen linked $[-As(O^-)-O]_\infty$ chains isoelectronic with SeO_2.

Other compounds

The **sulfides** of As and Sb are found in nature. As_2S_3 and Sb_2S_3 with the stoichiometries expected for As^{III} and Sb^{III} have polymeric structures. Compounds such as As_4S_4 (**6**) and P_4S_n (n = 3–10) are molecules based on P_4 or As_4 tetrahedra with bridging –S– groups inserted; some of the phosphorus compounds also have terminal P=S groups similar to P=O in **5**.

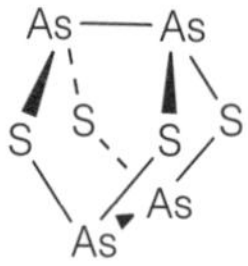

6

Phosphazines are compounds containing repeated $-PX_2N-$ units. For example, the reaction

$$nPCl_5 + nNH_4Cl \rightarrow (NPCl_2)_n + 4nHCl$$

gives rings and chains with a distribution of n values. The (PX_2N) unit has the same number of valence electrons as (Me_2SiO), which forms silicone polymers (see Topic F1, *Table 1*, and Topic F4). In the valence structure as drawn in **7** P and N carry formal charges, but there is probably some P=N double bonding.

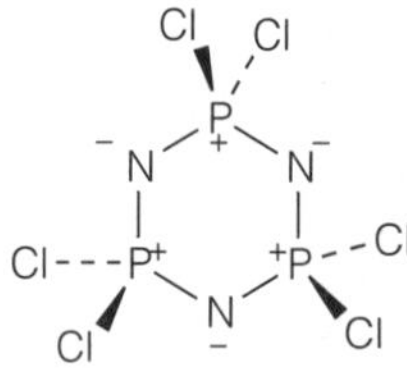

7

Binary compounds with metals are generally of low ionic character. Many of those with transition metals have the NiAs and related structures (see Topics D3 and D4) and show metallic properties. Some compounds appear to contain polyanionic species (e.g. P_2^{4-} isoelectronic with S_2^{2-} in Sr_2P_2, and P_7^{3-} in Na_3P_7), although the bonding is certainly not fully ionic.

F7 OXYGEN

Key Notes

The element

Oxygen compounds are extremely abundant on Earth. The element exists as dioxygen O_2 (which has two unpaired electrons) and the less stable allotrope ozone O_3. The strongly oxidizing properties of O_2 are moderated by the strength of the double bond.

Oxides

Nonmetallic elements form molecular or covalent polymeric structures and have acid properties, giving oxoacids with water. Many oxides of metallic elements have ionic structures and are basic. Intermediate bonding types and chemical properties are common, for example, with metals in high oxidation states.

Peroxides and superoxides

Ionic peroxides and superoxides contain O_2^{2-} and O_2^-, respectively. Hydrogen peroxide and other peroxo compounds contain O–O bonds, which are weak.

Positive oxidation states

Salts containing $[O_2]^+$ and some oxygen fluorides are known.

Related topics

Electronegativity and bond type (B1)
Chemical periodicity (B2)
Introduction to nonmetals (F1)
Sulfur, selenium and tellurium (F8)

The element

Oxygen is the second most electronegative element after fluorine, and forms thermodynamically stable compounds with nearly all elements. It rivals fluorine in the ability to stabilize the highest known oxidation states of many elements, examples where there is no corresponding fluoride being $Cl^{VII}O_4^-$ and $Os^{VIII}O_4$. Oxidation reactions with O_2 are often slow because of the strength of the O=O double bond (490 kJ mol^{-1}).

Oxygen is the most abundant element on Earth, making around 46% of the Earth's crust by mass. The commonest minerals are complex oxides such as silicates and carbonates. Oxygen is also a constituent of water, and of nearly all biological molecules. Atmospheric O_2 comes almost entirely from photosynthesis by green plants, and is not found on other known planets. Reactions involving dioxygen, both in photosynthesis and in respiration by air-breathing animals, are important in biological chemistry (see Topic J3).

Oxygen can be extracted from the atmosphere by liquefaction and fractional distillation. The liquid boils at –183°C (90 K) and is dangerous when mixed with combustible materials. The compressed gas is used in metallurgy (e.g. steel-making) and the liquid as an oxidizer for rocket propulsion.

Oxygen has two allotropes, the normal dioxygen O_2 form and **ozone O_3 (1)** formed by subjecting O_2 to an electric discharge. Ozone is a trace constituent of the atmosphere, where it plays an important role as an absorber of UV radiation.

$$^{-}O-\overset{+}{\ddot{O}}=O \longleftrightarrow O=\overset{+}{\ddot{O}}-O^{-}$$

1

As predicted by molecular orbital theory (see Topic C3) dioxygen has two unpaired electrons and some of its chemistry shows diradical characteristics; in particular, it reacts readily with other radicals. **Singlet oxygen** is an excited state in which the two electrons in the π antibonding orbitals have paired spins. It is produced in some chemical reactions and has different chemical reactivity.

Oxides

Oxygen forms binary compounds with nearly all elements. Most may be obtained by direct reaction, although other methods (such as the thermal decomposition of carbonates or hydroxides) are sometimes more convenient. Oxides may be broadly classified as **molecular**, **polymeric** or **ionic** (see Topics B1 and B2). Covalent oxides are formed with nonmetals, and may contain terminal (E=O) or bridging (E–O–E) oxygen. Especially strong double bonds are formed with C, N and S. Bridging is more common with heavier elements and leads to the formation of many polymeric structures such as SiO_2 (see Topics F1 and F4).

Water H_2O is the most abundant molecular substance on Earth. It is highly polar, with physical properties dominated by hydrogen bonding, and an excellent solvent for ionic substances and reactions (see Topics C9 and E1–E5). Many **hydrated salts** are known (e.g. $CuSO_4.5H_2O$), which contain water bound by coordination to metal ions and/or hydrogen bonding to anions. Autoprotolysis gives the ions H_3O^+ and OH^-, which are also known in solid salts, H_3O^+ with anions of strong acids (e.g. $[H_3O]^+[NO_3]^-$; hydrated species such as $[H_5O_2]^+$ are also known), and OH^- in **hydroxides**, which are formed by many metals.

Oxides of most metallic elements have structures that may be broadly classed as ionic (see Topics D3 and D4). The closed-shell O^{2-} ion is unknown in the gas phase, the reaction

$$O^-(g) + e^- \rightarrow O^{2-}(g)$$

being very endothermic. It is therefore only the large lattice energy obtained with the O^{2-} ion that stabilizes it in solids (see Topic D6). The variety of coordination numbers (CN) of oxide is large, examples being:

CN =	2	3	4	6	8
	ReO_3	TiO_2	ZnO	MgO	Li_2O

Oxide has a notable tendency for symmetrical coordination in ionic solids (linear, planar or tetrahedral with CN = 2, 3 or 4, respectively) and unlike sulfide rarely forms layer structures.

The distinction between ionic and polymeric solids is not absolute, and oxides of metals with low electropositive character (e.g. HgO) or in high oxidation states (e.g. CrO_3) are better described as having polar covalent bonds. A few metals in very high oxidation states form molecular oxides (e.g. Mn_2O_7, OsO_4).

Many ternary and more complex oxides are known. It is normal to distinguish **complex oxides** such as $CaCO_3$, which contain discrete oxoanions, and **mixed oxides** such as $CaTiO_3$, which do not (see Topic D5).

In water, the very basic O^{2-} ion reacts to form hydroxide:

$$O^{2-}(aq) + H_2O \rightarrow 2OH^-(aq)$$

Table 1. Some oxoacids, showing their anhydrides and the anions formed by them

Anhydride	Acid name	Acid formula	Anion formula
B_2O_3	Boric	$B(OH)_3$	$[B(OH)_4]^-$ [a]
CO_2	Carbonic	H_2CO_3	CO_3^{2-} [b]
SiO_2	Silicic	$Si(OH)_4$	SiO_4^{4-} [a,b]
N_2O_3	Nitrous	HNO_2	NO_2^-
N_2O_5	Nitric	HNO_3	NO_3^-
P_4O_{10}	Phosphoric	H_3PO_4	PO_4^{3-} [b]
SO_2	Sulfurous	H_2SO_3	SO_3^{2-} [b]
SO_3	Sulfuric	H_2SO_4	SO_4^{2-} [b]
Cl_2O	Hypochlorous	ClOH	ClO^-
(Cl_2O_5) [c]	Chloric	$HClO_3$	ClO_3^-
Cl_2O_7	Perchloric	$HClO_4$	ClO_4^-

[a] Anion with a strong tendency to polymerize and form complex structures.
[b] Polyprotic acid with intermediate states of ionization possible.
[c] Parent anhydride unknown.

and so ionic oxides are **basic** and either form alkaline solutions if soluble in water, or otherwise dissolve in acid solution. Covalent oxides (including those such as CrO_3 formed by metals in high oxidation states) are **acidic** and react with water to form **oxoacids**:

$$P_4O_{10} + 6H_2O \rightarrow 4H_3PO_4 \rightleftharpoons 4H_2PO_4^- + 4H^+ \rightleftharpoons \text{etc.}$$

(See Topic E2 for Pauling's rules on acid strength.) Such oxides may therefore be regarded as **acid anhydrides**. *Table 1* shows a selection of oxoacids with their anhydrides and illustrates the conventional nomenclature. For example, **sulfurous** and **sulfuric** acids display the lower (+4) and higher (+6) oxidation state, respectively, and their anions are called **sulfite** and **sulfate**.

Some oxides are **amphoteric** and have both acidic and basic properties; this often happens with a metal ion with a high charge/size ratio such as Be^{2+} or Al^{3+} (see examples in Topics E2 and G3–G5). A few nonmetallic oxides (e.g. CO) are **neutral** and have no appreciable acid or basic properties.

Peroxides and superoxides

Adding one or two electrons to dioxygen gives the **superoxide** O_2^- and **peroxide** O_2^{2-} ions. As the added electrons occupy the π antibonding orbital (see Topic C3) the bond becomes progressively weaker and longer. Superoxides MO_2, rather than simple oxides M_2O are the normal products of reacting the heavier alkali metals with oxygen; peroxides M_2O_2 are also formed. This may be explained by lattice energy arguments (see Topic D6). With most metal ions, the higher lattice energy obtained with O^{2-} forces the disproportionation of the larger O_2^- and O_2^{2-} ions. With large, low-charged cations, however, the lattice energy gain is insufficient to cause disproportionation. The peroxide ion can also be stabilized in **peroxo** complexes, where it acts as a ligand to transition metals, as in $[Cr^V(O_2)_4]^{3-}$.

The simplest covalent peroxide is **hydrogen peroxide** H_2O_2, which is normally encountered in aqueous solution. Although kinetically fairly stable, it can act as either an oxidizing agent (giving H_2O) or a reducing agent (giving O_2), and many transition metal ions catalyze its decomposition. Organic peroxides (R_2O_2) and peroxoacids (e.g. the percarbonate ion, **2**) contain the fairly weak peroxo O–O linkage. Some covalent peroxides can be unpredictably and dangerously explosive.

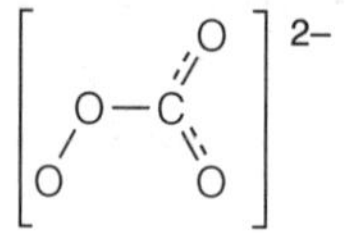

2

Positive oxidation states

Reaction with strong oxidizing agents gives the O_2^+ ion, which has a stronger and shorter bond than O_2 (see Topic C3):

$$O_2 + BF_3 + 1/2\ F_2 \rightarrow [O_2^+][BF_4]^-$$

Fluorides include F_2O and F_2O_2. The latter has a considerably shorter O–O bond than in peroxides, a fact that may indicate some contribution of ionic valence structures such as (**3**), which allow a degree of multiple bonding. All compounds in positive oxidation states are very strongly oxidizing. Compounds with heavier halogens are normally regarded as halogen oxides and are discussed in Topic F9.

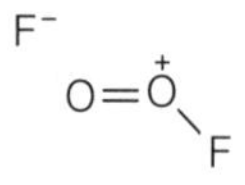

3

F8 SULFUR, SELENIUM AND TELLURIUM

Key Notes

The elements

The elements known as chalcogens show pronounced differences from oxygen in the same group, being much less electronegative. Sulfides are important minerals for some elements. Elemental structures are based on rings and chains with single bonds.

Chalcogenides

The hydrides are toxic gases. Metal chalcogenides are much less ionic than oxides, and often have different (e.g. layer) structures.

Halides

Many halides are known in oxidation states up to +6. Most are molecular compounds but some have polymeric structures.

Oxides and oxoacids

EO_2 and EO_3 compounds have structures that are increasingly polymeric for heavier elements. They form oxoacids, of which sulfuric acid is the most important.

Other compounds

Cationic species such as E_4^{2+} can be prepared. Sulfur and nitrogen form an interesting range of binary compounds.

Related topics

Introduction to nonmetals (F1)   Oxygen (F7)

The elements

The elements known collectively as the **chalcogens** are in the same group (16) as oxygen (Topic F7). They form some compounds similar to those of oxygen, but show many differences characteristic of other nonmetal groups (see Topic F1).

Sulfur is widespread in the Earth's crust, occurring as metal sulfides, sulfates, and **native** or elemental sulfur formed by bacterial oxidation of sulfides. Many less electropositive metals known as **chalcophiles** are found commonly as sulfide minerals (see Topic J2); some important examples are pyrites (FeS_2), sphalerite (zinc blende, ZnS), molybdenite (MoS_2), cinnabar (HgS) and galena (PbS). Volatile sulfur compounds such as H_2S and organic compounds are also found in petroleum and natural gas. The element is used in large amounts for the manufacture of **sulfuric acid** (see below). Selenium and tellurium are much rarer, found as minor components of sulfide minerals.

Sulfur has several allotropic forms, the most stable of which are molecular solids containing S_8 rings. The elemental forms of Se and Te have spiral chains and are semiconductors. In all of these solids each atom forms two single bonds to neighbors (see Topic D2). Sulfur combines directly with oxygen and halogens (except I), and with many less electronegative elements to form sulfides. The other elements show similar properties although reactivity declines down the group.

Chalcogenides

Molecular compounds include H_2S and its analogs, and many organic compounds. The hydrides are made by the action of Brønsted acids on metal chalcogenides. They are extremely toxic gases, weakly acidic in water (e.g. for H_2S, $pK_1 = 6.8$, $pK_2 = 14.2$). Many **polysulfanes** H_2S_n containing S–S bonds are also known.

Solid chalcogenides are formed by all metallic elements and by many non-metals. Only with the most electropositive metals do they commonly have the same structures as oxides (see Topics D3 and D4). With transition metals, compounds MX (which are frequently of variable stoichiometry) have the nickel arsenide or similar structures in which metal–metal bonding is present. MX_2 compounds either have layer structures (e.g. TiS_2, $TiSe_2$, $TiTe_2$, all CdI_2 types) or structures containing diatomic ions (e.g. FeS_2 has S_2^{2-} units and so is formally a compound of Fe^{II} not Fe^{IV}). Chalcogenides of electropositive metals are decomposed by water giving hydrides such as H_2S, but those of less electropositive elements (often the ones forming sulfide ores, see above) are insoluble in water.

Halides

A selection of the most important halides is show in *Table 1*. With sulfur the fluorides are most stable and numerous, but Se and Te show an increasing range of heavier halides. Compounds such as S_2Cl_2 and S_2F_{10} have S–S bonds; S_2F_2 has another isomer $S{=}SF_2$. Sulfur halides are molecular and monomeric with structures expected from VSEPR (e.g. SF_4 'see-saw', SF_6 octahedral; see Topic C2). With the heavier elements increasing polymerization is found, as in $(TeCl_4)_4$ (**1**) and related tetramers.

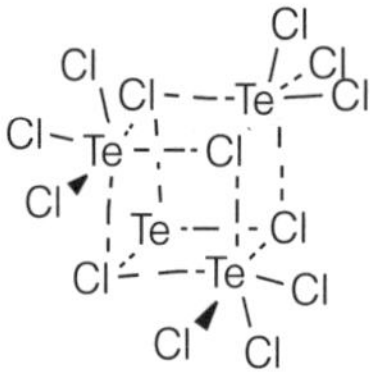

1

The hexahalides are kinetically inert, but most other halides are highly reactive and are hydrolyzed in water giving oxides and oxoacids. Intermediate hydrolysis products are **oxohalides** of which **thionyl chloride** $SOCl_2$ and **sulfuryl chloride** SO_2Cl_2 are industrially important compounds.

Some of the halides show donor and/or acceptor properties (see Topic C8). For example, SF_4 reacts with both Lewis acids (forming compounds such as $[SF_3]^+[BF_4]^-$) and bases (forming either simple adducts such as $C_5H_5N{:}SF_4$ with pyridine, or compounds containing the square pyramidal ion $[SF_5]^-$). The complex ions $[SeX_6]^{2-}$ and $[TeX_6]^{2-}$ (X = Cl, Br, I) are interesting as they appear to have regular octahedral structures in spite of the presence of a nonbonding electron pair on the central atom (see Topic C2).

Table 1. Principal halides of S, Se and Te

S	Se	Te
S_2F_2, SF_2, S_2F_4, SF_4, S_2F_{10}, SF_6	SeF_4, SeF_6	TeF_4, TeF_6
S_2Cl_2, SCl_2	Se_2Cl_2, $(SeCl_4)_4$	Te_2Cl, $(TeCl_4)_4$
S_nBr_2	Se_2Br_2, $(SeBr_4)_4$	Te_2Br, $(TeBr_4)_4$ Te_2I, Te_4I_4, $(TeI_4)_4$

Oxides and oxoacids

The major oxides of all three elements (E) are EO_2 and EO_3. Sulfur in addition forms many oxides of low thermodynamic stability, for example S_8O with a structure containing an S_8 ring. **Sulfur dioxide** SO_2 is the major product of burning sulfur and organic sulfur compounds in air, and is a serious air pollutant giving rise (after oxidation to H_2SO_4; see Topic J6) to acid rain. With one lone-pair, SO_2 is a bent molecule and has both Lewis acid and basic properties. The liquid is a good solvent for reactions with strong oxidizing agents. SO_2 dissolves in water giving acid solutions containing the pyramidal hydrogensulfite (HSO_3^-) and **sulfite** (SO_3^{2-}) ions. The expected sulfurous acid H_2SO_3, however, is present only in very low concentrations. SeO_2 and TeO_2 have polymeric structures and give oxoacid salts similar to those from sulfur.

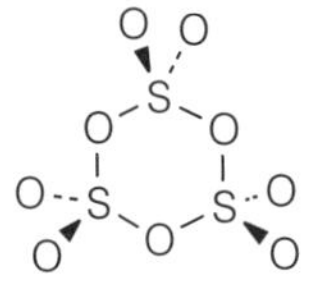

2

Sulfur trioxide SO_3 is made industrially as a route to sulfuric acid, by oxidizing SO_2 with oxygen using a vanadium oxide catalyst. It can exist as a monomeric planar molecule but readily gives cyclic S_3O_9 trimers and linear polymers with corner-sharing SO_4 units (see **2** and Topic D3). The highly exothermic reaction with water gives **sulfuric acid** H_2SO_4, which is the world's major industrial chemical, being used in many large-scale processes for making fertilizers, dyestuffs, soaps and detergents, and synthetic fibers (see Topic J4). Anhydrous sulfuric acid undergoes a series of acid–base equilibria such as

$$2H_2SO_4 \rightleftharpoons H_3SO_4^+ + HSO_4^-$$

$$2H_2SO_4 \rightleftharpoons H_3O^+ + HS_2O_7^-$$

(see Topic E1). It is a very strongly acid medium, in which HNO_3 (a strong acid in water) acts as a base:

$$HNO_3 + 2H_2SO_4 \rightarrow NO_2^+ + H_3O^+ + 2HSO_4^-$$

The resulting 'nitrating mixture' is used for preparing aromatic nitro compounds by electrophilic reactions of NO_2^+.

Reaction of HF with SO_3 gives **fluorosulfonic acid** HSO_3F, which is even more strongly acidic than sulfuric acid. In mixtures with SO_3 and powerful fluoride acceptors such as SbF_5 it gives **superacid media**, which are capable of protonating even most organic compounds.

SeO_3 and selenic acid H_2SeO_4 are similar to the sulfur analogs except that they are more strongly oxidizing. Tellurium behaves differently, as telluric acid has the octahedral $Te(OH)_6$ structure, which, as expected from Pauling's rules, is a very weak acid (see Topic E2).

There are many other oxoacids of sulfur, of which the most important are peroxodisulfate $S_2O_8^{2-}$, which has a peroxo (O–O) bond, and compounds with S–S bonds including thiosulfate $S_2O_3^{2-}$, dithionite $S_2O_4^{2-}$ and tetrathionate $S_4O_6^{2-}$. The reaction

$$2S_2O_3^{2-} + I_2 \rightarrow 2I^- + S_4O_6^{2-}$$

is used for the quantitative estimation of I_2 in aqueous solution.

Other compounds

Oxidation of the elements (e.g. by AsF_5) in a suitable solvent such as SO_2 or H_2SO_4 gives a series of **polyatomic cations** such as $[S_8]^{2+}$ and $[S_4]^{2+}$. The latter (and its Se and Te analogs) has a square-planar structure and can be regarded as a 6π-electron ring (see Topic C6).

Also of note are **sulfur–nitrogen compounds**. The cage-like S_4N_4 (see Topic C6) is formed by the reaction of S_2Cl_2 with ammonia or NH_4Cl. Passing the heated vapor over silver wool gives the planar S_2N_2 with the same valence electron count as $[S_4]^{2+}$. Polymerization forms polythiazyl $(SN)_x$, a linear polymer with metallic conductivity arising from delocalization of the one odd electron per SN unit.

F9 HALOGENS

Key Notes

The elements

The halogens are electronegative and oxidizing elements, fluorine exceptionally so. They occur in nature as halides, and form highly reactive diatomic molecules.

Halides and halide complexes

Molecular halides are formed with most nonmetals, ionic halides with metals. Some halides are good Lewis acids, and many halide complexes are known.

Oxides and oxoacids

Most halogen oxides are of low stability, but several oxoacids are known except for fluorine. Redox stability depends on pH, Cl_2 and Br_2 disproportionating in alkaline solution.

Interhalogen and polyhalogen compounds

Halogens form an extensive range of neutral and ionic compounds with each other, including some cationic species.

Related topics

Introduction to nonmetals (F1)
Binary compounds: simple structures (D3)
Binary compounds: factors influencing structure (D4)

The elements

The halogen group (17) is the most electronegative in the periodic table, and all elements readily form halide ions X^-. Trends in chemistry resemble those found in other groups (see Topic F1). Fluorine is limited to an octet of valence electrons. It is the most electronegative and reactive of all elements and often (as with oxygen) brings out the highest oxidation state in other elements: examples where no corresponding oxide is known include PtF_6 and AuF_5 (see Topic H5).

F and Cl are moderately abundant elements, principal sources being **fluorite** CaF_2 and **halite** NaCl, from which the very electronegative elements are obtained by electrolysis. Bromine is mainly obtained by oxidation of Br^- found in salt water; iodine occurs as iodates such as $Ca(IO_3)_2$. Astatine is radioactive and only minute amounts are found in nature. Chlorine is used (as ClO^- and ClO_2) in bleaches and is an important industrial chemical, other major uses (as with all the halogens) being in the manufacture of halogenated organic compounds (see Topic J4).

The elements form diatomic molecules, F_2 and Cl_2 being gases at normal temperature and pressure, Br_2 liquid and I_2 solid. They react directly with most other elements and are good oxidizing agents, although reactivity declines down the group. X–X bond strengths follow the sequence $F < Cl > Br > I$ (see Topic C7).

Halides and halide complexes

Nearly all elements form thermodynamically stable halides. The normal stability sequence is F > Cl > Br > I, which in covalent compounds follows the expected order of bond strengths, and in ionic compounds that of lattice energies (see Topics C7 and D6). The thermodynamic stability of fluorides (and the kinetic reactivity of F_2) is also aided by the weak F–F bond. Many halides can be made by direct combination, but fluorinating agents such as ClF_3 are sometimes used in preference to F_2, which is very difficult to handle.

The structural and bonding trends in halides follow similar patterns to those in oxides (see Topics B2 and F7). Most nonmetallic elements form simple **molecular compounds** in which halogen atoms each have a single bond to the other element. This is true also for metals in high oxidation states (e.g. $TiCl_4$ and UF_6). The compounds may be solids, liquids or gases, with volatility in the order F > Cl > Br > I as expected from the strength of van der Waals' forces. In the **hydrogen halides** HF is exceptional because of strong hydrogen bonding (see Topic C9). HF is a weak acid in water, the other HX compounds being strong acids (see Topic E2).

Covalent halides are less often polymeric in structure than oxides, a difference partly caused by the different stoichiometries (e.g. SiF_4 versus SiO_2), which provide a higher coordination number in the monomeric molecular halides. However, the halides of some metals (e.g. beryllium; Topic G3) may be better regarded as polymeric than ionic. Some molecular halides of both metallic and nonmetallic elements form halogen-bridged dimers and higher oligomers (e.g. Al_2Cl_6; Topic G4).

Most metallic elements form solid halides with structures expected for **ionic solids** (see Topics D3 and D4). Structural differences often occur with MX_2 and MX_3, fluorides more often having rutile, fluorite or rhenium trioxide structures, and the heavier halides layer structures. These differences reflect the more ionic nature of fluorides, and the higher polarizability of the larger halide ions. Many halides are very soluble in water, but low solubilities are often found with fluorides of M^{2+} and M^{3+} ions (e.g. CaF_2, AlF_3), and with heavier halides of less electropositive metals (e.g. AgCl, TlCl). These differences are related to lattice energy trends (see Topics D6 and E4).

Many halides of metals and nonmetals are good Lewis acids (see Topic C8). Such compounds are often hydrolyzed by water, and also form halide complexes (e.g. $AlCl_4^{2-}$, PF_6^-), which can make useful counterions in solids with large or strongly oxidizing cations. Both cationic and anionic complexes may be formed by halide transfer, for example, in solid PCl_5 (Topic F6) and in liquid BrF_3 (see below). Many metal ions also form halide complexes in aqueous solution. For a majority of elements the fluoride complexes are more stable but softer or class b metals form stronger complexes with heavier halides (see Topic E3).

Oxides and oxoacids

I_2O_5 is the only halogen oxide of moderate thermodynamic stability. Other compounds include X_2O (not I), X_2O_2 (F and Cl), the odd-electron XO_2 (Cl and Br), and Cl_2O_7. Most of these compounds are strongly oxidizing, have low thermal stability and can decompose explosively. ClO_2 is used as a bleaching agent.

Except for fluorine the elements have an extensive oxoacid chemistry. *Figure 1* shows Frost diagrams with the oxidation states found in acid and alkaline solution (see Topic E5). The sharp trend in oxidizing power of the elements (X_2/X^- potential) can be seen. As expected from Pauling's rules (see Topic E2) the **hypohalous acids** X(OH) and **chlorous acid** ClO(OH) are weak acids, but the **halic acids** XO_2(OH) and especially **perchloric acid** ClO_3(OH) and perbromic acid are strong. **Periodic acid** is exceptional, as, although periodates containing the tetra-

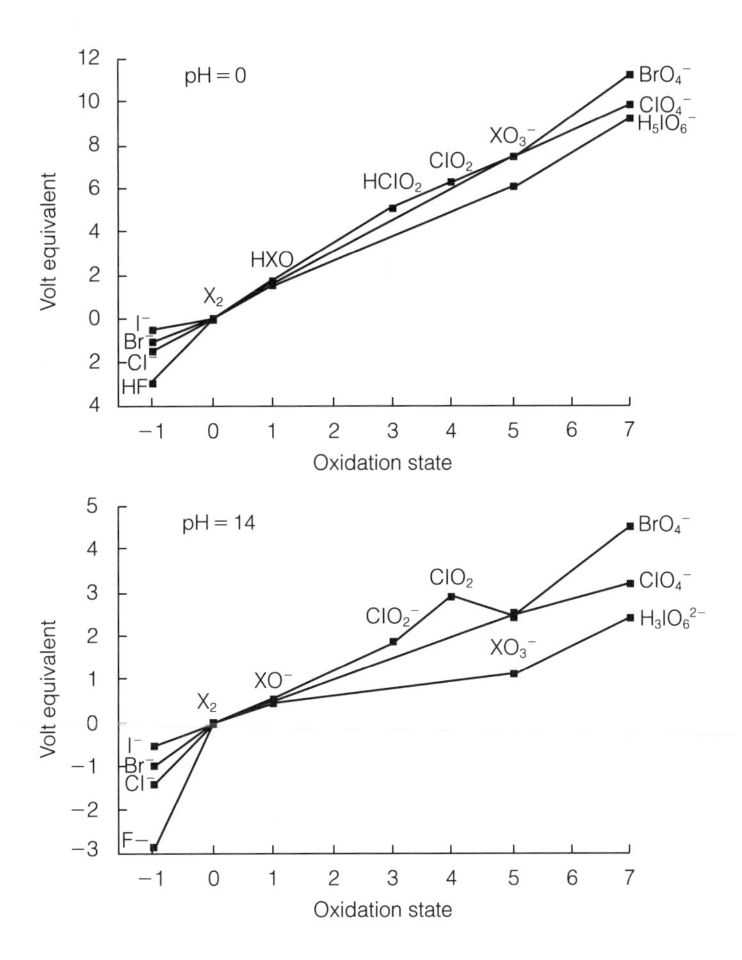

Fig. 1. Frost diagrams for the halogens in aqueous solution at pH = 0 (a) and pH = 14 (b). X represents any halogen, except F for positive oxidation states.

hedral IO_4^- ion are known, the predominant form in water is the octahedral $IO(OH)_5$, which, as expected, is a weak acid.

The redox behavior is strongly pH dependent but is also influenced by kinetic factors. From the pH = 14 diagram in *Fig. 1* it can be seen that Cl_2 and Br_2 disproportionate in alkaline solution. The thermodynamically expected products are X^- and XO_3^- but the hypochlorite ion ClO^- is formed in cold conditions, and further disproportionation occurs on heating.

The perhalic acids and their anions are strong oxidizing agents, especially BrO_4^-, which is not thermodynamically stable in aqueous solution. They do, however, have considerable kinetic stability. Perchlorates of organic or organometallic cations are very dangerous as they may appear stable, but can explode unpredictably with extreme force.

Interhalogen and polyhalogen compounds

Binary compounds known as **interhalogen compounds** with stoichiometry XY_n are found between every pair of halogens F–I. For neutral molecules n is an odd number and when $n > 1$ the terminal atom Y is always the lighter element. The maximum n found with a given pair increases with the difference in period number, some examples being IBr, ICl_3, BrF_5 and IF_7. Most interhalogen compounds are obtained by direct reaction. They are strongly oxidizing and the fluorides are good fluorinating agents.

Many interhalogen and polyhalogen anions and cations are also known, some forming easily. For example, aqueous solutions containing I^- dissolve I_2 to form I_3^-. In liquid BrF_3 the following equilibrium occurs:

$$2BrF_3 = BrF_2^+ + BrF_4^-$$

In accordance with the **solvent-system** concept (see Topic E1), fluoride donors such as NaF act as bases in this medium (giving Na^+ and BrF_4^-), and fluoride acceptors such as SbF_5 act as acids (giving BrF_2^+ and SbF_6^-).

Other **cationic species** can be prepared by strong oxidation of the elements (e.g. with AsF_5) in a suitable nonaqueous solvent. Examples include Cl_3^+, Br_2^+ and I_5^+, which are also known in solid salts with anions such as AsF_6^-.

Most species have the structures predicted by the VSEPR model (see Topic C2). Listed according to the steric number (SN) below, the geometries are

SN = 4: XY_2^+ (bent);

SN = 5: XY_2^- (linear), XY_3 (T-shaped), XF_4^+ ('see-saw');

SN = 6: XY_4^- (square-planar), XY_5 (square-pyramidal), XY_6^+ (octahedral);

SN = 7: XY_7 (pentagonal bipyramidal)

F10 NOBLE GASES

Key Notes

The elements — Noble gases occur as uncombined atoms in the atmosphere, and are uncommon except for argon. Helium has an exceptionally low boiling point and does not solidify except under pressure.

Xenon compounds — Xenon forms some binary fluorides and oxides, as well as fluoride complexes and oxoanions. All are very reactive compounds.

Krypton compounds — The only binary compound is a very unstable difluoride.

Related topic — Introduction to nonmetals (F1)

The elements

With their closed-shell electron configurations the noble gas elements of group 18 were long regarded as chemically inert. However, in 1962 Bartlett noted that the ionization energy of xenon was similar to that of O_2, and by reaction with PtF_6 attempted to prepare the compound analogous to $[O_2]^+[PtF_6]^-$ (see Topic F7). He obtained a complex product containing the ion $[XeF]^+$ (with a valence structure **1** isoelectronic to dihalogen molecules) rather than the expected Xe^+. Many compounds of xenon are now known, mostly with F and O, and few of krypton.

The gases are not generally abundant on Earth, although argon (formed by the radioactive decay of ^{40}K) makes up about 1 mol % of the atmosphere, and helium (formed by radioactive decay of uranium and thorium; see Topics A1 and I2) occurs in natural gas. Radon is radioactive, ^{222}Rn with a half-life of 3.8 days also being formed by radioactive decay from ^{238}U. The boiling points of the elements show the trend expected from van der Waals' forces (Topic C9), that of helium (4.2 K) being the lowest of any substance. Helium is also unique as it does not solidify except under pressure; the remaining elements form monatomic solids with close-packed structures (see Topic D2). Liquid helium is used for maintaining very low temperatures (e.g. for superconducting magnets), argon as an inert gas in some metallurgical processes, and all the elements in gas discharge tubes.

$$[:\ddot{X}e-\ddot{F}:]^+$$

1

Xenon compounds

The binary **fluorides** XeF_2, XeF_4 and XeF_6 are thermodynamically stable and can be prepared by direct reaction under appropriate conditions. They are reactive fluorinating agents. The bonding can be described by three-center molecular orbital pictures or by resonance structures (e.g. **2**; see Topic C5) in which no valence-shell expansion is required. The structures of XeF_2 (linear) and XeF_4

(square-planar) are those expected in the VSEPR model (see Topic C2) but that of gas-phase XeF_6 has proved elusive. It is believed that (as predicted for a molecule with a lone-pair) the shape is not a regular octahedron, but that **fluxional processes** lead to a rapid interchange between different distorted configurations. In the solid structure, some association between molecules occurs and the geometry around Xe is distorted, as expected in the VSEPR theory.

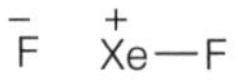

$\overset{-}{F} \quad \overset{+}{Xe}—F$

2

Compounds that appear to contain the $[XeF]^+$ (**1**) and bent $[Xe_2F_3]^+$ ions are known although the former is always strongly coordinated to a counterion such as SbF_6^-. Complex anions include XeF_5^-, XeF_7^- and XeF_8^{2-}, the first of which has a unique pentagonal planar structure (**3**), as expected from VSEPR.

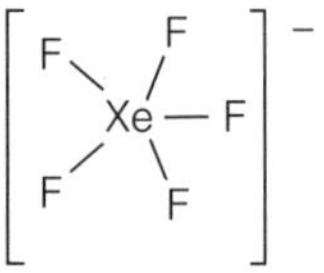

$[XeF_5]^-$

3

Oxohalides such as $XeOF_4$ are known. Hydrolysis of XeF_6 gives XeO_3, which disproportionates in alkaline solution:

$$2XeO_3 + 4OH^- \rightarrow 2HXeO_4^- + 2OH^- \rightarrow XeO_6^{4-} + Xe + O_2 + 2H_2O$$

Salts containing the octahedral Xe^{VIII} **perxenate ion** XeO_6^{4-} are known, and by the action of acid the tetrahedral **xenon tetroxide** XeO_4 is formed.

All xenon–oxygen compounds are very strongly oxidizing, and some decompose explosively. Compounds with Xe–O bonds attaching polyatomic groups are known, and weak Xe–N and Xe–C bonds can also be formed, as in $Xe(CF_3)_2$, which decomposes rapidly at room temperature.

Krypton compounds

No krypton compounds appear to be thermodynamically stable, but KrF_2 can be made from the elements in an electric discharge at very low temperatures, and a few compounds of the cationic species $[KrF]^+$ and $[Kr_2F_3]^+$ are also known. As the ionization energy of Kr is higher than that of Xe, the lower stability of krypton compounds is expected from the bonding models shown in structures **1** and **2**, where Xe carries a formal positive charge.

G1 INTRODUCTION TO NON-TRANSITION METALS

Key Notes

Scope

Non-transition metals include groups 1 and 2 of the *s*-block elements, group 12, and *p*-block elements in lower periods. Aluminum and the elements of groups 1 and 2 are classed as pre-transition metals, the remaining ones as post-transition metals.

Positive ions

Formation of compounds with positive ions depends on a balance between ionization energies and lattice or solvation energies. Post-transition metals have higher ionization energies and are less electropositive than pre-transition metals.

Group trends

Trends down groups 1 and 2 are dominated by increasing ionic size. In later groups the structural and bonding trends are less regular, and there is an increased tendency to lower oxidation states, especially in period 6.

Non-cationic chemistry

Many of the elements can form anionic species. Compounds with covalent bonding are also known: these include organometallic compounds and (especially with post-transition metals) compounds containing metal-metal bonds.

Related topics

The periodic table (A4)
Chemical periodicity (B2)
Trends in atomic properties (A5)
Lattice energies (D6)

Scope

The transition metals and the lanthanides and actinides have characteristic patterns of chemistry and are treated in Sections H and I. The remaining **non-transition metals** include the elements of group 12 although they are formally part of the *d*-block, as the *d* orbitals in these atoms are too tightly bound to be involved in chemical bonding and the elements do not show characteristic transition metal properties (see Topic G4).

Figure 1 shows the position of non-transition metals in the periodic table. They fall into two classes with significantly different chemistry. The **pre-transition metals** comprise groups 1 and 2 and aluminum in group 13. They are 'typical' metals, very electropositive in character and almost invariably found in oxidation states expected for ions in a noble-gas configuration (e.g. Na^+, Mg^{2+}, Al^{3+}). In nature they occur widely in silicate minerals, although weathering processes give rise to concentrated deposits of other compounds such as halides (e.g. NaCl, CaF_2) carbonates ($CaCO_3$) and hydroxides (AlO(OH)) (see Topic J2).

Metallic elements from periods 4–6 in groups following the transition series are **post-transition metals**. They are less electropositive than the pre-transition metals and are typically found in nature as sulfides rather than silicates. They form

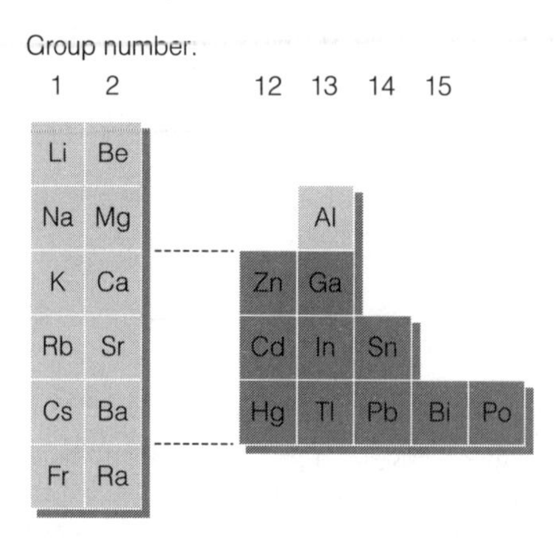

Fig. 1 Position of non-transition metals in the periodic table, with post-transition metals shaded.

compounds with oxidation states corresponding to d^{10} ions where *s* and *p* electrons have been ionized (e.g. Cd^{2+}, In^{3+}, Sn^{4+}) but these are less ionic in character than corresponding compounds of pre-transition metals. In solution, post-transition metals form stronger complexes than with pre-transition metals. Lower oxidation states (e.g. Tl^{+}, Sn^{2+}) are also common.

Positive ions

The formation of ionic compounds depends on a balance of energies as illustrated for NaCl in Topic D6, *Fig. 1*. Energy input required to form ions must be compensated by the lattice energy of the compound. For ions in solution, a similar cycle could be drawn, including the solvation energy rather than the lattice energy. For group 1 atoms with the $(ns)^1$ configuration, the second ionization energy involves an electron from an inner shell and is so large that the extra lattice or solvation energy obtainable with M^{2+} cannot compensate for it. For group 2 elements with the $(ns)^2$ configuration the second ionization energy is more than compensated by extra lattice energy. Thus M^{2+} compounds are expected, a solid such as CaF(s) having a strong tendency to disproportionate.

Figure 2 gives some data for groups 2 and 12 that are relevant in understanding the trends in pre- and post-transition metal groups. Ionization energies decrease, and ion sizes increase, down group 2 (see Topic A5). Increasing size gives smaller lattice energies, and so a decrease in ionization energy is also required if the electropositive character is to be retained. This happens in groups 1 and 2, and the electrode potentials shown in *Fig. 2* become slowly more negative for the lower elements.

Group 12 atoms have the electron configuration $((n-1)d)^{10}(ns)^2$ and also form positive ions M^{2+} by removal of the *s* electrons. Filling the *d* shell from Ca to Zn involves an increase of effective nuclear charge that raises the ionization energy and reduces the ionic radius. Lattice energies for Zn^{2+} are expected to be somewhat larger than for Ca^{2+}, and the formation of Zn^{2+} is also assisted by the slightly lower sublimation energy of metallic zinc. Nevertheless, these factors do not compensate fully for the increased ionization energy, and so zinc is less electropositive (less negative $E^{\ominus}$ value) than calcium. On descending group 12, ionization energies do not decrease to compensate for smaller lattice energies as they do in group 12, and $E^{\ominus}$ values increase down the group. This is particularly marked with

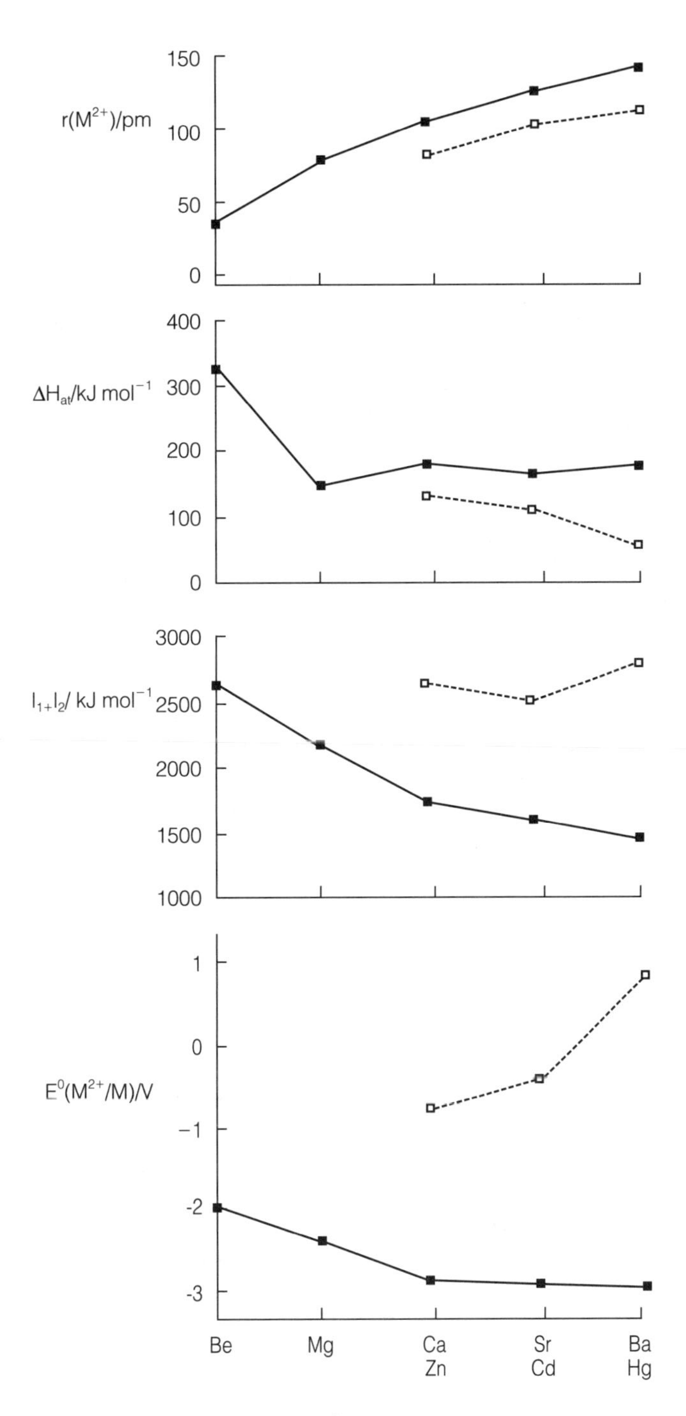

Fig. 2 Data for formation of M^{2+} ions of groups 2 and 12, showing (a) ionic radii, (b) sublimation enthalpies of the elements, (c) sum of the first two ionization energies, and (d) standard electrode potentials.

mercury, where especially high ionization energies result from the extra nuclear charge consequent on filling the 4*f* shell in the sixth period, combined with relativistic effects (see Topic A5).

Group trends

The above analysis shows how electropositive character remains strong throughout pre-transition groups. The major vertical trends in the stability and structure of compounds result from the changing ionic size. The small radius of Li^+ and Be^{2+} gives some peculiarities, which are sometimes described as **diagonal relationships**. Thus the solubilities and thermal stabilities of lithium compounds are often closer to those of magnesium than to those of other group 1 elements. Beryllium has even more marked differences from the rest of group 2, showing similarities with its diagonal neighbor aluminum. These relationships can be related to the size/charge ratio of ions. The small ion Li^+ gives lattice and solvation energies more similar to Mg^{2+} than to Na^+. The very small Be^{2+} is comparable with Al^{3+} in its polarizing power, which produces deviations from ionic character in solid-state and solution chemistry.

Size also increases down post-transition metal groups but the chemical trends are less regular. Solid compounds often have lower coordination numbers than expected by comparison with pre-transition metal ions of similar size, and have patterns of stability and solubility that suggest an appreciable degree of covalent bonding. The changing balance between ionization and lattice (or solvation) energies also has the consequence that lower oxidation states become more favorable. These tendencies are especially marked in period 6 (Hg, Tl, Pb, Bi). Thus many Tl^I and Pb^{II} compounds are known, the states Tl^{III} and Pb^{IV} being strongly oxidizing (see further discussion in Topic G5). The **inert-pair effect** is a somewhat misleading term for this phenomenon, implying the existence of an electron pair $(ns)^2$ too tightly bound to be involved in bonding. In fact, the 'inert pair' can have important structural consequences (see Topic G6). The discussion above also emphasizes that the relative stability of oxidation states always depends on a balance of factors, not on ionization energies alone.

Non-cationic chemistry

Although cationic chemistry has been emphasized above, other types of bonding are possible with the elements of all groups in this Section. These include the following.

- **Covalent compounds**. Compounds with predominantly covalent character include organometallic compounds.
- **Anionic compounds**. Under unusual conditions, group 1 elements can form anions such as Na^-. Some post-transition elements form polyatomic ions.
- **Metal-metal bonding**. This is especially a feature of post-transition groups and can accompany many 'unusual' oxidation states, of which Hg^I (in fact Hg_2^{2+}) is the commonest example.

G2 Group 1: alkali metals

Key Notes

The elements

All elements are found in silicates; sodium and potassium are more abundant and occur in chloride deposits. The elements are very electropositive and reactive.

Solution chemistry

M^+ aqua ions show only weak complexing properties except with macrocyclic ligands. The elements form strongly reducing solutions in liquid ammonia.

Solid compounds

Very ionic compounds are formed with halides, oxides and many complex ions. The heavier elements form superoxides, peroxides and some sub-oxides. Alkalides (containing M^- ions) and electrides can be made.

Organometallic compounds

Lithium alkyls such as $Li_4(CH_3)_4$ are oligomeric compounds with multicenter bonding. Organometallic compounds of the heavier elements are more ionic and less stable.

Related topic

Introduction to non-transition metals (G1)

The elements

The elements of group 1 are collectively known as **alkali metals** after the alkaline properties of their hydroxides such as NaOH. The atoms have the $(ns)^1$ electron configuration and the M^+ ions are therefore easily formed. Alkali metals are the most electropositive of all elements, and their compounds among the most ionic. Some group trends are shown in *Table 1*. Roughly constant electropositive character is maintained down the group by parallel fall in atomization, ionization, and lattice or hydration energies (see Topic G1). In some respects, lithium differs slightly from the rest of the series. The solubilities and the thermal stabilities of its compounds follow patterns that are more similar to those of group 2 elements than to those of the rest of group 1. This **diagonal relationship** can be understood from the small size of the Li^+ cation, which leads to trends in lattice energies and solvation energies more like those of the higher charged ions in group 2.

Only sodium and potassium are moderately abundant on Earth, and are major elements of life (see Topic J3). They occur in many silicates, but weathering reactions at the Earth's surface lead to the dissolution of the very soluble cations, which are common in sea water and are eventually deposited in halide minerals such as NaCl and KCl (see Topic J2). Li, Rb and Cs are of lower abundance, and obtained from silicate minerals. Francium is radioactive. Its longest-lived isotope ^{223}Fr has a half-life of only 22 min and occurs in exceedingly small amounts in uranium minerals (see Topic I2).

Table 1. Properties of alkali metals: melting and boiling points, atomization and ionization enthalpies, ionic radii and standard electrode potentials

Element	MP(°C)	BP(°C)	ΔH_{at} (kJ mol^{-1})	I (kJ mol^{-1})	$r(M^+)$ (pm)	E^0 (M^+/M) (V)
Li	180	1347	162	520	76	−3.03
Na	98	881	110	496	102	−2.71
K	63	765	90	419	138	−2.92
Rb	39	688	88	403	152	−2.93
Cs	28	705	79	376	167	−2.92

The elements are soft low-melting metals and are very strong reducing agents, reacting violently with many substances. Their major applications are as compounds (especially sodium chloride, hydroxide and carbonate) but the elements can be made by electrolysis of fused halides, and sodium metal is used in industrial processes such as the production of metallic Ti (see Topic J4).

Solution chemistry

Aqueous chemistry is entirely dominated by the M^+ ions. The M^+/M electrode potentials are all extremely negative (see *Table 1*), that of Li being slightly more so than the others because of the large solvation energy as a result of its small size. The higher solvation of lithium can be seen in the **ionic mobilities** determined from the ionic conductivities of dissolved salts. It might be expected that the smallest ion would be the most mobile, but in fact Li^+ is the least mobile and it appears that the smallest 'bare' ion becomes the largest on solvation.

The M^+ ions have only weak complexing tendencies, but these can be enhanced by suitably sized **macrocyclic ligands** (see Topic E3). Ligands with different cavity sizes can be used to discriminate between alkali ions.

The metallic elements dissolve in **liquid ammonia** (see Topic F5) and related amines (e.g. ethylamine $C_2H_5NH_2$) to give solutions which contain **solvated electrons** in addition to cations. In some solvents there is evidence for equilibria involving alkali anions M^-. The solutions are useful reducing agents for the preparation of unusually low oxidation states (e.g. $[Ni^0(CN)_4]^{4-}$) including anionic compounds of the alkali elements themselves (see below).

Solid compounds

The alkali metals react with many other elements directly to make binary solids. The **alkali halides** are often regarded as the most 'typical' ionic solids (see Topics D3–D6). Their lattice energies agree closely with calculations although their structures do not all conform to the simple radius ratio rules, as all have the rocksalt (NaCl) structure at normal temperature and pressure, except CsCl, CsBr and CsI, which have the eight-coordinate CsCl structure. The alkali halides are all moderately soluble in water, LiF being the least so. (The influence of ionic radius on solubility is discussed in Topic E4.)

The elements also form **hydrides** by direct interaction between the elements. LiH is the most stable and is a useful precursor for other hydrides (see Topic F2). Lithium also reacts with N_2 to form the nitride Li_3N.

The elements form **oxides** M_2O, which have the antifluorite structure for Li–Rb. Cs_2O has the very unusual anti-CdI_2 structure with adjacent layers of Cs^+ (see Topic D4). All compounds are very basic and react with water and CO_2 to produce hydroxides and carbonates, respectively. Except for Li, however, the simple oxides are not the normal products of burning the elements in air. K, Rb and Cs form **superoxides** MO_2 containing the O_2^- ion, and sodium the **peroxide** Na_2O_2 with O_2^{2-}. The relative stability of these compounds with large cations of low charge can be understood by lattice energy arguments (see Topics D6 and F7). Rb and Cs also form **suboxides** when oxygen supply is very deficient, for example, Rb_9O_2 (**1**) and $Cs_{11}O_3$; the structure of the former compound is based on two face-sharing octahedra with direct Rb–Rb bonding giving distances shorter than in the metallic element.

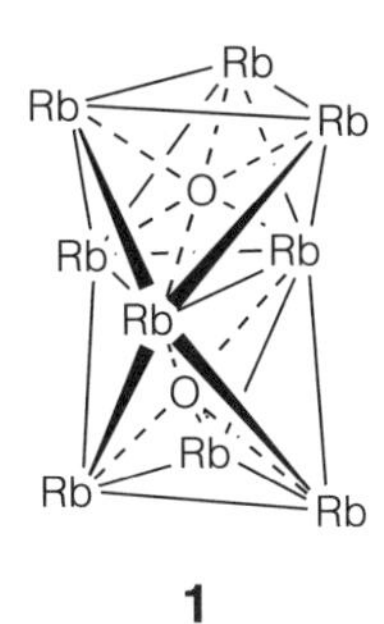

1

Hydroxides MOH are very important compounds for all the alkali metals, being easily formed by reaction of oxides with water (or atmospheric moisture), and soluble in water giving classic **strong base** behavior (see Topic E2). Compounds of oxoacids are commonly encountered, such as carbonate, nitrate, sulfate, etc. As these anions are fairly large, lithium compounds tend to be the most soluble in the series (see Topic E4). Many of these compounds crystallize in a variety of hydrated forms (e.g. $Na_2CO_3.nH_2O$ with $n = 1, 7$ or 10).

The combination of the reducing power of alkali metal–ammonia solutions with the strong complexing power of macrocyclic ligands allows compounds to be made containing unusual anions, such as $[Sn_9]^{4-}$ (see Topics C6 and G6). Among the unexpected products of such reactions are **alkalide** and **electride** salts. An example of an alkalide is $[Na(2.2.2.crypt)]^+Na^-$, where crypt is the cryptand ligand **2**. The crystal structure shows that the Na^- ion is larger than I^-. In electrides such as $[Cs(18\text{-crown-}6)_2]^+e^-$ there is a 'bare' electron trapped in a cavity in the lattice.

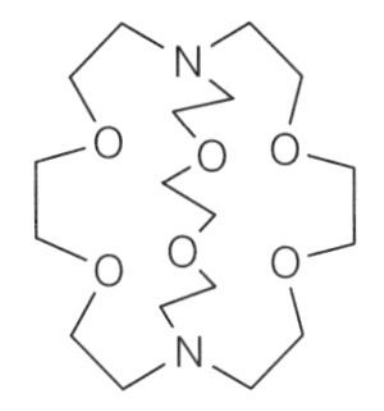

2

Organometallic compounds

Lithium is exceptional in forming molecular alkyls with oligomeric structures, for example, the tetrameric $Li_4(CH_3)_4$ (**3**). Bonding in the 'cubane'-like framework is provided by delocalized electrons. These compounds may be prepared by direct reaction between Li metal and alkyl halides and are useful reagents for preparing organometallic compounds of other elements, and as alternatives to Grignard reagents in organic synthesis (see Topic G3). Organometallic compounds of the other elements form solids with somewhat more ionic character.

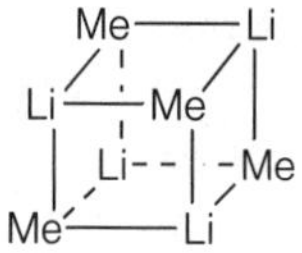

3

G3 Group 2: alkaline earths

Key Notes

The elements

Beryllium is a rare element; the others form many minerals such as carbonates and sulfates. All elements are highly electropositive and reactive, with chemistry dominated by the +2 oxidation state.

Solution and coordination chemistry

Be^{2+} is amphoteric, the other M^{2+} aqua ions basic. They form complexes with electronegative (and especially chelating) ligands, stability generally declining down the group.

Solid compounds

Be is normally four-coordinate and its compounds are more polymeric than ionic. The other elements form ionic oxides and halides with coordination numbers ranging from six to eight. Thermal stability of oxoanion salts increases with cation size.

Organometallic compounds

Beryllium alkyls are polymeric. Magnesium forms Grignard reagents, which are useful in organic and organometallic synthesis.

Related topic

Introduction to non-transition metals (G1)

The elements

The elements known commonly as **alkaline earths** have atoms with the $(ns)^2$ configuration and almost always have the +2 oxidation state in their compounds. Molecules such as MgH can be detected at high temperatures in the gas phase, the instability of the +1 state under normal conditions being due to the much greater lattice energies obtained with M^{2+} (see Topic D6). Some data illustrating the factors underlying group trends are discussed in Topic G1. Beryllium is distinct, as the very small and polarizing Be^{2+} ion forms compounds with more covalent character than with the other elements, where a high degree of ionic character is normal. Be shows some similarities both with its diagonal neighbor aluminum, and with the group 12 element zinc (see Topics G4 and G5).

Calcium and magnesium are very abundant elements, being common in silicate minerals and occurring in major deposits of $CaCO_3$, $CaMg(CO_3)_2$ (dolomite) and $MgKCl_3.3H_2O$ (carnallite). Calcium fluoride and phosphate minerals are the major sources of the elements F and P, respectively (see Topic J2). The moderately abundant heavier elements are found principally as sulfates $SrSO_4$ and $BaSO_4$, whereas beryllium is rather rare and occurs in beryl $Be_3Al_2Si_6O_{18}$. Radium is radioactive, its longest-lived isotope ^{226}Ra having a half-life of 1600 years and being found in uranium minerals (see Topic I2). Calcium and magnesium are major elements in life but beryllium and its compounds are very toxic (see Topic J3).

The metallic elements are all potentially very reactive towards air, water and most elements, but Be and Mg form passivating oxide films. Elemental magnesium is manufactured in large quantities either by electrolysis of molten $MgCl_2$ or by reduction of MgO, and is used in lightweight alloys and as a reducing agent. The other elements are used mainly as compounds.

Solution and coordination chemistry

The properties of the M^{2+} aqueous ions show trends expected from their increasing size down the group. Be^{2+} (like Al^{3+}) is **amphoteric** (see Topic E2). The insoluble hydroxide dissolves in both acid solution:

$$Be(OH)_2 + 2H_2O + 2H^+ \rightarrow [Be(H_2O)_4]^{2+}$$

and in alkaline conditions:

$$Be(OH)_2 + 2OH^- \rightarrow [Be(OH)_4]^{2-}$$

The simple aqua cation is present only in strongly acidic conditions. As the pH increases, successive protolysis and polymerization reactions first give soluble species with Be–OH–Be bridges, and then the solid hydroxide. The other M^{2+} ions are basic. As the hydroxide $M(OH)_2$ becomes more soluble in the series Mg < Ca < Sr < Ba precipitation requires increasingly high pH (see Topic E4).

Complex formation is dominated by class a or 'hard' behavior (see Topic E3) and is generally most favorable for the smaller ions. Beryllium forms $[BeF_4]^{2-}$ and strong complexes with some bidentate ligands such as oxalate $C_2O_4^{2-}$. From carboxylic acids unusual complexes such as $[Be_4O(O_2CCH_3)_6]$ can be obtained; the structure (**1**) has a central oxygen atom surrounded by a Be_4 tetrahedron with acetate groups bridging the edges (only one shown). The larger ions form complexes with chelating ligands such as EDTA. Complexes with ammonia such as $[Mg(NH_3)_6]^{2+}$ can be made in nonaqueous conditions but are not stable in water. However, **chlorophylls**, which are essential for photosynthesis in all green plants, have magnesium coordinated by nitrogen in macrocyclic porphine derivatives: **2** shows the basic framework, which has other organic groups attached; Mg^{2+} normally has one water molecule also coordinated.

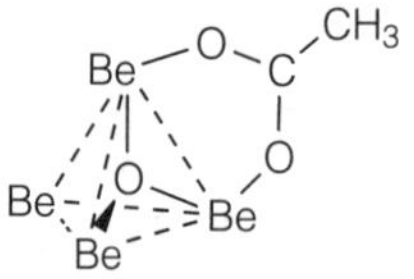

Be—O—C—CH3, C—O, O, Be, Be, Be

1

N, N, Mg, N, N

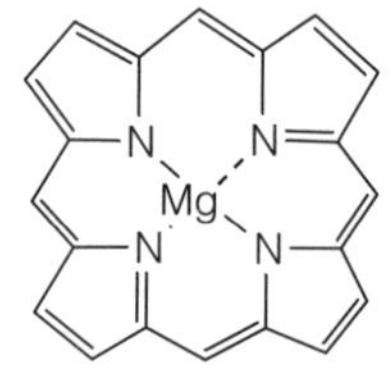

2

Solid compounds

Binary compounds are formed with all nonmetallic elements, many by direct combination. Beryllium is exceptional as its coordination is almost always tetrahedral, giving structures that may be regarded as polymeric rather than highly ionic. Thus BeO has the wurtzite structure (see Topic D3), BeF_2 is similar to SiO_2, and $BeCl_2$ (like SiS_2) has a chain structure (**3**) based on edge-sharing tetrahedra. BeH_2 is similar, with bridging hydrogens forming three-center bonds as in B_2H_6 (see Topic C5).

Compounds of the remaining elements have structures more in line with the expectations of the ionic model (see Topics D3 and D4). **Oxides** MO all have the rocksalt structure; as the cation size increases they become increasingly basic and reactive towards water and CO_2, giving $M(OH)_2$ or MCO_3, respectively. **Peroxides** such as BaO_2 are formed by the heavier elements in the group (see Topic F7). **Halides** show increasing coordination with size, six for Mg and seven or eight for the larger ions. MgF_2 has the rutile structure and the other MF_2 compounds the fluorite structure. Heavier halides of Mg give layer structures ($CdCl_2$ and CdI_2) whereas for the larger cations somewhat distorted structures are formed (e.g. distorted rutile for $CaCl_2$); these appear to be dictated by the tendency to asymmetrical coordination of the halide ion, with cations too large to form normal layer structures. Fluorides (especially CaF_2) have low solubility in water, but other halides are extremely soluble.

Binary compounds with less electronegative elements include hydrides, nitrides, sulfides and phosphides. They are decomposed by water and can provide convenient routes for the preparation of nonmetal hydrides (see Topic F2). The anions may be polyatomic or polymerized, as with CaC_2, which contains C_2^{2-} and reacts with water to give acetylene (ethyne) C_2H_2.

The elements form an enormous range of compounds with **oxoanions**, many of those with calcium (carbonate, silicate, phosphate, sulfate) being common minerals in the Earth's crust. Hydrated forms are common. Their thermal stability towards decomposition to the oxide is less than that for the alkali metals, and increases with cation size. Thus Be (like Al) does not form a stable carbonate; the decomposition temperatures for the others range from 400°C for $MgCO_3$ to 1400°C for $BaCO_3$. These trends can be understood using lattice energy arguments, as discussed in Topic D6.

Organometallic compounds

Be and Mg form an extensive range of organometallic compounds, those of Ca, Sr and Ba being much more reactive and difficult to characterize. Beryllium alkyls such as $Be(CH_3)_2$ have chain structures (see **3** with X = CH_3) with multicenter bonding similar to that in $Li_4(CH_3)_4$ and $Al_2(CH_3)_6$ (see Topics G2 and G5). Be and Mg form biscyclopentadienyl compounds $M(C_5H_5)_2$; the Mg compound has an η^5 sandwich structure like that of ferrocene (Topic H10, Structure 3) but is more reactive and at least partially ionic: $M^{2+}(C_5H_5^-)_2$. The Be compound is less symmetrical with one ring displaced sideways, presumably because of the small size of Be.

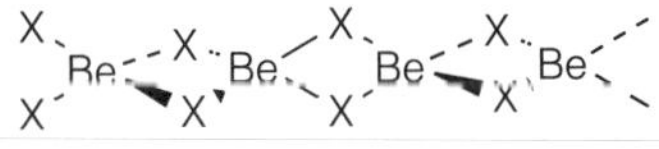

3

By far the most commonly encountered organometallic compounds in group 2 are the **Grignard reagents** RMgX, formed by reaction of Mg metal with an alkyl or aryl halide RX in an ether solvent. Solid compounds with additional ether molecules coordinated to Mg can be obtained, but the reagents are generally used in solution. They are very useful for alkylation and arylation reactions, either for forming C–C bonds in organic chemistry, or for preparing organometallic compounds of other elements.

G4 GROUP 12: ZINC, CADMIUM, AND MERCURY

Key Notes

The elements

Group 12 elements are found in nature as sulfides. Reactivity and electropositive character is much less than in group 2, especially for mercury.

M^{II} solution chemistry

Zn^{2+} is amphoteric. All elements form strong complexes, Hg having an exceptional affinity for soft ligands.

M^{II} solids

Most compounds show a marked deviation from ionic character, both in structures and properties. Typical coordination numbers are four for Zn, six for Cd, and two or four for Hg.

Lower oxidation states

Hg^{I} compounds contain Hg_2^{2+} with ligands coordinated. Similar species can be formed with Zn and Cd but are much less stable.

Organometallic compounds

Compounds R_2M and RMX are least reactive with Hg, and are very toxic.

Related topics

Introduction to non-transition metals (G1)

Complex formation (E3)

The elements

Group 12 elements have the electron configurations $(n-1)d^{10}ns^2$ with $n = 4, 5$ and 6 for Zn, Cd and Hg, respectively. They are formally part of the *d* block but the electrons of the $(n-1)d$ shell are too tightly bound to be involved directly in chemical bonding, and these elements show typical post-transition metal behavior. The extra nuclear charge associated with filling the *d* orbitals leads to high ionization energies in comparison with group 2 elements and hence reduced electropositive character (see Topic G1). This is especially pronounced with mercury, which forms few compounds that can be regarded as ionic.

The elements are found in nature as sulfides, especially ZnS (zinc blende or sphalerite) and HgS (cinnabar). Overall abundances in the crust are low. Zinc is an important element of life; Cd and Hg are not essential and are very toxic (see Topics J3 and J6).

The elements may be obtained by reduction of sulfides or oxides (e.g. ZnO with C). Zinc and cadmium are used for corrosion-resistant coatings. The metals have melting and boiling points that are lower than for group 2 elements, especially with Hg, which is one of two elements (Br being the other) existing as a liquid at

25°C. Zn and Cd are more reactive than Hg, dissolving in non-oxidizing acids and forming oxide films in air. Mercury oxidizes at room temperature but HgO decomposes readily on heating, a reaction historically important in the discovery of oxygen. Mercury dissolves many metallic elements to form **amalgams**, which can be useful reagents (e.g. sodium amalgam as a reducing agent, being much easier to handle than elemental sodium).

M^{II} solution chemistry

The aqua ions M^{2+} are more acidic than those in the same periods in group 2 (see Topic E2). Zinc (like Be) is amphoteric, dissolving at high pH to form $[Zn(OH)_4]^{2-}$. The other ions are not amphoteric as they have little tendency to complex with the hard ion OH^-, but Hg^{2+} is very strongly protolyzed and readily precipitates as HgO unless complexing ligands are present.

All the ions can form strong complexes, the overall formation constants for tetrahedral $[ML_4]$ species (e.g. $[HgCl_4]^{2-}$) with a selection of ligands L being shown in *Table 1* (see Topic E3). There is an increasing tendency to 'soft' class b behavior in the order Zn < Cd << Hg. Complexes with Hg^{2+} are among the strongest known with any element. In addition to the $[ML_4]$ complexes, mercury can form linear $[HgL_2]$ and sometimes $[HgL_3]$. As in the solid compounds, these trends indicate a pronounced tendency to covalent bonding; on grounds of size alone, the large Hg^{2+} ion could support a coordination number of six or even eight.

M^{II} solid compounds

Only the **fluorides** have structures and properties expected for ionic compounds with cations of the appropriate size (ZnF_2 rutile, the others fluorite; see Topics D3 and D4). In other compounds the characteristic coordination numbers are four for Zn, four or six for Cd, and two or four for Hg.

Zn and Cd **halides** (apart from fluorides) are based on close-packed lattices of halide ions, with Zn occupying tetrahedral holes and Cd octahedral ones. The Zn compounds are best regarded as polymeric, whereas those of Cd are prototypes of the important $CdCl_2$ and CdI_2 **layer structures**. Both sets of compounds are soluble in water, but solutions of Cd halides contain a variety of complex ions $[CdX_n]$ in equilibrium. Hg halides have varying coordination, with two close neighbors in $HgCl_2$ making this compound essentially molecular, the others being more polymeric. Solubility in water is low but increases markedly with rise in temperature, giving undissociated HgX_2 molecules.

Among the **oxides and sulfides**, only CdO adopts the octahedral rocksalt structure found with group 2 elements, although the solid is normally very deficient in oxygen and the electrons not used in bonding give rise to metallic properties. ZnO and ZnS are prototypes of the tetrahedrally coordinated **wurtzite** and **zinc blende** (or **sphalerite**) structures; in fact, ZnS can adopt either structure, as can CdS and CdSe. HgO and HgS have chain structures with linear two-coordination of Hg.

Table 1. Overall equilibrium constants ($log_{10}\beta_4$) for the formation of some $[ML_4]$ complexes

	$log_{10}\beta_4$		
ligand (L)	Zn^{2+}	Cd^{2+}	Hg^{2+}
Cl^-	0	3	15
Br^-	−1	4	21
I^-	−2	6	30
NH_3	9	7	19
CN^-	21	19	41

Many of these compounds are colored and show electronic properties characteristic of small bandgaps and nonstoichiometry (see Topic D7).

Lower oxidation states

The +1 oxidation state is fairly stable for mercury, and invariably involves the dimeric $[Hg-Hg]^{2+}$ ion. Evidence for this comes from solid-state structures, and in solution from many sources:

- Hg^{I} species are diamagnetic whereas Hg^{+} would have an unpaired electron;
- Raman spectra of solutions show a band from the Hg–Hg stretching vibration similar to that seen in solids;
- Equilibrium studies (e.g. by electrochemistry) are consistent with

$$Hg_2^{2+}(aq) \rightleftharpoons Hg(l) + Hg^{2+}(aq)$$

with an equilibrium constant $[Hg^{2+}]/[Hg_2^{2+}] = 0.011$ at 25°C; the equilibrium expression involving Hg^{+} would have a different form.

Uncomplexed Hg_2^{2+} is marginally stable in aqueous solution, but the disproportionation equilibrium can be upset by any ligand for which the Hg^{II} compound is more stable. Thus addition of sulfide, cyanide and many other ligands causes disproportionation. In solid compounds the Hg_2^{2+} ion always has two ligands strongly bonded. For example, Hg_2Cl_2 has linear Cl–Hg–Hg–Cl molecules, and salts with noncomplexing anions such as nitrate contain the hydrated ion $[H_2O\text{–}Hg\text{–}Hg\text{–}H_2O]^{2+}$.

Oxidation of Hg with AsF_5 gives species containing linear Hg_3^{2+} and Hg_4^{2+} ions, culminating in a metallic compound $Hg_{0.33}AsF_6$, which contains linear chains of mercury atoms.

Zn and Cd analogs of Hg_2^{2+} are much less stable, principally because the larger lattice energies obtained with the smaller M^{2+} ions tend to force disproportionation (see Topic D6). Zn_2^{2+} and Cd_2^{2+} can both be identified spectroscopically when the elements react with melts of the corresponding chloride. Adding $AlCl_3$ gives the solid compound $[Cd_2^{2+}][AlCl_4^-]_2$ but no solid zinc (I) compounds have been prepared.

Organometallic compounds

The elements form compounds R_2M and RMX, where R is an alkyl or aryl group and X a halide. M–C bond strengths are in the order Zn > Cd > Hg but nevertheless the mercury compounds are the most easily formed; for example, from

$$RH + HgX_2 \rightarrow RHgX + HX$$

The Hg compounds are also the least reactive towards air or water, partly because the competing Hg–O bond is so much weaker than with Zn or Cd. They are useful for preparing organometallic compounds of other elements. Water-soluble ions can be obtained, such as $[CH_3Hg]^{+}$, which has been used as a prototype 'soft' acid in the hard and soft acid and base (HSAB) classification (see Topic C8). All organomercury compounds are **extremely toxic**, as they pass through cell membranes much more easily than inorganic forms.

G5 GROUP 13: ALUMINUM TO THALLIUM

Key Notes

The elements

Aluminum is the commonest metallic element on Earth, occurring widely in aluminosilicate minerals and in deposits of the hydroxide bauxite. It is very electropositive and potentially very reactive, but forms a stable oxide film. Gallium, indium and thallium are rarer and less electropositive.

M^{III} aqueous chemistry

Al^{3+} is amphoteric and complexes strongly with hard ligands. Ga^{3+} and In^{3+} are similar, but Tl^{3+} is a strong oxidizing agent and shows soft complexing properties.

M^{III} compounds

Al^{III} is octahedral in fluorides and in most oxides (including many complex oxides), and tetrahedral with larger or less electronegative atoms (and sometimes also in oxides). Heavier halides form molecular dimers Al_2X_6. The tetrahedral $[AlH_4]^-$ complex and dimeric alkyls are also formed. The heavier elements of the group form less stable compounds, especially Tl^{III}.

Lower oxidation states

Stability of M^I increases down the group. Tl^+ is strongly basic and shows some resemblance to group 1 cations in its solution and solid-state chemistry. Mixed valence and metal-metal bonded compounds of Ga, In and Tl are also known.

Related topic

Introduction to non-transition metals (G1)

The elements

The elements aluminum, gallium, indium and thallium have valence electron configurations $(ns)^2(np)^1$ and for the lighter elements their chemistry is dominated by the +3 oxidation state. The group trends are very different, however, from those in groups 1 and 2. The Al^{3+} ion has a large charge/radius ratio and is strongly polarizing, so that significant deviations from simple ionic behavior are often observed. The filling of the *d* shells (and 4*f* in period 6) leads to decreased electropositive character for Ga, In and Tl similar to that shown in group 12 (see Topics G1 and G4). There is also a progressive stabilization of lower oxidation states down the group.

Aluminum is the commonest metallic element in the Earth's crust, being a constituent of almost all silicate minerals (see Topic J2). Weathering leaves deposits of the very insoluble aluminum minerals AlO(OH) and $Al(OH)_3$, known together as **bauxite**, which forms the principal source of the element. The metal is extracted by electrolysis of fused cryolite $Na_3[AlF_6]$. Although reactive when clean, the metal easily forms a very resistant oxide film, which allows widespread applications as a lightweight construction material and in cooking and other vessels.

Ga, In and Tl are much less common elements, obtained in small amounts from sulfide minerals of other elements and used only in specialized applications. The metals are less reactive than aluminum; *Fig. 1* shows a Frost diagram in which the much larger negative slope (negative electrode potential; see Topic E5) of Al is apparent. Thallium compounds are extremely toxic but do not normally pose an environmental hazard because they are little used.

M^{III} aqueous chemistry

Al^{3+} is **amphoteric** and will dissolve in acidic and alkaline solutions (see Topic E2). The $[Al(H_2O)_6]^{3+}$ ion is formed at low pH but undergoes increasing protolysis as the pH increases above four, and polymeric species such as $[Al_{13}O_4(OH)_{24}(H_2O)_{12}]^{7+}$ can be identified. The very insoluble $Al(OH)_3$ is formed at neutral pH but redissolves above pH 10:

$$Al(OH)_3 + OH^- \rightarrow [Al(OH)_4]^-$$

Al^{3+} shows typically 'hard' complexing behavior and has a particularly strong affinity for negative charged and/or chelating ligands, such as oxalate ($C_2O_4^{2-}$) and EDTA (see Topic E3).

Ga^{3+} is similar to Al^{3+} but In^{3+} is more basic. Tl^{3+} differs as it is a strong oxidizing agent, readily forming Tl^+ (see below and *Fig. 1*). It also shows strong class b or 'soft' complexing behavior, although not so marked as that of Hg^{2+} (see Topic G4).

M^{III} compounds

All **aluminum halides** can be made by direct reaction, but AlF_3 is best produced by reaction with anhydrous HF. It has a structure based on corner-sharing AlF_6 octahedra (similar to ReO_3; see Topic D3). Solid $AlCl_3$ has a polymeric layer structure, but in the gas phase or nonpolar solvents is molecular and dimeric Al_2Cl_6 (see Topic C8, Structure 1). The bromine and iodide have the molecular dimeric form in the solid state. Aluminum halides are strong Lewis acids (see Topic C8)

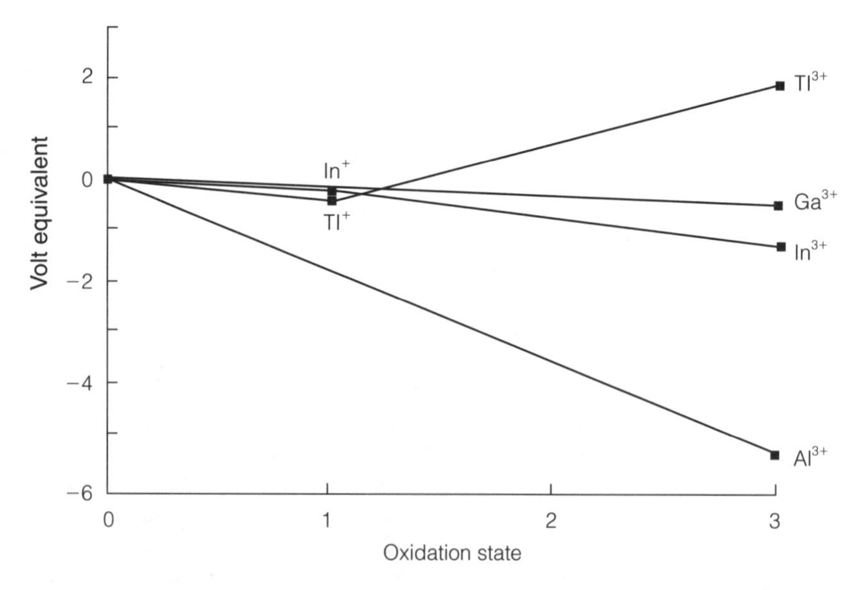

Fig. 1 Frost diagram showing the oxidation states of Al, Ga, In, Tl in aqueous solution at pH=0.

and $AlCl_3$ is frequently used as an acid catalyst, for example, in organic Friedel-Crafts reactions. Complex halides containing the ions $[AlF_6]^{3-}$ and $[AlCl_4]^-$ are easily formed and can be useful for the preparation of compounds containing unusual cations such as Cd_2^{2+} (see Topic G4).

The most stable form of the **oxide Al_2O_3** is α-alumina with the **corundum** structure where Al^{3+} ions occupy two-thirds of the octahedral holes in a hexagonal close-packed oxide lattice. Another form γ-Al_2O_3 has a defect spinel structure (see below). So-called **β-alumina** is in fact a mixed oxide of aluminum of approximate formula $NaAl_{11}O_{17}$ with a disordered arrangement of Na^+ ions, and is a good ionic conductor (see Topic D7).

Aluminum forms many mixed oxides of which the **aluminosilicates** are major constituents of minerals (see Topic J2). In these compounds aluminum sometimes replaces a portion of the silicon present as corner-sharing SiO_4 groups (see, e.g. **zeolites**, Topic D5). The mixed oxide mineral **spinel** $MgAl_2O_4$ gives its name to an important structure type. One-half of the octahedral holes and one-eighth of the tetrahedral holes are filled in a cubic close-packed array of oxide ions. In the **normal spinel** form adopted by $MgAl_2O_4$ the divalent Mg^{2+} ion is in tetrahedral sites and the trivalent Al^{3+} is octahedral. (See Topic H4 for other examples.) In the defect spinel structure of γ-Al_2O_3 a fraction of the cation sites are occupied at random.

Halides and oxides of Ga and In are fairly similar to those of Al, but have less negative enthalpies of formation and (with In) a tendency to higher coordination. Tl^{III} is more strongly oxidizing; for example, there is no Tl^{III} iodide, and the compound of stoichiometry TlI_3 in fact contains Tl^I with the linear tri-iodide ion I_3^- (see Topic F9).

Al, Ga and In form tetrahedrally coordinated solids with elements of group 15, which are part of the series of **III–V semiconductors** (i.e. groups 13–15, III–V in old nomenclature; see Topic A4). The mixed compounds gallium aluminum phosphide $Ga_{1-x}Al_xP$ and the arsenide $Ga_{1-x}Al_xAs$ are used for light-emitting diode (LED) displays and semiconductor lasers.

Aluminum hydride AlH_3 has a structure similar to that of AlF_3. The **tetrahydroaluminate ion** $[AlH_4]^-$ is a powerful reducing and hydride transfer agent, generally used in the form of 'lithium aluminum hydride' $LiAlH_4$ made by reaction of LiH with $AlCl_3$. Stability of hydrides decreases down the group but $[GaH_4]^-$ is fairly stable and the unstable **digallane** molecule Ga_2H_6 has been identified with a structure like that of diborane (see Topic C1, Structure 16).

Organoaluminum compounds are dimeric but the bonding is different from that of halides as the bridging methyl groups in $Al_2(CH_3)_6$ (**1**) must be held by three-center two-electron bonds similar to those in diborane (see Topic C6). Organometallic compounds of Ga, In and Tl are less stable than for Al and do not dimerize.

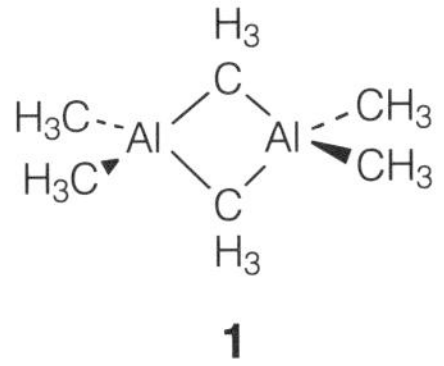

H3
C
H3C Al Al CH3
H3C CH3
C
H3

1

Lower oxidation states

Gas-phase molecules such as AlH, AlCl and AlO are known at high temperatures and low pressures but, as in group 2, disproportionation occurs under normal conditions because of the much higher lattice or solvation energies associated with M^{3+} (see Topics D6 and G1). As these energies decrease with ion size down the group, the tendency to disproportionation also declines, and lower oxidation states become commoner. *Figure 1* shows the possibility of forming In^+ and Tl^+, the former prone to disproportionation, the latter much more stable. The increasing stability of ions with the $(ns)^2$ configuration in lower periods is often called the **inert-pair effect**. It is particularly marked in period 6 because of the high ionization energies of these elements (see Pb^{II}, Topic G6) but it is important to remember that it depends not on ionization energies alone but on a balance of different energy trends.

Like K^+, which has a very similar size, Tl^+ is very basic in solution, and forms some compounds with similar structures to those of alkali metals (e.g. TlCl has the CsCl structure). It has a greater affinity for soft ligands, however, and sometimes its solid structures show an irregular coordination suggesting the influence of a lone-pair of electrons as with Sn^{II} (see Topic G6).

Ga and In form +1 compounds with large low-charged anions, and also some in which the oxidation state is apparently +2 (or sometimes even fractional). The gas-phase M^{2+} ions have the $(ns)^1$ configuration with one unpaired electron, and in chemical situations always either disproportionate or form metal-metal bonds. The former possibility leads to **mixed valence compounds** such as '$GaCl_2$' (in fact, $Ga^+[Ga^{III}Cl_4]^-$). The alternative gives ions $[M-M]^{4+}$ (isoelectronic to Hg_2^{2+}), although they are never found on their own but are always strongly bonded to ligands, as in $[Ga_2Cl_6]^{2-}$ (**2**). (Note the difference between this structure and that of Ga_2Cl_6 (like Al_2Cl_6), where there are no electrons available for direct Ga–Ga bonding.)

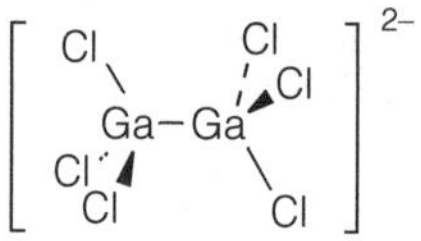

$$\left[Cl_3Ga-GaCl_3 \right]^{2-}$$

2

All elements of the group form **Zintl compounds** with electropositive metals (see Topic D5). Continuous networks of covalently bonded atoms are generally found, rather than the clusters common with group 14. For example, NaAl and NaTl have tetrahedral diamond-like networks of Al or Tl, which can be understood on the basis that Al^- and Tl^- have the same valence electron count as carbon.

G6 Group 14: tin and lead

Key Notes

The elements — Found in the minerals SnO_2 and PbS, the elements are commoner than other heavy metals. They have rather low electropositive character. Lead compounds are very toxic.

M^{IV} chemistry — Many Sn^{IV} compounds are known, some with molecular structures. Pb^{IV} is strongly oxidizing and binary compounds are limited to oxide and fluoride, although complex ions and covalent compounds are known. There is no simple M^{IV} aqueous chemistry for either element.

M^{II} chemistry — Most Sn^{II} compounds have structures influenced by the pair of nonbonding electrons. Pb^{II} compounds more often have regular ionic structures. Aqueous Sn^{2+} is amphoteric; Pb^{2+} forms strong complexes.

Other compounds — Organometallic compounds are known in both M^{II} and M^{IV} states. Polyatomic anions can be made.

Related topics — Carbon, silicon and germanium (F4); Introduction to non-transition metals (G1)

The elements

Tin and lead show some resemblance to the lighter elements in group 14, especially Ge (see Topic F4). Although they are distinctly more metallic in their chemical and physical characteristics, simple cationic chemistry is the exception rather than the rule. As with group 13 (see Topic G5), two oxidation states M^{II} and M^{IV} are found, the M^{II} form becoming more stable for lead.

Both elements have rather low abundance, but are commoner than other heavy metals. They occur in the minerals **cassiterite** SnO_2 and **galena** PbS. They each have several stable isotopes, Sn more than any other element (10). Some Pb isotopes are derived from the radioactive decay of uranium and thorium (see Topic I2). The isotopic composition of Pb (and thus its atomic mass) varies detectably according to the source, and such variations have been used to estimate the age of rocks and of the Earth.

The elements are readily produced by reduction of their ores and are soft, low-melting, somewhat unreactive metals. Tin is used for plating, and both elements in low-melting alloys (e.g. solder) and as many compounds. Applications of lead, however, are declining as its compounds are very toxic (see Topics J3 and J6). A continuing major use is in **lead-acid batteries**, which depend on two reactions involving the Pb^0, Pb^{II} and Pb^{IV} states:

$$PbO_2(s) + SO_4^{2-}(aq) + 4H^+(aq) + 2e^- = PbSO_4(s) + 2H_2O(l)$$

$$Pb(s) + SO_4^{2-}(aq) = PbSO_4(s) + 2e^-$$

Occurring at different electrodes, these give a cell potential of 2 V, larger than can be obtained easily from any other pair of electrode reactions in aqueous solution.

M^{IV} chemistry

Many binary Sn^{IV} compounds are known. SnO_2 has the rutile structure, and SnX_2 with X = S, Se, Te the CdI_2 layer structure (see Topic D3). SnF_4 has a layer structure constructed from corner-sharing octahedra, but other tetrahalides form tetrahedral molecules. The halides are good Lewis acids, especially SnF_4, which forms complexes such as $[SnF_6]^{2-}$.

The Pb^{IV} state is strongly oxidizing and only oxides and fluorides form stable binary compounds. PbO_2 and PbF_4 have the same structures as with tin, and mixed-valency oxides such as Pb_3O_4 (containing Pb^{IV} and Pb^{II}) are known. Other Pb^{IV} compounds include salts containing the $[PbCl_6]^{2-}$ ion as well as some molecular covalent compounds, such the tetraacetate $Pb(CH_3CO_2)_4$ and organometallic compounds (see below).

Neither element shows any simple aqueous chemistry in the M^{IV} state, as the oxides MO_2 are insoluble in water at all pH values. Reaction of SnO_2 in molten KOH gives the octahedral hydroxoanion $[Sn(OH)_6]^{2-}$, in contrast to the normal tetrahedral silicates and germanates, but in parallel with isoelectronic compounds such as $Te(OH)_6$ also found in period 5 (see Topics F1 and F8). Other 'stannates' are mixed oxides without discrete oxoanions (e.g. $CaSnO_3$ with the perovskite structure; see Topic D5).

M^{II} chemistry

The structural chemistry of Sn^{II} and Pb^{II} compounds is extremely complex and varied. The M^{2+} ions have the $(ns)^2$ configuration and hence a nonbonding electron pair which can have a stereochemical influence analogous to that in molecules (see Topic C2). Thus the structure of SnO (**1**) shows tin with four oxygen neighbors on one side and a 'vacant' coordination site apparently occupied by the lone-pair. Sn^{II} sulfide and halides have polymeric structures with similar stereochemical features, but Pb^{II} compounds appear to be more ionic, and less influenced by the nonbonding electrons. One form of PbO has the same structure as SnO, but the structures of many other compounds are similar to those found with the larger M^{2+} ions in group 2 (see Topic G3), examples being PbS (rocksalt) and PbF_2 (fluorite). Solubility patterns of some Pb^{II} salts also parallel those found in group 2 (e.g. insoluble sulfate and carbonate) but differences appear with softer anions: thus PbS is insoluble in water, the heavier halides insoluble in cold water but more soluble in hot.

Sn
O O O
O

1

The aqueous M^{2+} ions are fairly acidic, Sn^{2+} especially so and shows typical amphoteric behavior, undergoing strong protolysis to form polymeric hydroxo species, which dissolve in alkali to form the pyramidal $[Sn(OH)_3]^-$. Pb^{2+} forms complexes with a class b pattern of stability analogous to that of Cd^{2+} (see Topics E3 and G4) although it does not complex with NH_3 in aqueous solution.

Other compounds

M^{II} **organometallic compounds** are found with cyclopentadienyl. $Sn(C_5H_5)_2$ has a 'bent sandwich' structure **2**, where the stereochemical influence of the lone-pair is apparent (compare ferrocene, Topic H10, Structure 3). M^{IV} organometallic compounds with M–C σ bonding are extremely varied and include simple tetraalkyls MR_4 and compounds with Sn–Sn bonds similar to those of Si and Ge (see Topic F4). Tetraethyl lead has been widely used as a gasoline additive to improve combustion but is being phased out because of the toxic hazard associated with all lead compounds.

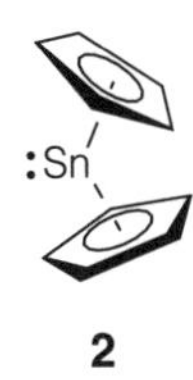

2

Reaction of alloys such as $NaSn_x$ with macrocyclic ligands in amine solvents gives compounds containing anionic clusters such as $[Sn_5]^{2-}$, $[Sn_9]^{4-}$ and $[Pb_5]^{2-}$. These have multicenter metal-metal bonding, which can often be rationalized by Wade's rules (see Topic C6, Structures 4 and 7).

H1 INTRODUCTION TO TRANSITION METALS

Key Notes

Scope

Transition elements form groups 3–11 in the *d* block. They have distinct chemical characteristics resulting from the progressive filling of the *d* shells. These include the occurrence of variable oxidation states, and compounds with structures and physical properties resulting from partially filled *d* orbitals.

Vertical trends

Elements of the 3*d* series are chemically very different from those in the 4*d* and 5*d* series, showing weaker metallic and covalent bonding, stronger oxidizing properties in high oxidation states, and the occurrence of many more compounds with unpaired electrons.

Horizontal trends

Electropositive character declines towards the right of each series. Elements become less reactive and their compounds show a tendency towards 'softer' behavior. Later elements in the 4*d* and 5*d* series are relatively more inert.

Electron configurations

Neutral atoms have both *s* and *d* valence electrons, but in chemically important states are often regarded as having purely d^n configurations.

Related topics

Many-electron atoms (A3)
The periodic table (A4)
Trends in atomic properties (A5)
Introduction to non-transition metals (G1)

Scope

Transition metals are elements of the *d* block that form compounds where electrons from *d* orbitals are ionized or otherwise involved in bonding. Typical transition metal characteristics include: the possibility of variable oxidation states; compounds with spectroscopic, magnetic or structural features resulting from partially occupied *d* orbitals; an extensive range of complexes and organometallic compounds including ones with very low oxidation state (zero or even negative); and useful catalytic properties shown by metals and by solid or molecular compounds. Different transition metals display these features to different degrees, but together the properties form a sufficiently coherent pattern that the elements are best dealt with by themed Topics rather than individually or as groups.

Although formally part of the *d* block, the elements of group 12 do not show typical transition metal characteristics, as the *d* orbitals are too tightly bound to be involved in chemical bonding. These elements are better regarded as post-transition metals, and are dealt with in Section G (Topic G4).

Vertical trends

The smaller size of the 3*d* orbitals compared with 4*d* and 5*d* has some important consequences.

- Electron repulsion is large between electrons in 3*d* orbitals. Exchange energy effects are more significant (see Topic A3); also, successive ionization energies (IEs) rise more sharply compared with later series (see Topic A5).
- 3*d* orbitals are not much larger than the 3*p* orbitals of the argon core $(3p)^6$. Good overlap with other atoms is hard to achieve, and covalent or metallic bonding involving 3*d* orbitals is weak compared with 4*d* and 5*d*.

One consequence of the IE trend is that higher oxidation states are less stable (more strongly oxidizing) compared with the 4*d* and 5*d* series. For example, in group 7 MnO_4^- is much more strongly oxidizing than ReO_4^-, and in group 8 FeO_4 is unknown although RuO_4 and OsO_4 are stable compounds.

The bond-strength trend $3d \ll 4d < 5d$ is the reverse of that normally found in main groups (see Topic C7). Its influence can be seen in the atomization enthalpies of the elements, reflecting the strength of bonding in the metallic state, and shown in *Fig. 1* for elements of the three series. The very high atomization energies of elements such as tungsten (5*d* group 6) are reflected in their extremely high melting and boiling points, a property important in applications such as electric light bulb filaments. Sublimation energies in the middle of the 3*d* series are much less, partly because the relatively poor overlap of 3*d* orbitals gives weaker bonding, and partly because of the exchange energy stabilization of the free atoms, which have several unpaired electrons (e.g. six with Cr). Compounds with unpaired electrons in *d* orbitals are also much commoner in the 3*d* series, those of the 4*d* series more often forming low-spin configurations or having *d* electrons involved in metal–metal bonds (see Topics H2 and H5).

Between the 4*d* and 5*d* series the expected decrease of IEs and increase of radius is counteracted by the increase of nuclear charge involved in filling the 4*f* shell before 5*d* (see Topic A4). 5*d* elements in early groups are very similar to the corresponding 4*d* ones, although this feature is less marked in later groups.

Horizontal trends

The chemical trends along each series are dominated by the increase in nuclear charge and in the number of valence electrons. Earlier elements can achieve the

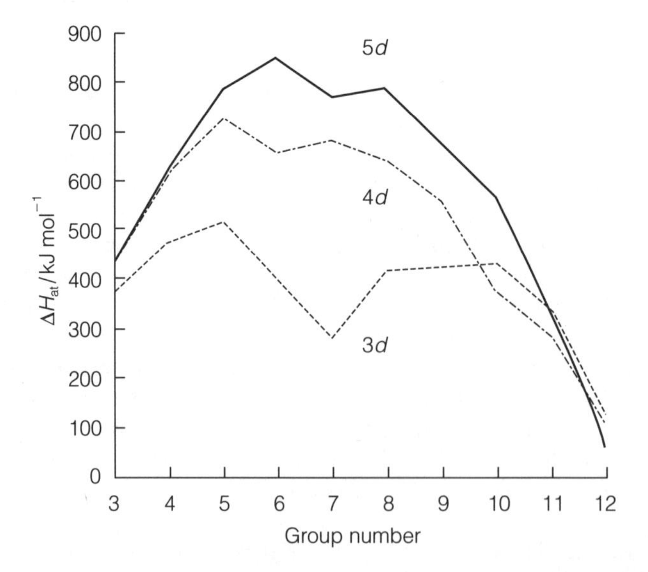

Fig. 1. Standard enthalpies of atomization for elements of the three series.

group oxidation state corresponding formally to ions with a noble gas configuration (up to Mn^{VII} in the 3*d* series and Ru^{VIII} and Os^{VIII} in 4*d* and 5*d*). Increasing effective nuclear charge brings an increase in IEs as shown for the 3*d* elements in *Fig. 2*. Not only does the group oxidation state become very strongly oxidizing for later elements, but redox potentials for any given states (e.g. M^{3+}/M^{2+}) also increase along the series, as the extra lattice or solvation energies of the higher state become less able to compensate for the higher IE values (see Topics D6 and G1).

With increasing IEs comes also a general decline in electropositive character. Early elements in each series are thermodynamically extremely reactive towards oxygen and other electronegative elements (although the formation of an inert oxide film may kinetically prevent the solid elements from further oxidation). Later elements are less reactive, a trend that culminates in the 'noble' or 'coinage' metals Cu, Ag and Au of group 11. The trend is exacerbated in the later 4*d* and 5*d* elements by the high atomization energies, and the elements Ru, Rh, Pd, Os, Ir and Pt form a group known as the **platinum metals**, often occurring together in nature, sometimes as metallic alloys. The change in electronegativity is also shown by different patterns of chemical stability: whereas earlier elements of both series generally form more stable compounds with 'harder' anions such as oxide and fluoride (and are found in nature in oxide minerals), the later ones are 'softer' in character and are more often found as sulfides. The trend along the series thus provides a link between the chemical characteristics of the pre-transition and post-transition metals (see Topic G1).

A general decline in atomic size is another consequence of increasing effective nuclear charge. *Figure 2* also shows the ionic radii of M^{2+} ions of the 3*d* series.

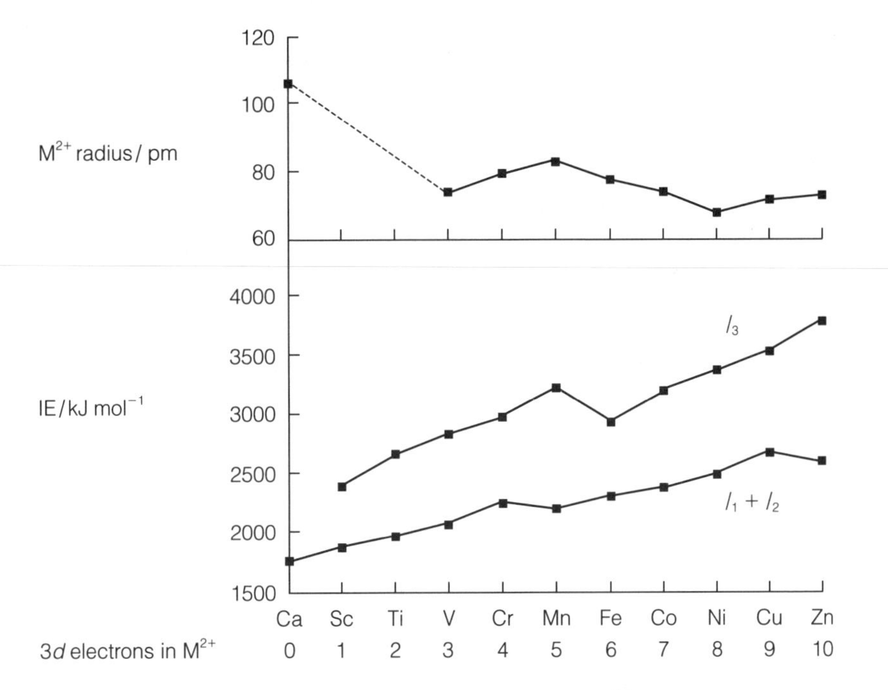

Fig. 2. Data for ions of the elements Ca–Zn showing: radii of M^{2+} ions, third IE and the sum of first and second IEs, and the $(3d)^n$ configurations of M^{2+}.

The expected decrease across the series is modulated by **ligand field** effects (see Topic H2).

Electron configurations

Electron configurations of the neutral atoms are complex and have both *d* and *s* electrons in outer shells. For example, in the 3*d* series most atoms have the configuration $(3d)^n(4s)^2$, where *n* increases from one to 10; chromium and copper are, however, exceptions with $(3d)^5(4s)^1$ and $(3d)^{10}(4s)^1$, respectively. The configurations depend on a balance of two factors:

(i) 3*d* orbitals are progressively stabilized relative to 4*s* across the series;
(ii) repulsion between electrons is large in the small 3*d* orbitals, and so minimum energy in the neutral atom is achieved in spite of (i) by putting one or two electrons in the 4*s* orbitals.

An important consequence of this balance is that in forming positive ions, 4*s* electrons are always removed first. Thus for M^{2+} ions and ones of higher charge, outer-shell electrons are left only in the 3*d* orbitals. *Figure 2* lists the value of *n* in the configuration $(3d)^n$ for M^{2+}; values for higher charges may be found from these by subtracting the appropriate number of electrons (e.g. Ti^{4+} $(3d)^0$ and Fe^{3+} $(3d)^5$). These numbers can be used to interpret the IE trends shown: whereas I_1 and I_2 rise fairly steadily (with small irregularities resulting from the exceptional configurations of Cr and Cu), the I_3 plot shows a pronounced break after manganese. With six or more electrons in the *d* shell some must pair up, thus giving greater electron repulsion and a lower IE than expected from the previous trend (see Topic A5).

Ligand field theory deals with the important consequences of the progressive filling of the *d* shell. It is normal to specify the *d* electron number associated with the appropriate transition metal ion, even though the bonding is not assumed to be completely ionic. For example, any Fe^{III} compound is assigned the configuration $(3d)^5$, a Pt^{II} compound $(5d)^8$ (corresponding to Ni^{2+} in the same group). In compounds with very low oxidation states, or with ligands such as organic groups where bonding is largely covalent, a different electron counting scheme is often used (see Topics H9 and H10). In applying the **18-electron rule** one needs to count the total number of valence electrons in a neutral atom, irrespective of whether they are *d* or *s*. This is simply the group number, thus eight for Fe and 10 for Pt. If ligand field arguments are used for very low oxidation states the electrons in the appropriate ion are assigned entirely to *d* orbitals. For example, a Co^I compound would be regarded as $(3d)^8$ even though the free Co^+ ion has the configuration $(3d)^7(4s)^1$. The justification for this procedure is that the energy balance between *d* and *s* orbitals changes on compound formation; what were *s* orbitals in the free ion become strongly antibonding molecular orbitals in a complex and are no longer occupied in the ground state.

H2 LIGAND FIELD THEORY

Key Notes

Octahedral splittings

Ligand field splitting of the d orbitals arises from a combination of σ and π bonding interactions with ligands. In octahedral geometry two orbitals (e_g) are at higher energy than the other three (t_{2g}). The spectrochemical series puts ligands in order of field strength. High-field ligands are strong σ donors and π acceptors.

High and low spin

High-spin complexes have as many d electrons unpaired as possible, and are common with 3d series elements. Low-spin complexes have as many electrons as possible in the lower set of orbitals, and are common in the 4d and 5d series.

Ligand field stabilization energy

Ligand field stabilization energy (LFSE) is calculated relative to the average of all d orbital energies. Large values are found for octahedral d^3, d^8 and low-spin d^6 complexes. A large LFSE leads to smaller ions and higher lattice or solvation energies.

Other geometries

Different ligand geometries give characteristic patterns of ligand field splitting. The square-planar geometry is common for some d^8 ions, and a Jahn–Teller distortion from a regular octahedron is normal for d^9 and high-spin d^4 ions.

Related topics

Atomic orbitals (A2)
Lewis acids and bases (C8)
Complexes: electronic spectra and magnetism (H8)

Octahedral splittings

The five d orbitals with different values of the magnetic quantum number (m) have the same energy in a free atom or ion (see Topic A2). In any compound, however, they interact differently with the surrounding ligands and a **ligand field splitting** is produced. The commonest coordination is octahedral with six surrounding ligands (see *Fig. 1*). Then two of the d orbitals (d_{z^2} and $d_{x^2-y^2}$ known together as the e_g set) are found at higher energy than the other three (d_{xy}, d_{xz} and d_{yz}, known as t_{2g}). Such a splitting (denoted Δ_o) occurs in any transition metal

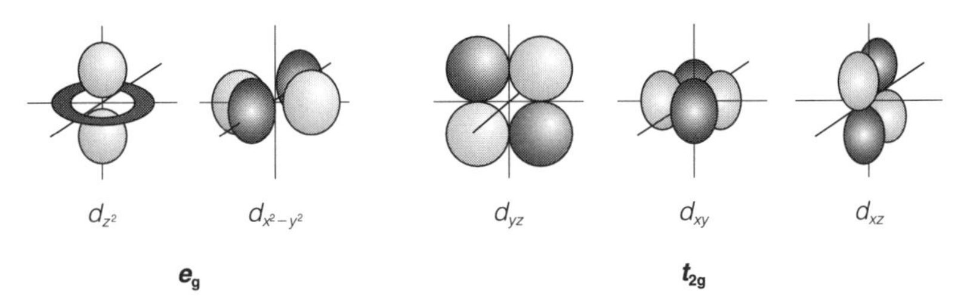

Fig. 1. The five d orbitals, showing e_g and t_{2g} sets in an octahedral complex, with ligands along the x, y and z axes.

compound with octahedral coordination, including aqua ions and many solids. Electronic transitions between t_{2g} and e_g orbitals give rise to colors, which are a familiar feature of transition metal complexes, and allow Δ_o to be measured experimentally (see Topic H8).

Although originally explained in terms of electrostatic repulsion between *d* electrons and the ligands, it is now recognized that ligand field splittings come from the same type of orbital overlap effects as donor–acceptor interactions (see Topic C8). Most ligands coordinate to the metal ion using nonbonding electrons (see Topic C1). A ligand lone-pair orbital pointing directly towards the metal overlaps with the e_g orbitals (**1**) but has the wrong symmetry to interact with t_{2g}. The overlap gives rise to σ bonding and antibonding molecular orbitals (see *Fig. 2*, and Topics C3 and C4). The bonding orbitals are occupied by the electrons from the ligand, and it is the σ antibonding levels that form the 'metal' e_g set, available for the *d* electrons of the metal ion. A strong **σ-donor** ligand will produce a large splitting Δ_o by raising the e_g energy. π bonding arises when ligands have orbitals directed perpendicular to the metal–ligand axis, which can interact with the metal t_{2g} orbitals (**2**). Ligands such as halide ions have occupied *p*π orbitals and act as **π-donors**. This interaction raises the energy of the metal t_{2g} orbitals, and decreases Δ_o. On the other hand, **π-acceptor** ligands such as CO have empty antibonding π orbitals (see Topic H9). Overlap with the metal in this case causes the t_{2g} orbitals to be lowered in energy so that Δ_o is increased (see *Fig. 2b and c*).

The order of Δ_o values produced by different ligands is known as the **spectrochemical series**. A partial series in order of increasing splitting is:

$$I^- < Br^- < Cl^- < F^- < OH^- < H_2O < NH_3 < PPh_3 < CN^- < CO$$

As expected, strong σ donors are generally high in the series, π donors are low, and π-acceptor ligands such as CN^- and CO are among the highest, and known as **strong field ligands**. The major trends with different metal ions are (i) Δ_o increases with charge on the ion, and (ii) splittings are larger for 4*d* and 5*d* series elements than in the 3*d* series.

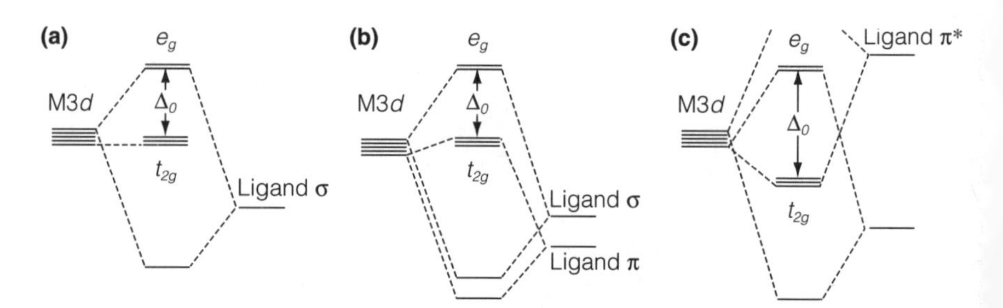

Fig. 2. Partial MO diagram showing an octahedral complex with (a) σ-donor only, (b) π-donor, and (c) π-acceptor ligands.

(a) e_g / t_{2g} (b) e_g / t_{2g}

Fig. 3. Electron configurations for d^5 in (a) high-spin and (b) low-spin octahedral complex.

High and low spin

Assignment of the electron configuration of an octahedral complex involves (i) finding the *d* electron number of the ion (see Topic H1) and (ii) determining the occupation of the t_{2g} and e_g orbitals. Electron repulsion effects are important, and other things being equal the ground state will be formed with the maximum number of electrons in different orbitals and with parallel spin (see Topic A3). Two and three *d* electrons will occupy the t_{2g} orbitals with parallel spin, but with four or more there are different possibilities. If the extra repulsion coming from **spin-pairing** is large enough, the ground state will be of the **high-spin** type formed by keeping electrons in separate orbitals as far as possible. On the other hand, if Δ_o is larger than the spin-pairing energy, the favored configuration will be **low-spin** formed by placing as many electrons as possible in t_{2g} even though they must be paired. As shown in *Fig. 3*, the high- and low-spin configurations for d^5 are $(t_{2g})^3(e_g)^2$ (five unpaired electrons) and $(t_{2g})^5$ (one unpaired electron), respectively.

The spin state of a transition metal ion can often be measured from the paramagnetic susceptibility (see Topic H8). For ions of the 3*d* series it is found that most complexes with ligands such as halides, water or ammonia are high-spin compounds, the notable exception being Co^{3+}, a d^6 ion that normally forms low-spin compounds. Low-spin complexes are found with strong field ligands such as CN^-, and nearly always with 4*d* and 5*d* elements whatever the ligand.

Ligand field stabilization energy

The **ligand field stabilization energy (LFSE)** of an ion is calculated by summing the orbital energies of the *d* electrons present, measured relative to the average energy of all five *d* levels. In octahedral coordination, each electron in a t_{2g} orbital is assigned an energy $-(2/5)\Delta_o$, and each e_g electron an energy $+(3/5)\Delta_o$. LFSE values in terms of Δ_o are shown for high- and low-spin configurations in *Table 1*. LFSE is zero for ions with the d^{10} and high-spin d^5 configurations where all *d* orbitals are equally occupied.

Table 1. Electron configurations for d^n high- and low-spin octahedral complexes, with corresponding ligand field stabilization energies

n	High spin		Low spin	
	Configuration	LFSE	Configuration	LFSE
0	–	0	–	–
1	$(t_{2g})^1$	$-2/5\ \Delta_o$	–	–
2	$(t_{2g})^2$	$-4/5\ \Delta_o$	–	–
3	$(t_{2g})^3$	$-6/5\ \Delta_o$	–	–
4	$(t_{2g})^3(e_g)^1$ [a]	$-3/5\ \Delta_o$	$(t_{2g})^4$	$-8/5\ \Delta_o$
5	$(t_{2g})^3(e_g)^2$	0	$(t_{2g})^5$	$-10/5\ \Delta_o$
6	$(t_{2g})^4(e_g)^2$	$-2/5\ \Delta_o$	$(t_{2g})^6$	$-12/5\ \Delta_o$
7	$(t_{2g})^5(e_g)^2$	$-4/5\ \Delta_o$	$(t_{2g})^6(e_g)^1$ [a]	$-9/5\ \Delta_o$
8	$(t_{2g})^6(e_g)^2$	$-6/5\ \Delta_o$	–	–
9	$(t_{2g})^5(e_g)^3$ [a]	$-3/5\ \Delta_o$	–	–
10	$(t_{2g})^5(e_g)^4$	0	–	–

[a] Configurations susceptible to Jahn–Teller distortion.

Maximum values of octahedral LFSE in high-spin states occur with the d^3 and d^8 configurations, and for low-spin with d^6. These patterns of LFSE influence thermodynamic, structural and kinetic aspects of complex formation (see Topics H6 and H7). They also have an effect on ionic radii and on lattice and solvation energies. Superimposed on a general decrease of radius along the 3*d* series, the ions with the largest LFSE have smaller radii (and also larger lattice or solvation energies) than otherwise expected (see Topic H1, *Fig. 2*). One interpretation of this effect is that in an ion with large LFSE, the repulsion between closed shells is decreased by the predominance of metal electrons in t_{2g} orbitals that do not point directly towards the ligands.

Other geometries

The pattern of ligand field splitting depends on the coordination geometry; generally those *d* orbitals that point most strongly towards the ligands are raised in energy relative to the others. *Figure 4* shows the splittings produced by some other ligand coordination geometries. **Tetrahedral** coordination gives a splitting in the opposite direction (and about half the magnitude) to that found with octahedral. **Tetragonally distorted octahedral** coordination arises where two opposite ligands are further from the metal than the other four. In this and in **square-planar** coordination, the *d* orbital pointing towards ligands in the *xy* plane is higher in energy than the others. (The main difference from the octahedral case is the lowering in energy of d_{z^2} as this interacts less strongly with the ligands).

Ligand field splitting is sometimes important in understanding the geometrical preferences of an ion, although other factors may play a part. The splitting in tetrahedral coordination is only about half that for octahedral, and so in competition between octahedral and tetrahedral geometry the octahedral LFSE is more important; thus ions such as Cr^{3+} (d^3) and Co^{3+} (d^6 low-spin) are nearly always found in octahedral coordination and are notably resistant to forming tetrahedral complexes. Square-planar complexes are found for d^8 ions when the ligand field splitting is large enough for the electrons to pair in the four lowest orbitals (see *Fig. 4c*); examples are Ni^{2+} with strong-field ligands, and Pd^{2+} and Pt^{2+} in nearly all situations (see Topics H5 and H6).

The geometry of *Fig. 4b* arises from a so-called **Jahn–Teller** distortion of the octahedron. The e_g orbitals are split in energy, and this allows stabilization of a complex if these two orbitals are unequally occupied. Thus in d^9 (Cu^{2+}) two electrons occupy the d_{z^2} and one the $d_{x^2-y^2}$. Nearly all Cu^{2+} compounds show this type of distortion, as do many high-spin d^4 ions such as Cr^{2+}.

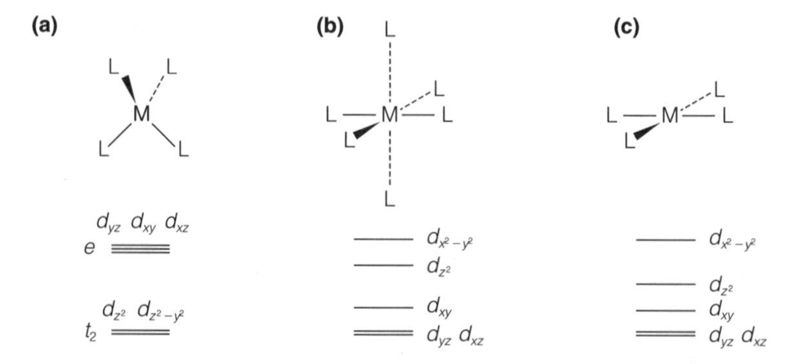

Fig. 4. Ligand field splitting patterns for (a) tetrahedral, (b) tetragonally distorted octahedral, and (c) square-planar complexes.

H3 3D SERIES: AQUEOUS IONS

Key Notes

Oxidation states

Elements from Sc to Mn can form oxidation states up to the group number (e.g. MnO_4^-). These become increasingly strongly oxidizing, and lower states are more stable for later elements. The M^{3+}/M^{2+} couple shows a trend related to ionization energies and ligand field stabilization energy.

Effect of pH

High pH stabilizes some high oxidation states and reduces the tendency of others to disproportionate. The species present change according to the acidic, basic or amphoteric character of the metal ions.

Complex formation

Early elements show hard complexing behavior; later ones have an increasing affinity for ligands such as NH_3. Complexing can alter redox potentials, and stabilizes some states such as Co^{III} and Cu^{I}.

Related topics

Complex formation (E3)
Electrode potentials (E5)
Introduction to transition metals (H1)
3*d* series: solid compounds (H4)

Oxidation states

Figure 1 shows a **Frost diagram** for the elements Sc–Zn with the electrode potentials for aqueous species appropriate to acid solution (pH = 0). Lines with positive (negative) slopes on this diagram indicate potentials that are more (less) strongly oxidizing with respect to the standard couple (H^+/H_2). The diagram displays clearly the major trend to less electropositive character across the series. (See Topic E5 for the construction and interpretation of these diagrams.)

The negative slopes for the M^{2+}/M or M^{3+}/M couples show that metals early in the series are strong reducing agents. This tendency decreases across the series, and copper has a positive Cu^{2+}/Cu slope, showing that copper metal does not react with acids to give hydrogen. It will, however, dissolve in strongly oxidizing acids such as HNO_3 or in the presence of some complexing agents.

Higher oxidation states are also much more accessible for elements early in the series. As far as Mn (group 7), elements can attain the **group oxidation state** (corresponding to the formal d^0 electron configuration; see Topic H1). This becomes more oxidizing along the series $Ti^{IV} < V^{V} < Cr^{VI} < Mn^{VII}$, and permanganate MnO_4^- is used widely as a strong oxidizing agent in acid solution.

M^{2+} ions are stable for all elements except Sc and Ti, where these oxidation states are too strongly reducing to exist in water. M^{3+} ions are formed by elements up to Co, although for Mn and Co the uncomplexed ions are very strongly oxidizing in acid solution. The ionization energy (IE) trend is the most important factor controlling the change in redox stability, and the discontinuity in the trend of third IE values after Mn (see Topic H1, *Fig. 2*) is reflected in a similar break in

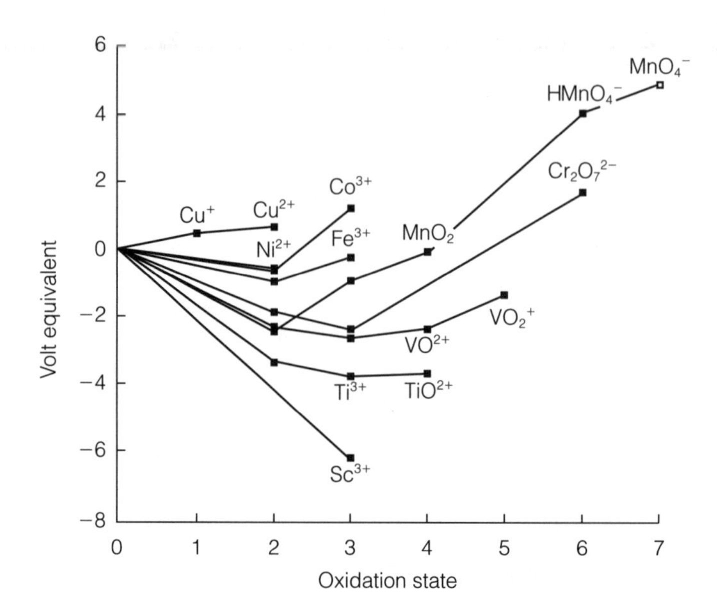

Fig. 1. Frost diagram for elements of the 3d series in aqueous solution at pH = 0.

the M^{3+}/M^{2+} redox potentials, Fe^{3+} being less strongly oxidizing than Mn^{3+}. Changing solvation energies also have an effect, and these are in turn influenced by ligand field stabilization energies (LFSE, see Topic H2). The Cr^{3+}/Cr^{2+} couple is more reducing than expected from its third IE as a consequence of the large reduction of LFSE between Cr^{3+} (d^3) and Cr^{2+} (d^4).

Also seen in the diagram is that some intermediate oxidation states are prone to **disproportionation**. Thus Mn^{VI} undergoes the following reaction in acid solution:

$$3MnO_4^{2-} + 4H^+ \rightarrow MnO_2 + 2MnO_4^- + 2H_2O$$

Mn^{3+} and Cu^+ also disproportionate although all these reactions can be influenced by pH or complexing.

Effect of pH

In a half-cell reaction such as

$$FeO_4^{2-} + 8H^+ + 4e^- \rightarrow Fe^{2+} + 4H_2O$$

increasing pH (and hence decreasing H^+ concentration) will favor the left-hand side and so lower the redox potential. Thus some high oxidation states such as FeO_4^{2-} are more accessible in alkaline than in acid solution. Changing pH can also alter the tendency to disproportionate. For example, Mn^{III}, which is unstable in alkaline solution, is nevertheless readily formed as $Mn(OH)_3$ by air oxidation of $Mn(OH)_2$. Mn^{VI} also resists disproportionation in alkaline solution (see Topic E5, *Fig. 2*, for data in alkaline solution).

The species present may change with pH in a way that depends on the oxidation state. Low oxidation states (+2) are always cationic and as pH increases an insoluble hydroxide is eventually precipitated. As the oxidation state increases so does the acidic character of the hydrated cation (see Topic E2). Thus M^{3+} ions undergo protolysis even at pH values as low as 1 or 2; deprotonation can be a first step in the formation of oxygen-bridged dimers as with

$$2[Fe(H_2O)_6]^{3+} \rightleftharpoons [(H_2O)_5Fe{-}O{-}Fe(H_2O)_5]^{4+} + 2H^+$$

These may undergo further polymerization before precipitating as $Fe(OH)_3$.

High oxidation states (+6, +7) are acidic and always present as anionic species (CrO_4^{2-}, MnO_4^-), although with CrO_4^{2-} dimerization occurs at low pH:

$$2CrO_4^{2-} + 2H+ \rightleftharpoons Cr_2O_7^{2-} + H_2O$$

With intermediate oxidation states more complex amphoteric and polymeric behavior is observed. Thus V^V forms hydrated VO_2^+ in acid solution below pH 2, and the anionic species VO_4^{3-} at high pH. Over an intermediate pH range complex **polyvanadates** are formed, most prominent being the decavanadate ion $[V_{10}O_{28}]^{6-}$ (normally present in protonated forms).

Complex formation

Complexing behavior depends on the oxidation state and position in the series. higher oxidation states tend to form stronger complexes with the 'hard' anionic ligands F^- and chelating agents such as EDTA (see Topic E3). Thus we have complexes such as $[TiF_6]^{2-}$, $[VF_6]^-$ and $[FeF_6]^{3-}$. Later elements, especially in low oxidation states, have more affinity for softer ligands such as heavier halides or ammonia. The stabilities of complexes found with many ligands, especially ammonia or amines, follow a trend known as the **Irving–Williams series**:

$$Mn^{2+} < Fe^{2+} < Co^{2+} < Ni^{2+} < Cu^{2+} > Zn^{2+}$$

Two contributions to this trend are (a) the general decrease in electropositive character resulting from increased effective nuclear charge (see Topic H1) and (b) ligand field stabilization energies (see Topic H2), which increase the stability of complexes with ligands higher in the spectrochemical series than water in all ions except Mn^{2+} (d^5) and Zn^{2+} (d^{10}).

Complexing can have a strong effect on redox chemistry, the general rule being that a ligand stabilizes whichever oxidation state it complexes with most strongly (see Topic E5). Two important examples are the following.

(i) Cu^+ forms strong complexes with ligands such as CN^- and I^- so that the Cu^I/Cu potential becomes negative and copper metal will react with acids to form hydrogen; these ligands also stabilize the Cu^I state against disproportionation.
(ii) Many ligands (e.g. NH_3) complex strongly with Co^{3+}, giving a low-spin d^6 state with a large LFSE. The resulting complexes such as $[Co(NH_3)_6]^{3+}$ are much less strongly oxidizing than aqua Co^{3+} ion, which itself oxidizes water.

Generally speaking, negatively charged ligands complex more strongly with ions of higher oxidation state and so reduce the redox potential, whereas neutral π-acceptor ligands, being electron withdrawing, tend to stabilize cations of lower charge and so raise the potential. Ligand field stabilization and other effects cause many complications, however, which can upset these simple generalizations.

H4 3D SERIES: SOLID COMPOUNDS

Key Notes

Oxidation states

Compounds are formed with elements in the group oxidation state up to Mn, higher states being found mostly with oxides and fluorides. Lower oxidation states are more stable for later elements. Many mixed-valency and nonstoichiometric compounds are known.

Halide and oxide structures

Transition metal ions are most often found in octahedral coordination. Tetrahedral coordination is commoner for high oxidation states. Ligand field stabilization energies sometimes influence the coordination. Ions with d^4 and d^9 configurations have distorted coordination geometries.

Other binary compounds

Sulfides have different structures from oxides (often NiAs or layer types), and some contain S_2^{2-} ions. Compounds with N and C are often described as interstitial.

Elements: occurrence and extraction

Elements early in the series occur in oxide minerals that cannot be reduced easily. Later elements occur in sulfides and are easier to extract.

Related topics

Binary compounds: simple structures (D3)
Lattice energies (D6)
3*d* series: aqueous ions (H3)

Oxidation states

Table 1 shows some oxides and halides of the 3*d* series elements, selected to show the range of stable oxidation states. These follow the same trends as found in aqueous chemistry (see Topic H3). Elements early in the series form compounds up to the **group oxidation state**, for example, TiO_2, VF_5 and CrO_3. With increasing group number the higher oxidation states become increasingly hard to form, and can be found only with oxides and/or fluorides, and sometimes only in ternary but not binary compounds. For example, with V^V we can make VF_5 and V_2O_5 but not VCl_5. With Mn^{VII} the only binary compound is Mn_2O_7 but this is much less stable than ternary permanganates such as $KMnO_4$.

The stabilization of high oxidation states by O and F may be attributed at least partly to their small size, which gives the large lattice energies necessary according to the ionic model to compensate for ionization energies (see Topic D6). Additional lattice stabilization is possible in ternary structures, as in compounds such as K_2FeO_4 and K_2CoF_6 where no binary compounds with the corresponding oxidation state are stable. It should be recognized that many of the compounds in high oxidation states are not very ionic, and arguments based on the high bond strengths formed by O and F to more electropositive elements may be more satisfactory than using the ionic model.

Low oxidation states (e.g. +2) are of limited stability for the early elements. The unusual **metal-rich** compound Sc_2Cl_3 has a structure with extensive Sc–Sc

Table 1. A selection of oxides and halides of the elements Sc–Cu. X represents any halogen unless specified. Oxidation states are shown in mixed-valency and ternary compounds.

Element	Oxides	Halides
Sc	Sc_2O_3	Sc_2Cl_3, ScX_3
Ti	TiO_x, Ti_2O_3, $Ti^{III}_2Ti^{IV}_2O_7$, TiO_2	TiX_2, TiX_3, TiX_4
V	VO_x, V_2O_3, VO_2, V_2O_5	VX_2, VX_3, VX_4, VF_5
Cr	Cr_2O_3, CrO_2, CrO_3, $K_2Cr^{VI}_2O_7$	CrX_2, CrX_3, CrF_4, CrF_5, CrF_6
Mn	MnO, $Mn^{II}Mn^{III}_2O_4$, Mn_2O_3, MnO_2, Mn_2O_7, $KMn^{VIII}O_4$	MnX_2, MnF_3, MnF_4
Fe	Fe_xO, $Fe^{II}Fe^{III}_2O_4$, Fe_2O_3, $Sr_2Fe^{IV}O_4$, $K_2Fe^{VI}O_4$	FeX_2, FeX_3 (*not* I)
Co	CoO, $Co^{II}Co^{III}_2O_4$, $LaCo^{III}O_3$, $Na_4Co^{IV}O_4$	CoX_2, CoF_3, $Cs_2Co^{IV}F_6$
Ni	NiO, $NaNi^{III}O_2$	NiX_2, $K_2Ni^{IV}F_6$
Cu	Cu_2O, CuO, $LaCu^{III}O_3$	CuX (*not* F), CuX_2 (*not* I), $K_3Cu^{III}F_6$, $Cs_2Cu^{IV}F_6$

bonds. Compounds such as TiO_x and VO_x are nonstoichiometric (see below) and are also stabilized by metal–metal bonding using *d* electrons. With Cu the +1 oxidation state is stable in compounds such as Cu_2O and CuCl, but CuF is not known, presumably because the larger lattice energy of fluorides makes this unstable with respect to disproportionation to Cu and CuF_2. The differential stability of oxidation states with different halogens is also shown by the existence of CuI but not CuI_2.

The existence of several stable oxidation states gives rise to the possibility of **mixed valency compounds** where an element is present in different oxidation states. Thus the compounds M_3O_4 with M = Mn, Fe, Co, have both M^{II} and M^{III} states present. Many oxides also show **nonstoichiometry** where a continuous range of composition is possible. For example, 'TiO' is really TiO_x where x can vary continuously over a wide range, and 'FeO' does not actually exist but is approximately $Fe_{0.9}O$ (and thermodynamically unstable below 550°C). Such nonstoichiometric compounds are better described by phase diagrams than by simple stoichiometric formulae, which can be misleading.

Halide and oxide structures

A majority of halides and oxides have the structures expected for largely ionic compounds, with the metal in **octahedral coordination** (see Topic D3, especially *Fig. 1*). Common oxide structures are rocksalt (e.g. MnO, NiO), corundum (see Topic G5, e.g. Cr_2O_3, Fe_2O_3) and rutile (e.g. TiO_2, CrO_2). Most MF_2 compounds have the rutile structure, other dihalides forming layer ($CdCl_2$ and CdI_2) types. Many ternary oxides and halides also follow this pattern; for example, the $LaMO_3$ compounds formed by all elements of the series (M = Sc–Cu) have the perovskite structure (see Topic D5).

The $3d^4$ ions Cr^{2+} and Mn^{3+} and the $3d^9$ ion Cu^{2+} are subject to **Jahn-Teller distortions** (see Topic H2). For example, CuO does not have the rocksalt structure, but one with four close Cu–O neighbors and two at longer distance; similar tetragonally distorted coordination is found in most other simple compounds of Cr^{2+} and Cu^{2+}. (Note that CrO is unknown.)

Tetrahedral coordination is also sometimes found. In high oxidation states (e.g. molecular $TiCl_4$, polymeric CrO_3 and in complex ions such as VO_4^{3-}, CrO_4^{2-} and MnO_4^-) this can be understood in terms of the small size of the transition metal ion. However, tetrahedral (zinc blende) structures are also found in Cu^I halides such as CuCl. As Cu^+ has the $3d^{10}$ configuration this appears to be typical post-transition metal behavior as seen, for example, with Zn^{2+}, and must involve some degree of covalent bonding (see Topics D4 and G4).

Some ternary and mixed-valency oxides have the **spinel structure** where metal ions occupy a proportion of tetrahedral and octahedral holes in a cubic close-packed lattice (see Topic G5). Examples include M_3O_4 with M = Mn, Fe, Co. The distribution of M^{2+} and M^{3+} ions between the tetrahedral and octahedral sites shows the influence of ligand field stabilization energies (see Topic H2). In Fe_3O_4, Fe^{2+} ($3d^6$) has an octahedral preference whereas Fe^{3+} ($3d^5$) has none, and this compound has the **inverse spinel** structure where Fe^{2+} is octahedral and Fe^{3+} is present in both octahedral and tetrahedral sites. In Co_3O_4 the low-spin $3d^6$ ion Co^{3+} has a very strong octahedral preference and the **normal spinel** structure is found with all Co^{3+} in octahedral sites and Co^{2+} tetrahedral. Mn_3O_4 is also based on the normal spinel structure, but with a tetragonal distortion as expected for the sites occupied by Mn^{3+} ($3d^4$).

Other binary compounds

Sulfides are formed by all elements and have structures different from oxides. Many MS compounds (which are generally nonstoichiometric) have the NiAs structure. TiS_2 and VS_2 have layer (CdI_2) structures, but later disulfides contain S_2^{2-} ions (e.g. FeS_2 with the **pyrites** and **marcasite** structures; this is a compound of Fe^{II} not Fe^{IV}). The compound CuS is particularly complicated, having apparently Cu^{I} and Cu^{II} present as well as S^{2-} and S_2^{2-}.

Hydrides, nitride and carbides are known for some of the elements. Some have simple stoichiometry and structure, such as TiN and TiC with the rocksalt structure. Many are nonstoichiometric with metallic properties, and some can be regarded as **interstitial compounds** with the nonmetal atom occupying sites between metallic atoms in the normal elemental structure.

Elements: occurrence and extraction

The decreasing electropositive character of the elements across the series is shown in the typical minerals they form, and in the methods required to extract them (see Topics B4 and J2). Early elements are found in oxide or complex oxide minerals (e.g. TiO_2, $CrFeO_3$) and are known as **lithophilic**, whereas later elements are found mainly in sulfides (e.g. NiS) and are called **chalcophilic**. Iron forms the dividing line in this trend, and is found both as Fe_2O_3 and FeS_2. Reduction of later elements is relatively easy, as sulfides may be roasted to form oxides and then reduced with carbon. For example, iron, a major structural metal, is produced in blast furnaces by reduction of Fe_2O_3:

$$2Fe_2O_3 + 3C \rightarrow 4Fe + 3CO_2$$

However, early transition metal oxides cannot be reduced in this way, because they form stable carbides (e.g. TiC) and/or because the temperature required for reduction by carbon is too high. The **Kroll process** for manufacture of Ti involves first making $TiCl_4$,

$$TiO_2 + 2Cl_2 + 4C \rightarrow TiCl_4 + 4CO$$

which is then reduced by metallic magnesium. Titanium is widely used as a lightweight structural metal; although potentially very reactive towards water and air it forms a very inert protective TiO_2 film.

H5 4D AND 5D SERIES

Key Notes

Oxidation states

Higher oxidation states are more stable than in the 3*d* series, and lower ones less common. The group oxidation state is found up to group 8. 4*d* and 5*d* elements of early groups are very similar; in later groups higher oxidation states occur in the 5*d* series.

Aqueous chemistry

Very few simple aqua cations are found, but many complexes are known, increasingly dominated by softer ligands for later elements. High oxidation states form oxoanions that are less strongly oxidizing than corresponding 3*d* species, and that form extensively polymerized structures.

Solid structures

Larger ions formed by early elements have high coordination numbers. Many compounds show extensive metal–metal bonding. Later ions have low coordination numbers related to specific electron configurations.

Related topics

Introduction to transition metals (H1)
3*d* series: aqueous ions (H3)
3*d* series: solid compounds (H4)

Oxidation states

Table 1 shows the main binary oxides and halides formed by transition elements of the 4*d* and 5*d* series. Comparison with the corresponding information for the 3*d* series (Topic H4, *Table 1*) shows a similar pattern, with early elements in the series forming states up to the group maximum (Zr^{IV}, Nb^{V}, etc.) where all valence electrons are involved in bonding. The principal difference is that this trend persists further in the lower series, the compounds RuO_4 and OsO_4 having no counterpart with the 3*d* element iron. Following group 8, the highest oxidation state shown in *Table 1* remains higher than ones in 3*d* elements, as in RuF_6, IrF_6, PtF_6 and AuF_5. In addition to oxides and halides, high oxidation states are sometimes found with surprising ligands, such as in the ion $[ReH_9]^{2-}$ (**1**), which is formally a hydride complex of Re^{VII}.

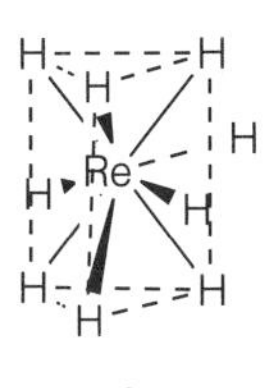

1

A counterpart to the stability of higher oxidation states is that lower ones (+2, +3) are less often found than in the 3*d* series.

For the earlier groups the patterns of 4*d* and 5*d* behavior are so similar that the corresponding elements (Zr, Hf, etc.) are placed together in *Table 1*, but in later groups high oxidation states become slowly less stable in the 4*d* compared with the 5*d* series. This tendency is especially marked with Pd, Pt, Ag and Au. The factors underlying the differences from 3*d* elements, and the general similarity of the two lower series, are discussed in Topic H1. The slow divergence between 4*d* and 5*d* series arises because increasing nuclear charge across the series has more effect on ionization energies of 4*d* orbitals than on the larger 5*d*.

Another trend apparent from *Table 1* is the preponderance of oxidation states with even rather than odd electron configurations in later groups; these include Pt^{IV}, Au^{V} (d^6), Pd^{II}, Au^{III} (d^8) and Ag^{I} (d^{10}). AgF_2 (d^9) is an exception although sometimes the stoichiometry is misleading, AgO being a mixed valency compound, $Ag^{I}Ag^{III}O_2$. Even electron configurations are favored by the large ligand field splittings found in these series, giving low spin states, with the d^6 octahedral and d^8 square-planar arrangements being particularly favorable (see Topic H2).

Aqueous chemistry

Unlike elements of the 3*d* series, 4*d* and 5*d* elements have little simple aqueous cationic chemistry. The main exceptions are Y^{3+} and La^{3+} (see lanthanides, Topic I1), and Ag^+, which forms some soluble salts (AgF, $AgNO_3$). The aqua Ag^+ ion shows strong class b complexing behavior, with an affinity for ligands such as NH_3, I^- and CN^- comparable with Cd^{2+} in the next group (see Topics E3 and G4). Some other aqua cations can be made, but they are extensively hydrolyzed and polymerized (e.g. Zr^{4+}, Hf^{4+}), strongly reducing (e.g. Mo^{3+}), or have a very high affinity for other ligands (e.g. Pd^{2+}) and are difficult to prepare in uncomplexed form.

Numerous complexes are, however, formed, the most stable with early groups being ones with F^- and oxygen donor ligands, and in later groups ones with softer ligands such as heavy halides and nitrogen donors. This trend is similar to that found in the 3*d* series but is more marked. The most commonly encountered solu-

Table 1. A selection of oxides and halides of elements from the 4d and 5d series. M represents either of the two elements from the corresponding group, and X any halogen unless exceptions are specified.

Element	Oxides	Halides
Y, La	M_2O_3	MX_3
Zr, Hf	MO_2	ZrCl, MX_4
Nb, Ta	NbO, MO_2, M_2O_5	M_6X_{14} (*not* F), MX_3, MX_4, MX_5
Mo, W	MO_2, MO_3, Na_xMO_3	MX_2 (*not* F), MX_3, MX_4, MoF_5, $MoCl_5$, WX_5, (*not* I), MoF_6, WF_6, WCl_6
Tc	TcO_2, Tc_2O_7	$TcCl_4$, TcF_5, TcF_6
Re	ReO_2, Re_2O_5, ReO_2, Re_2O_7	Re_3X_9 (*not F*), ReX_4 (*not* I), ReX_5 (*not* I), ReF_6, $ReCl_6$, ReF_7
Ru	RuO_2, RuO_4	RuX_2 (*not* F), RuX_3, RuF_4, RuF_5, RuF_6
Os	OsO_2, OsO_4	OsX_3 (*not* F), OsX_4 (*not* I), OsF_5, $OsCl_5$, OxF_6
Rh	Rh_2O_3, RhO_2	RhX_3, RhF_4, RhF_5, RhF_6
Ir	IrO_2	IrX_3, IrX_4, IrF_4, IrF_5, IrF_6
Pd	PdO	PdX_2, PdF_4
Pt	PtO_2	PtX_2 (*not* F), PtX_4, PtF_5, PtF_6
Ag	Ag_2O, $Ag^{I}Ag^{III}O_2$	Ag_2F, AgX, AgF_2
Au	Au_2O_3	AuX (*not* F), AuX_3 (*not* I), AuF_5

tion species for later elements are chloride complexes such as $[PdCl_4]^{2-}$, $[PtCl_6]^{4-}$ and $[AuCl_4]^{-}$.

Oxoanions are formed by elements of groups 5–8, examples being MoO_4^{2-}, ReO_4^{-} and RuO_4^{2-}. They are invariably less strongly oxidizing than their counterparts in the 3*d* series. Mo^{VI} and W^{VI}, and to a lesser extent Nb^{V} and Ta^{V}, form extensive series of polymeric oxoanions: **isopolymetallates** such as $[Mo_6O_{19}]^{2-}$ and $[Ta_6O_{19}]^{8-}$ are mostly based on metal oxygen octahedra sharing corners and edges (see Topic D3); **heteropolymetallates** such as the phosphopolymolybdate ion $[PMo_{12}O_{40}]^{3-}$ incorporate other elements, in this case as a tetrahedral PO_4 group.

Solid structures

Larger ionic radii compared with the 3*d* series elements often lead to higher coordination numbers (see Topics D3 and D4). ZrO_2 and HfO_2 can adopt the eight-coordinate fluorite structure as well as a unique seven-coordinate structure known as baddeleyite (cf. TiO_2, rutile). ReO_3 is the prototype of a structure with six-coordination, and is adopted also (in slightly distorted form) by WO_3, in contrast to Cr^{VI}, which is tetrahedral. MoO_3 and WO_3 form extensive series of insertion compounds known as **oxide bronzes** (see discussion of Na_xWO_3 in Topics D5 and D7). In halides, higher coordination often leads to polymeric forms for compounds MX_4 and MX_5 where the corresponding 3*d* compounds are molecular.

Compounds of elements in low oxidation states very frequently have extensive **metal–metal bonding**. Sometimes this acts to modify an otherwise normal structure, as in NbO_2, MoO_2 and WO_2, which have the rutile form distorted by the formation of pairs of metal atoms. Often the structures are unique. For example, $MoCl_2$ contains $[Mo_6Cl_8]^{4+}$ clusters formed by metal–metal bonded octahedra with chlorine in the face positions (see **2**; only one of eight Cl shown). Complex halides often show metal–metal bonding, such as in $[Re_2Cl_8]^{2-}$ (**3**) where all four *d* electrons of Re^{III} are paired to form a quadruple bond.

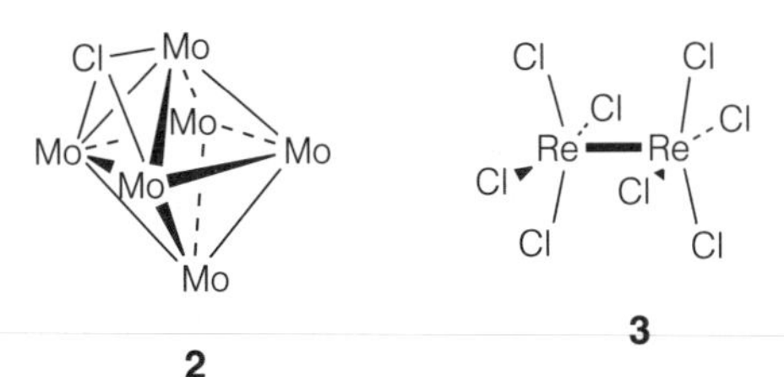

2

3

Later elements tend to show coordination geometries that are specific to certain low-spin electron configurations (see above and Topic H2). d^6 compounds are invariably octahedral, d^8 nearly always square planar (e.g. in $PdCl_2$ **4** and PdO; a rare exception is PdF_2, which, like NiF_2, has the octahedral rutile structure with two unpaired electrons per Pd). The d^{10} configuration often has a tendency to linear two-coordination (cf. Hg^{II}, Topic G4). Although AgF, AgCl and AgBr have the rocksalt structure some other Ag^{I} compounds such as Ag_2O have two-coordination, and it is normal for Au^{I}; for example, AuCl has a chain structure with a linear Cl–Au–Cl arrangement.

Cl Cl Cl
Pd Pd Pd
Cl Cl Cl

4

H6 COMPLEXES: STRUCTURE AND ISOMERISM

Key Notes

Coordinate number and geometry

Classical or Werner complexes have a metal in a positive oxidation state coordinated by donor ligands. The coordination number and geometry are determined by size and bonding factors, octahedral and tetrahedral being common for 3*d* ions, and square-planar coordination for some d^8 ions. Polynuclear complexes can have bridging ligands and/or metal–metal bonding.

Nomenclature

H_2O and NH_3 ligands are called aqua and ammine respectively. The names of anionic ligands ends in -o, and of anionic complexes in -ate. Either the oxidation state or the overall charge on the complex is specified.

Isomerism

The study of isomerism depends on kinetic factors limiting the rate of interconversion, and in the 3*d* series is confined to complexes of Cr^{III} and Co^{III}. Coordination, linkage, geometrical, and optical isomerism are possible.

Related topics

Describing inorganic compounds (B5)

Complex formation (E3)

Ligand field theory (H2)

Coordination number and geometry

Transition metal complexes are cationic, neutral or anionic species in which a transition metal is coordinated by ligands. A **classical** or **Werner complex** is one formed by a metal in a positive oxidation state with donor ligands such as H_2O, NH_3 or halide ions. Ligands with strong π-acceptor properties are discussed in Topic H9.

The **coordination numbers** (CN) observed in complexes range from two (e.g. $[Ag(NH_3)_2]^+$) to nine (e.g. $[ReH_9]^{2-}$; see Topic H5, Structure 1). The commonest geometries for 3*d* ions are **octahedral** (CN = 6, e.g. $[M(H_2O)_6]^{2+}$) and **tetrahedral** (CN = 4, e.g. $[MCl_4]^{2-}$). As in solid compounds, higher coordination numbers are often found with the larger 4*d* and 5*d* ions. Other coordination geometries may be dictated by bonding arrangements depending on the *d* electron number (see Topics H2 and H5).

The relative preference for octahedral or tetrahedral coordination is partly steric, but ligand field effects can also play a role. Ions with the d^3 and low-spin d^6 configurations (e.g. Cr^{3+} and Co^{3+}, respectively) have a large octahedral ligand field stabilization energy and are notably resistant to forming tetrahedral complexes. **Square-planar** complexes would never be predicted in preference to tetrahedra on steric grounds alone. They are commonly found, however, with $4d^8$ and $5d^8$ ions such as Pd^{2+} and Pt^{2+} where the pattern of ligand field splitting is favorable if its magnitude is large enough for spin-pairing to occur. The corresponding $3d^8$ ion Ni^{2+} gives square-planar complexes only with strong-field ligands such as CN^-; otherwise octahedral or sometimes tetrahedral coordination is found. With the d^9 or high-spin d^4 configuration a distorted octahedral geometry is often found with only four ligands strongly attached. This is common with Cu^{2+}, as in $[Cu(NH_3)_4]^{2+}$, where two weakly bound water molecules are also present.

Low coordination numbers are often found with post-transition metal ions having the d^{10} configuration (see Topic G4). This is also true for the d^{10} ions Cu^+, Ag^+ and Au^+, which form many linear complexes with CN = 2 (e.g. $[AuCl_2]^-$, isoelectronic to $HgCl_2$).

Polynuclear complexes contain more than one metal atom. Sometimes these may be held by **bridging ligands**, as in $[(RuCl_5)_2O]^{4-}$ (**1**). In other cases **metal–metal bonds** may be present, as in $[Re_2Cl_8]^{2-}$ (see Topic H5, Structure 3). Metal–metal bonding is commoner in the 4*d* and 5*d* series than with 3*d* elements, although binuclear compounds of Cr^{II} are known; for example, $[Cr_2(CH_3CO_2)_4]$ (**2**), which has bridging acetate groups (only one shown explicitly) and a quadruple Cr–Cr bond formed by all remaining valence electrons of the $3d^4$ ions.

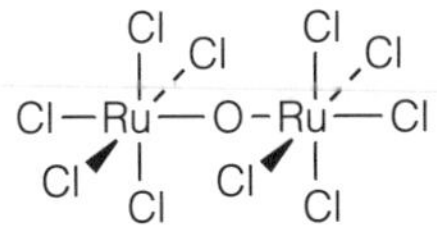

Cl–Ru(Cl)(Cl)(Cl)(Cl)–O–Ru(Cl)(Cl)(Cl)(Cl)–Cl

1

CH₃–C(O)(O) bridging Cr–Cr

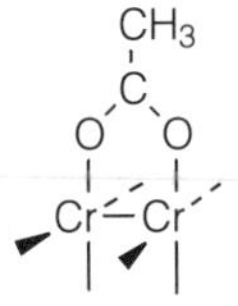

2

Nomenclature

The naming of coordination compounds is introduced in Topic B5. Some further examples will illustrate the principles involved.

- $[Ni(H_2O)_6]^{2+}$, hexaaquanickel(II) ion; $[Cu(NH_3)_4]^{2+}$, tetraamminecopper(II) ion. The terms **aqua** and **ammine** are used for water and ammonia ligands. Other neutral ligands are referred to by their normal (molecular) name. Sometimes the prefixes **bis**, **tris**, . . . are used where normal form (bi, tri, . . .) could cause confusion with the ligand name; for example, $[Co(H_2O)_3(CH_3NH_2)_3]^{3+}$, tris(methylamine)triaquacobalt(III) ion.
- $[CoCl_4]^{2-}$ tetrachlorocobaltate(II), $[Fe(CN)_6]^{3-}$ hexacyanoferrate(3−). For anionic ligands the normal ending -ide is replaced by -o. Names of anionic complexes end in -ate, and are sometimes based on Latin rather than English names of the metallic element (see Topic B5, *Table 1*). Either the oxidation state of the metal atom or the total charge on the complex is specified.
- $[CoCl(NH_3)_5]Cl_2$, pentaamminechlorocobalt(III) chloride. Coordinated ligands are shown in square brackets, others are assumed to be separate in the structure. Anionic ligands are usually written before neutral ones in the formula, but after them in the name.
- $[(RuCl_5)_2O]^{4-}$ (**1**), μ-oxo-bis(pentachlororuthenate)(4−). The Greek letter μ ('mu') is used to denote bridging ligands.

Isomerism

Isomers are compounds with the same (molecular) formula but different structure. When several isomers exist, one may be thermodynamically more stable than the others, or there may be an equilibrium between them (see Topic B3). Thus the isolation and study of individual isomers depends on **kinetic factors** that limit the rate of interconversion. Such kinetic inertness is associated with only a few ions (see Topic H7) and most examples of isomerism involve complexes of Cr^{III}, Co^{III} and Pt^{II}.

Ionization isomerism

This is best shown by an example. '$CrCl_3.6H_2O$' exists in four solid forms, which dissolve in water to give different species:

$[Cr(H_2O)_6]Cl_3 \rightarrow [Cr(H_2O)_6]^{3+} + 3Cl^-$
$[Cr(H_2O)_5Cl]Cl_2.H_2O \rightarrow [Cr(H_2O)_5Cl]^{2+} + 2Cl^- + H_2O$
$[Cr(H_2O)_4Cl_2]Cl.2H_2O \rightarrow [Cr(H_2O)_4Cl_2]^+ + Cl^- + 2H_2O$
$[Cr(H_2O)_3Cl_3].3H_2O \rightarrow [Cr(H_2O)_3Cl_3] + 3H_2O$

The different isomers all contain an octahedral Cr^{III} complex but the coordinated ligands are different; for example, in the first case the three Cl^- ions are present in the crystal lattice of the solid compound but are not directly bound to the metal.

Linkage isomerism

A few ligands are **ambidentate**, meaning that they can coordinate through alternative ligand atoms. Examples are nitrite NO_2^- (which can bind through N or O) and thiocyanate SCN^- (S or N). The nomenclature **N-nitrito** and **O-nitrito** is recommended for complexes where in formulae the ligand atom is underlined, $M–\underline{N}O_2$ and $M–\underline{O}NO$, respectively (although the nonsystematic names **nitro** and **nitrito** are also used for NO_2^- complexes).

Geometrical isomerism

The fact that a tetrahedrally coordinated compound MX_2Y_2 has only one possible isomer was historically important in establishing the structure of carbon compounds. When the coordination is square planar there are two possibilities, known as the ***cis*** (**3**) and ***trans*** (**4**) forms. Geometrical isomers occur also in octahedral complexes: with MX_2Y_4 the two isomers are also called *cis* (**5**) and *trans* (**6**), and for MX_3Y_3 the terms ***mer*** (**7** from 'meridional') and ***fac*** (**8** from 'facial') are used.

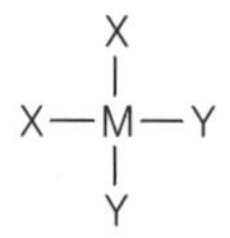

3 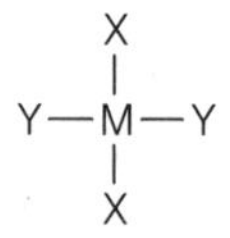 **4**  **5**

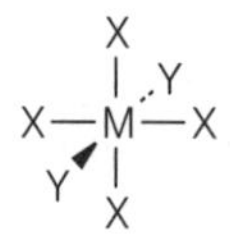

6 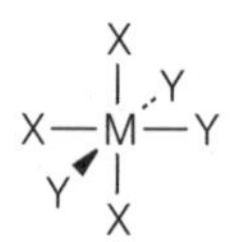 **7**  **8**

Geometrical isomerism can also refer to the possibility of different coordination geometries, although these are rather rare. Square-planar or tetrahedral coordination is, in principle, possible with CN = 4, and an example with CN = 5 occurs with $[Ni(CN)_5]^{3-}$, which can adopt shapes approximating either to a trigonal biyramid (the normally expected shape; see Topic C2) or a square pyramid (**9**).

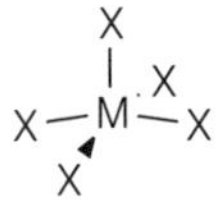

9  **10**  **11**

Optical isomerism

When a species cannot be superimposed on its mirror image the two forms are known as **enantiomers** or **optical isomers**. Most examples with coordination compounds have chelating (e.g. bidentate) ligands (see Topic E3). Structures **10** and **11** show respectively the ***delta*** and ***lambda*** isomers of a tris(chelate) complex, with the bidentate ligands each denoted by a simple bond framework.

H7 Complexes: kinetics and mechanism

Key Notes

Ligand exchange

Ligand exchange mechanisms may be associative (A), dissociative (D) or interchange (I_a or I_d) in nature. Kinetically inert complexes are formed by Cr^{3+} and Co^{3+}, and by 4*d* and 5*d* ions with d^6 and d^8 configurations.

Octahedral complexes

The mechanism becomes more dissociative for 3*d* ions later in the series. Substitution rates may be increased by the conjugate base mechanism.

Square - planar complexes

Exchange in square-planar complexes is associative, and is influenced by the *trans* effect, whereby some ligands facilitate substitution of *trans* ligands.

Electron transfer

Inner sphere mechanisms involve a bridging ligand, whereas in outer sphere mechanisms ligand coordination remains intact. The reorganization of metal–ligand distances is important in determining outer sphere electron transfer rates. These can be very slow for redox reactions involving Co^{III} complexes.

Related topics

Complex formation (E3)
Ligand field theory (H2)
Complexes: structure and isomerism (H6)

Ligand exchange

Ligand exchange reactions are of the kind

$$ML_nX + Y \rightarrow ML_nY + X$$

and are effectively nucleophilic substitutions. The possible mechanisms are classified as **associative** (A) or **dissociative** (D) according to whether the new bond is formed before or after the old one is broken, or **interchange** (I), the intermediate case, which can be subdivided into I_a or I_d according to the degree of associative or dissociative character. Kinetic studies of ligand exchange can sometimes distinguish between the mechanisms although these results can be misleading. Determination of the volume or entropy of activation (i.e. the volume or entropy change in the transition state) can often give guidance.

For many metal ions ligand exchange is an extremely fast reaction, with rate constants close to the limit of diffusion control (around 10^{10} M^{-1} s^{-1} in water). There is a correlation with the charge and size, and outside the transition series Be^{2+} and Al^{3+}, which have large charge/size ratio, have significantly slower exchange. With transition metals the influence of ligand field effects is apparent (see Topic H2). Complexes of Cr^{3+} (d^3) and Co^{3+} (d^6) and of many d^6 and d^8 ions in the 4*d* and 5*d* series are **kinetically inert** and undergo ligand substitution many orders of magnitude more slowly than comparable nontransition ions. These ions have a ligand field stabilization energy (LFSE) that contributes a barrier to the geometrical change required in the transition state. A large LFSE value also gives shorter bond lengths, which enhance other contributions (electrostatic, etc.) to the metal–ligand bond strength.

The existence of kinetically inert complexes is useful in mechanistic studies, and important for the separation of different isomers (see Topic H6).

Octahedral complexes

Most M^{2+} ions of the 3*d* series undergo ligand exchange at a rate comparable with that for nontransition metal ions of similar size. V^{2+} (d^3) and Ni^{2+} (d^8) are somewhat slower, these being the electron configurations that give maximum octahedral LFSE for high-spin ions. Entropies and volumes of activation suggest a change from predominantly I_a mechanisms early in the series (e.g. V^{2+}) to I_d towards the end (e.g. Ni^{2+}). Both decreasing size and increasing *d* orbital occupancy may contribute to this trend. Incoming ligands in the I_a mechanism must approach an octahedral complex along directions where the t_{2g} orbitals normally point (see Topic H2, *Fig. 1*). Filling these orbitals will tend to inhibit the approach of ligands and favor the dissociative pathway.

For the kinetically inert low-spin Co^{III} complexes the mechanism of exchange is certainly dissociative although kinetic studies can give results that are superficially misleading. For example, the base hydrolysis reaction

$$[Co(NH_3)_5Cl]^{2+} + OH^- \rightarrow [Co(NH_3)_5OH]^{2+} + Cl^-$$

has a rate proportional to the concentrations of both complex and OH^-. This is not indicative of an associative mechanism, but of a **conjugate base mechanism** where the first reversible step is deprotonation of the complex:

$$[Co(NH_3)_5Cl]^{2+} + OH^- \rightleftharpoons [Co(NH_3)_4(NH_2)Cl]^+ + H_2O$$

Deprotonation *trans* to the leaving group is especially effective at promoting the dissociation step. The conjugate base mechanism cannot operate if a tertiary amine with no ionizable proton is placed *trans* to the leaving group; as expected the rate of substitution is then slower and does not depend on $[OH^-]$.

Square-planar complexes

Kinetically inert square-planar complexes are formed by d^8 low-spin ions, especially Pt^{2+}. Ligand substitution is associative and correlated with the ease of forming a five-coordinate transition state (or intermediate). Substitution is much faster with Ni^{2+} where five-coordinate complexes such as $[Ni(CN)_5]^{3-}$ are more stable than for Pt. For a given metal, the rate of substitution is controlled by:

- the nature of the incoming and leaving ligands; more polarizable groups are generally faster in both bond-making and breaking processes;
- the ***trans* effect**, which is the ability of some ligands to facilitate the substitution of the ligand *trans* to them in the complex. Some ligands in order of increasing effectiveness are:

$$NH_3 < Cl^- < Br^- < H^-, PR_3 < CN^-, CO$$

The *trans* effect is a kinetic phenomenon and is influenced by different factors that operate either in the ground state or in the five-coordinate transition state. Some ligands weaken the bond *trans* to them in the original complex. This ground-state phenomenon is called the ***trans* influence**, and depends mostly on the σ bonding capability and the polarizability of the ligand. Some ligands such as CN^- do not show much *trans* influence but nevertheless have a large kinetic *trans* effect, because their π-acceptor properties help in the stabilization of the transition state.

The *trans* effect is useful in synthesis. For example, different isomers are formed in the reactions below by the greater *trans* directing ability of Cl^- compared with NH_3:

$$[Pt(NH_3)Cl_3]^- + NH_3 \rightarrow cis\text{-}[Pt(NH_3)_2Cl_2] + Cl^-$$
$$[Pt(NH_3)_3Cl]^+ + Cl^- \rightarrow trans\text{-}[Pt(NH_3)_2Cl_2] + NH_3$$

Electron transfer reactions

Electron transfer is the simplest type of redox process, an example being
$V^{2+} + Fe^{3+} \rightarrow V^{3+} + Fe^{2+}$
A majority of reactions of this type are very fast, but oxidation by some complexes (especially of Co^{III}) is much slower.

In an **inner sphere** process, the coordination sphere of one complex is substituted by a ligand bound to the other complex, which then acts as a bridge and may be transferred during the redox process. For example, isotopic labeling studies show that the oxidation of aqueous Cr^{2+} with $[Co^{III}(NH_3)_5Cl]^{2+}$ proceeds via a bridged species Cr–Cl–Co, the chlorine not exchanging with free labeled Cl^- in solution but remaining attached to the kinetically inert Cr^{III} product. An inner sphere mechanism requires one of the reactants to be substitutionally labile, and a ligand that can act as a bridge. One test is to compare the rates of reaction with the ligands azide N_3^- and (N bonded) thiocynanate NCS^-; azide is generally better at bridging and so gives faster rates if the inner sphere route is operating.

The **outer sphere** mechanism involves no disruption of the coordination of either complex, and is always available as a route to electron transfer unless the inner sphere rate is faster. The **Marcus theory** shows that the rate of outer sphere transfer depends on:

(i) the orbital interaction between the two metal centers involved, a factor that decreases roughly exponentially with the distance between them;
(ii) the change in metal–ligand distances resulting from electron transfer, the effect that provides most of the activation energy for the reaction;
(iii) an enhancement term, which depends on the difference of redox potentials of the two couples involved.

Reactions of complexes containing unsaturated ligands such as bipyridyl are generally fast because the π system facilitates transfer, and because the change in geometry is small (as significant charge is distributed over the ligand). On the other hand, oxidation by $[Co(NH_3)_6]^{3+}$ is often very slow. The orbital interaction term is small because the reaction is 'spin forbidden', the ground state of Co changing from low-spin d^6 with no unpaired electrons to high-spin Co^{2+} d^7 with three. The activation energy is also large because the number of e_g electrons increases by two, which gives a significant change of LFSE and so causes a large increase in the metal–ligand distances. The inner sphere route is unavailable as NH_3 does not normally act as a bridging ligand.

H8 COMPLEXES: ELECTRONIC SPECTRA AND MAGNETISM

Key Notes

Electronic transitions

Electronic absorptions, in which an electron is excited to a higher energy orbital, occur in the visible and neighboring parts of the spectrum. Transitions are classified as *d–d*, ligand-to-metal charge transfer (LMCT), metal-to-ligand charge transfer (MLCT), or ligand based.

d–d spectra

d–d transitions are weak, especially in centro-symmetric complexes. The number of transitions depends on the *d*-electron configuration. The energies provide information about ligand field splittings and electron repulsions.

Charge transfer spectra

Charge transfer energies may be correlated with redox potentials. LMCT is at low energy if the metal ion is easily reduced or the ligand easily oxidized.

Paramagnetism

Paramagnetism depends on the number of unpaired electrons and can provide information about spin states and metal–metal bonding.

Related topic

Ligand field theory (H2)

Electronic transitions

In an electronic transition an electron is excited from an occupied to an empty molecular orbital (MO). The energy of such transitions normally corresponds to photons in the near IR, visible or UV region of the electromagnetic spectrum. Electronic **absorption bands** give rise to the colors of compounds, including ones without transition metals (see Topic D7).

In *d*-block complexes various types of MO can be involved. In ***d–d*** **transitions** both the lower and upper MOs are those based on the *d* atomic orbitals, split by interaction with the ligands (see Topic H2). **Charge transfer transitions** involve ligand-based MOs as well, and may be divided into **ligand-to-metal charge transfer** (LMCT, the commonest type) or **metal-to-ligand charge transfer** (MLCT). There may also be transitions between two ligand MOs (e.g. π to π^* in unsaturated ligands). Charge transfer and ligand-based transitions often appear at higher energy than *d–d* transitions, and are generally also more intense. *Figure 1* shows the absorption spectrum of $[Ti(H_2O)_6]^{3+}$. The *d–d* transition peaks at around 20 000 cm^{-1} (500 nm) corresponding to green light, giving a violet color to the complex (transmitting red and violet light). The strong absorption rising beyond 25 000 cm^{-1} is due to LMCT.

d–d spectra

d–d transitions are weak because of atomic **selection rules,** which make transitions between *d* orbitals **forbidden**. They remain forbidden in complexes with a center of symmetry (e.g. octahedral or square planar), and appear only because of vibrational motions that break this symmetry. In complexes without a center of symmetry (e.g. tetrahedral) the transitions are stronger but are still weak compared with charge transfer. There are also **spin selection rules**, the strongest transitions being **spin-allowed** ones where there is no change in the number of unpaired electrons.

In a d^1 octahedral complex such as $[Ti(H_2O)_6]^{3+}$ excitation of an electron from t_{2g} to e_g gives a single absorption band at an energy equal to the ligand field splitting Δ_o (see Topic H2). The theory is more complicated for ions with many *d* electrons because the energy of a state is now determined by the repulsion between electrons as well as the occupancy of t_{2g} and e_g orbitals. The predicted number of spin-allowed *d–d* transitions in high-spin octahedral or tetrahedral complexes is shown below. Not all transitions may be visible in all cases, because bands may overlap or some may be obscured by charge transfer:

one for d^1, d^4, d^6 and d^9;
three for d^2, d^3, d^7 and d^8;
none for d^0, d^5 and d^{10}.

The absence of spin-allowed transitions for high-spin d^5 can be understood from the fact that in ground state all *d* orbitals are singly occupied by electrons having parallel spin (see Topic H2, *Fig. 3*). This is the only possible state with five unpaired electrons, and any *d–d* transition must involve a change of spin. *d–d* transitions in high-spin Mn^{2+} and Fe^{3+} complexes are indeed very weak compared with other ions, which have spin-allowed transitions.

A mathematical analysis of the transition energies in d^n ions allows Δ_o to be determined as well as **electron repulsion parameters**. Electron repulsion between *d* electrons in complexes is found to be less than in the free gas-phase d^n ions. This reduction is called the **nephelauxetic effect** (meaning 'cloud expanding') and arises because '*d* orbitals' in complexes are really MOs with some ligand as well

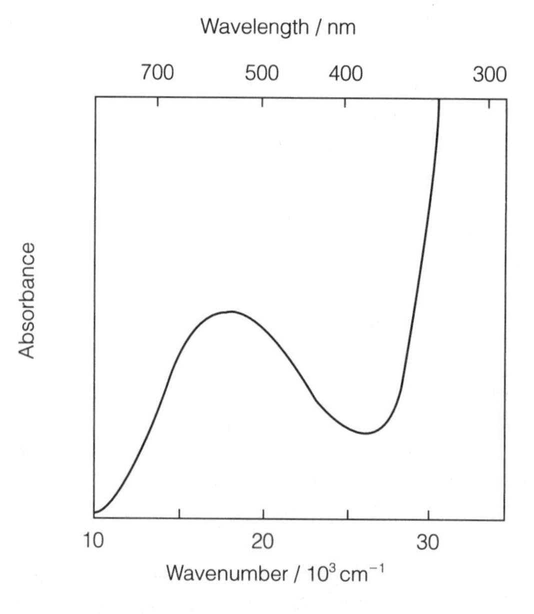

Fig. 1. Absorption spectrum of $[Ti(H_2O)_6]^{3+}$.

as metal contribution, so that electrons are on average further apart than in the pure *d* orbitals of the uncombined ions. Larger nephelauxetic reductions are observed in complexes with 'soft' ligands such as I^- than with 'hard' ones such F^-, reflecting the greater degree of covalent bonding in the former case.

Charge transfer spectra

Charge transfer is analogous to an **internal redox reaction**, and the absorption energies can be correlated with trends in redox properties (see Topics E5 and H3). In LMCT an electron is transferred to the metal, which is therefore reduced in the excited state. The more positive the redox potential concerned, the easier such reduction will be, and so the lower the LMCT energy. LMCT transitions in the visible region of the spectrum give intense color, as is found with permanganate MnO_4^-, a d^0 complex, which therefore has no *d–d* transitions. The energy trends in some d^0 species are:

(i) $MnO_4^- < CrO_4^{2-} < VO_4^{3-}$;
(ii) $MnO_4^- < TcO_4^- < ReO_4^-$;

which follow the trends towards less strongly oxidizing compounds, (i) towards the left in the 3*d* series (see Topic H3) and (ii) down each transition metal group (see Topic H5). The above orders of LMCT energy are reflected in the changing colors of the ions (e.g. MnO_4^- deep purple, CrO_4^{2-} deep yellow, VO_4^{3-} pale yellow, as the transition moves progressively to higher energy out of the visible spectrum into the UV).

LMCT energies also follow expected trends as the ligand is changed, for example, O > S, and F > Cl > Br, as the heavier ions in each group are more easily oxidized (see Topic F1). With different metal ions, there is a general decrease in energy towards the right in each series. For ions in lower oxidation states, LMCT often occurs in the UV rather than the visible part of the spectrum.

MLCT is less common, as it requires the existence of empty ligand orbitals of suitable energy. Many of these ligands are π acceptors (see Topics H2 and H9). With changing metal ions and oxidation states, MLCT bands often follow the reverse of the trends found with LMCT.

Paramagnetism

In **diamagnetism** substances are repelled by a magnetic field: this property is associated with all closed electron shells. **Paramagnetic** substances are attracted into a magnetic field, the force being related to the **magnetic susceptibility**. Paramagnetism normally arises from the spin of **unpaired electrons** (see Topic A3). The **Curie law** for the susceptibility per mole (χ_m) is

$$\chi_m = \frac{N_A \mu_0 \mu_{eff}^2}{3kT}$$

where N_A is Avogadro's number, μ_0 the magnetic permeability of free space, μ_{eff} the **effective magnetic moment** of the paramagnetic species, k is Boltzmann's constant and T the temperature in kelvin. The inverse temperature dependence arises because thermal agitation acts against the alignment of moments in an applied field. For many *d*-block compounds the **spin-only formula** is a fairly good approximation to the effective magnetic moment:

$$\mu_{eff} = 2[S(S+1)]^{1/2}\mu_B = [n(n+2)]^{1/2}\mu_B,$$

where S is the spin quantum number, equal to half the number of unpaired electrons n, and μ_B the **Bohr magneton**, equal to approximately 9.274×10^{-24} J T^{-1}. The most straightforward application of magnetic measurements is therefore to establish the number of unpaired electrons, and so to distinguish between high- and low-spin states. For example, most Co^{3+} complexes have $\mu_{eff} = 0$ as expected for low-spin d^6; however, $[CoF_6]^{3-}$ has μ_{eff} around $5\mu_B$, corresponding to four unpaired electrons and a high-spin state (see Topic H2).

Magnetic measurements are sometimes used to give information about **metal–metal bonding**. For example, dimeric Cr^{2+} complexes such as $[Cr_2(CH_3CO_2)_4]$ (see Topic H6, Structure 2) have $\mu_{eff} = 0$, suggesting that all four *d* electrons of Cr^{2+} are paired to form a quadruple bond. However, there are many other factors that can complicate magnetic properties. The oxygen-bridged complex $[(RuCl_5)_2O]^{4-}$ (Topic H6, Structure 1) also has $\mu_{eff} = 0$. In this case, there is no metal–metal bond and the electrons are paired as a consequence of the Ru–O bonding.

H9 COMPLEXES: π ACCEPTOR LIGANDS

Key Notes

Definition and evidence — π-acceptor ligands such as CO have empty π antibonding orbitals that can accept electron density from filled metal *d* orbitals. The CO bond is weakened as a result. Other π-acceptor ligands include NO and phosphines.

Binary carbonyls — Many transition metals form carbonyl compounds where the oxidation state of the metal is zero. Polynuclear compounds are also known with metal–metal bonds, and sometimes with bridging CO groups.

The 18-electron rule — In many carbonyls and related compounds, the metal atoms have a total valence count of 18 electrons. This rule can break down for steric reasons with early transition metals, and is less often obeyed in later groups.

16-electron complexes — Elements of groups 9–11 form many 16-electron square-planar complexes. These undergo various reactions including oxidative addition.

Related topics — Ligand field theory (H2); Organometallic compounds (H10)

Definition and evidence

Most ligands have a nonbonding electron pair that can act as a donor to empty orbitals on the metal atom (see Topics C8 and H2). In ligands known as π **acceptors** or π **acids** a donor–acceptor interaction also happens in the reverse direction. If a ligand has empty orbitals of π type symmetry with respect to the bond axis (see Topic C3) these may act as acceptors for electrons in filled metal orbitals of the correct symmetry. This is known as **back donation**. The simplest and commonest π acid ligand is carbon monoxide CO. This acts as a σ donor in the normal way, through the occupied lone-pair orbital centered on carbon (the 3σ MO; see Topic C4). The π antibonding orbital can also interact with filled *d* orbitals to give the π-acceptor interaction (*Fig. 1*). The combination of σ-donor and π-acceptor interaction is sometimes described as **synergic**, as the electron flows in opposite directions facilitate each other.

Evidence for the π-acceptor interaction comes from various sources.

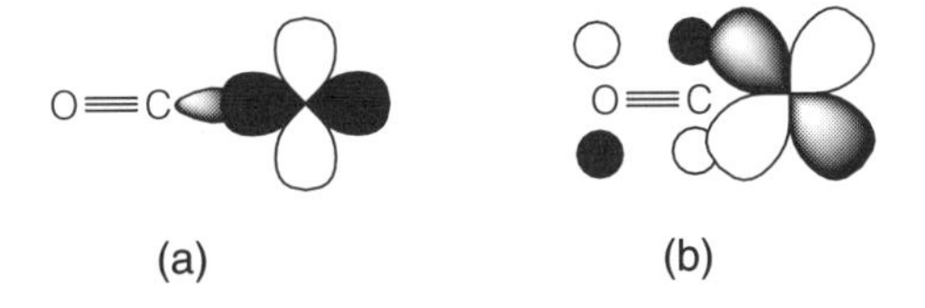

Fig. 1. Bonding in CO complexes showing (a) σ overlap of CO lone-pair with empty metal d orbital, and (b) overlap of CO π with occupied metal d orbital.*

- CO and related ligands stabilize **very low oxidation states** of transition elements, often zero (see below). π-acceptor interactions remove electron density from a metal atom and make possible a lower oxidation state than is commonly found with ligands such as water and ammonia.
- Partial occupation of the π antibonding orbital in CO weakens the bond. This is most easily seen from the **bond stretching frequency** measured by IR spectroscopy (see Topic C7). CO stretching frequencies in carbonyl compounds are nearly always less than in free CO, and also decrease in a sequence such as

  $[Mn(CO)_6]^+ > [Cr(CO)_6] > [V(CO)_6]^-$

  where the availability of metal electrons for back donation is increasing. (A few CO complexes, e.g. BH_3CO (Topic F3) and Au(Cl)CO, have stretching frequencies slightly higher than in the free molecule, indicating that little or no π interaction is taking place in these cases.)

π-acceptor properties in other ligands may be judged by their ability to stabilize low oxidation states in a similar way to CO, or by their effect on the CO stretching frequency when placed in the same complex. Two π-acceptor ligands in a *trans* configuration (see Topic H6) will compete for the same *d* orbitals. Placing a strong π acceptor *trans* to CO will therefore lessen the availability of electrons for back-bonding and so the CO stretching frequency will be higher than otherwise. On this basis the following **order of π-acceptor strength** has been deduced for some ligands:

$NO > CO > RNC > PF_3 > P(OR)_3 > PR_3 > RCN > NR_3$

π back-bonding with phosphines is generally assumed to involve valence expansion on the phosphorus. As expected, the strength increases with the electronegativity of the attached groups. Although nitrogen ligands such as pyridine (where N is part of an aromatic ring system) are π acceptors, amines R_3N are not, as nitrogen cannot expand its valence shell (see Topic F1).

Binary carbonyls

CO forms binary neutral compounds with most transition metals, and some anionic and cationic species. *Table 1* shows compounds from the 3*d* series. Some of these compounds can be obtained by direct reaction of the metal and CO at high pressure. The **Mond process** for the purification of nickel depends on the formation of nickel tetracarbonyl $Ni(CO)_4$ in this way, followed by its thermal decomposition to deposit metallic nickel. For earlier elements in the series **reductive carbonylation** is required, with a compound (generally a halide) reduced in the presence of CO at high pressure. Polynuclear carbonyls are formed naturally for some elements (Mn, Co); in other cases, such as Fe where the mononuclear carbonyl $Fe(CO)_5$ is stable, polynuclear compounds can be made from it by photolysis or controlled pyrolysis. Binary carbonyls are volatile compounds, often very toxic, and thermodynamically not stable in the presence of oxygen but often with considerable kinetic stability, especially for metals later in the series.

Table 1. Binary carbonyls and ions formed by 3d series elements

$V(CO)_6$	$Cr(CO)_6$	$Mn_2(CO)_{10}$	$Fe(CO)_5$	$Co_2(CO)_8$	$Ni(CO)_4$
$[V(CO)_6]^-$		$[Mn(CO)_6]^+$	$[Fe(CO)_4]^{2-}$	$[Co(CO)_4]^-$	
			$Fe_2(CO)_9$		
			$Fe_3(CO)_{12}$		

In mononuclear carbonyls CO is invariably attached to the metal through carbon giving a linear M–C–O arrangement. Polynuclear carbonyls have relatively short distances between metal atoms indicative of **metal–metal bonds**. CO can then bond in either a **terminal** or a **bridging** mode, the former bonded to one metal as in $Mn_2(CO)_{10}$ (**1**) and the latter attached to more than one metal as in $Co_2(CO)_8$ (**2**). In larger clusters formed by some elements, triply bridging CO is also possible. Terminal and bridging CO may be distinguished by IR spectroscopy, as bridging groups show a characteristically lower stretching frequency.

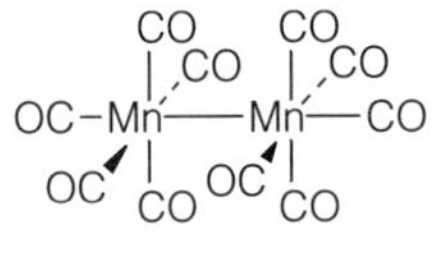

1

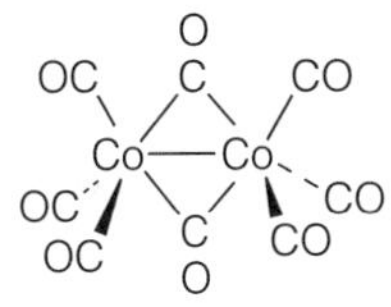

2

Many compounds are known containing CO in conjunction with other ligands, which may include π acceptors such as phosphines, and/or σ bonding ligands. For example, there is a series of compounds $Mn(CO)_5X$, where X = H, halogen or an alkyl group.

The 18-electron rule

A great majority of stable carbonyls obey the **18-electron rule** (sometimes called the **effective atomic number (EAN) rule**). To use this rule one first counts the number of valence electrons in the neutral atom, equal to the group number (thus both *s* and *d* electrons are included; see Topic H1), then adds two electrons for the lone-pair of each attached CO. For example, in $Fe(CO)_5$, the group number of Fe is eight; five COs make 18. The EAN calculation starts with the actual atomic number (Fe = 26). Adding two electrons for each CO makes an EAN = 36, which is the noble gas configuration of Kr. The only difference between the 18-electron and the EAN count is that the latter includes core electrons and so gives a different count for the three series: 36 (Kr core) for 3*d*, 54 (Xe core) for 4*d* and 86 (Ra core) for 5*d* (see Topic A4).

All the mononuclear species except $V(CO)_6$ in *Table 1* satisfy the 18-electron rule. The bi- and tri-nuclear species do also if (i) the two electrons in a metal–metal bond are counted as contributing to the valence shells of both metal atoms concerned, and (ii) a bridging CO contributes one electron to each metal. Monomeric Mn and Co carbonyls would have an odd number of electrons and dimerize in consequence. $V(CO)_6$ is exceptional as a stable radical with 17 valence-shell electrons, presumably because it is sterically impossible for it to dimerize without losing one CO ligand. It does, however, readily form the 18-electron anion $[V(CO)_6]^-$.

When other ligands are present it is normal in 18-electron counting to assume covalent rather than ionic bonding. In $Mn(CO)_5X$, where X = H, Cl or CH_3, Mn and X therefore contribute one electron each to the Mn–X bond, and X is regarded as a one-electron ligand even if it is a halogen.

One can make a connection between the 18-electron rule and ligand field theory by noting that a d^6 octahedral complex has 18 valence electrons. π-acceptor ligands provide strong fields and hence low-spin configurations (see Topic H2) thus making the d^6 octahedral combination extremely favorable. In general, the 18-electron configuration with π-acceptor ligands provides a large gap between the highest occupied MO (HOMO) and the lowest unoccupied MO (LUMO). Without the stabilization of the lower-energy set of *d* orbitals provided by a π-acceptor ligand the HOMO–LUMO gap is not so large, and the 18-electron rule does not generally apply to complexes with weak-field ligands. Even with π-acceptor ligands it can break down under some circumstances.

- With elements early in the transition series that contribute few electrons themselves it may be sterically impossible to coordinate enough ligands to achieve the 18-electron count. $V(CO)_6$ is an example.
- For later elements (group 9 onwards) there is a tendency towards lower electron counts (see below).

16-electron complexes

A square-planar complex of a d^8 ion, such as $[Ni(CN)_4]^{2-}$, has a valence electron count of 16 rather than 18. Similar **16-electron complexes** are formed by other elements in groups 9, 10 and 11, for example **Vaska's compound** $Ir(CO)(PPh_3)_2Cl$ (**3**). Some 16-electron complexes (especially in the 3*d* series) can readily add another ligand to form a five-coordinate 18-electron complex such as $[Ni(CN)_5]^{3-}$. Another important reaction is known as **oxidative addition**, where a molecule X–Y adds by cleavage of the bond to form an 18-electron complex that can be regarded as d^6 octahedral:

$$Ir^{I}(CO)(PPh_3)_2Cl + X\text{–}Y \rightarrow Ir^{III}(CO)(PPh_3)_2(X)(Y)Cl$$

X–Y can be a simple molecule such as H_2 or HCl, or an organic compound. Vaska's compound also reacts with O_2 to form $Ir(CO)(PPh_3)_2(O_2)Cl$ (**4**). In this case, O_2 remains intact on coordination, although the bond lengthens, suggesting that **4** can be regarded as a complex with a bidentate peroxo ligand (O_2^{2-}; see Topic F7).

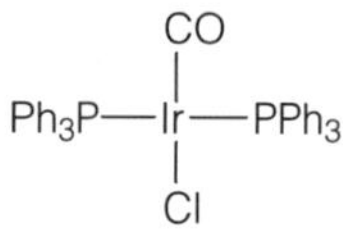

CO
Ph3P—Ir—PPh3
Cl

3

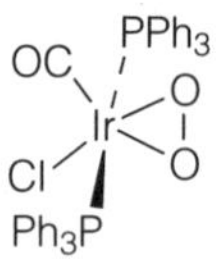

PPh3
OC
Ir
O
O
Cl
Ph3P

4

The reverse of oxidative addition is **reductive elimination**. Such reversible processes are important in many catalytic cycles involving transition metal compounds (see Topic H10, *Fig. 2*, and Topic J5).

H10 ORGANOMETALLIC COMPOUNDS

Key Notes

Ligand classification	Organic ligands for transition metals are classified by their hapticity (the number of bonded atoms) and by the number of electrons they provide in bonding. Sometimes but not always these numbers are equal.
Structure and bonding	Compounds with metal–carbon σ bonds may be unstable to elimination reactions; some have unexpected structures. π complexes including sandwich compounds are formed by interaction of metal *d* orbitals with π electrons in the ligand. The 18-electron rule can be useful for rationalizing differences of structure or stability.
Preparative methods	Methods include reduction of metal compound in the presence of the ligand, reaction with a main-group organometallic compound, and metal vapor synthesis.
Insertion and elimination	Carbonyl and alkene groups may insert into metal–carbon bonds; the reverse process gives elimination of a ligand. Together with oxidative addition and reductive elimination steps, these reactions form the basis for many catalytic applications.
Related topics	Complexes: π acceptor ligands (H9) Industrial chemistry: catalysts (J5)

Ligand classification

Organometallic compounds with metal–carbon bonds are formed by nearly all metallic elements, but those of transition metals show a diversity without parallel in main groups. Carbonyl and cyanide ligands are not considered organic, although they may also be present in organometallic compounds along with other π-acceptor ligands such as phosphines. *Table 1* shows a selection of the ligands found in organometallic compounds of transition metals, classified according to two properties.

- The **hapticity** is the number of carbon atoms bonded directly to the metal. With some ligands this can vary; for example, cyclopentadienyl can be η^1–C_5H_5, η^3–C_5H_5 or (most often) η^5–C_5H_5 (pronounced 'monohapto', 'trihapto', etc.).
- The **electron number** is the number of electrons the ligand contributes to the metal–carbon bonding. This is useful for applying the 18-electron (EAN) rule (see Topic H9). Ligands are taken to be neutral species even if they are known as stable anions (e.g. C_5H_5, not $C_5H_5^-$). For ligands of variable hapticity the electron number often varies accordingly, but electron number is **not** always equal to the hapticity, as can be seen with η^1 ligands, where the electron number can vary from one to three.

Table 1. Some organic ligands, classified according to hapticity and electron number

Ligand name	Ligand formula	Hapticity	Electron number
Methyl, akyl	CH_3, RCH_2	η^1	1
Alkylidene	R_2C	η^1	2
Alkylidyne	RC	η^1	3
Ethylene (ethene)	C_2H_4	η^2	2
Allyl (propenyl)	CH_2CHCH_2	η^1	1[a]
		η^3	3
Cyclopentadienyl	C_5H_5	η^1	1[a]
		η^3	3[a]
		η^5	5
Benzene	C_6H_6	η^6	6

[a] Uncommon bonding arrangements.

Structure and bonding

Alkyl ligands form metal–carbon σ bonds. Often they occur in conjunction with other organic ligands or CO, but can be found on their own, as in tungsten hexamethyl (**1**), and in $[Ti(CH_2SiMe_3)_4]$ where the bulky groups are helpful in stabilizing the compound. Compounds with H attached to β carbons (the nomenclature being $M–C_\alpha–C_\beta–C_\gamma$) tend to be unstable to **β-hydride elimination** of an alkene fragment, discussed below. The surprising structure of (**1**), trigonal prismatic rather than octahedral as found in WCl_6, has been attributed to the orientation of *d* orbitals available for σ bonding. In an octahedral complex only two *d* orbitals (the e_g set) can be involved, but four in the trigonal prismatic structure. (Unlike WMe_6, WCl_6 also has some degree of W–Cl π bonding, which can involve the other *d* orbitals (t_{2g}) in octahedral geometry; see Topic H2.)

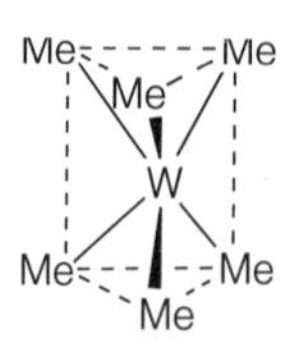

1

Alkylidene and alkylidyne ligands require metal–carbon π bonding in addition to σ (see Topic C5). This is different, however, from π **complexes** where bonding involves only the π orbitals of alkene or aromatic ligands. Examples are the ethene complex $[(\eta^2–C_2H_4)PtCl_3]^-$ (**2**) found in **Zeise's salt**, and the 'sandwich compound' **ferrocene** $[Fe(\eta^5–C_5H_5)_2]$ (**3**). The **Dewar–Chatt–Duncanson** model of bonding in ethene complexes is shown in *Fig. 1*, and is analogous to the σ-donor–π-acceptor description of the bonding in carbonyl complexes (see Topic H9, *Fig. 1*). In the present case the 'σ-donor' character comes from the occupied bonding π MO of ethene (*Fig. 1a*), back donation (*Fig. 1b*) involving the empty π* antibonding MO. The relative degrees of donor or acceptor behavior depend on the compound.

$[Cl_3Pt(CH_2{=}CH_2)]^-$

2

Fe

3

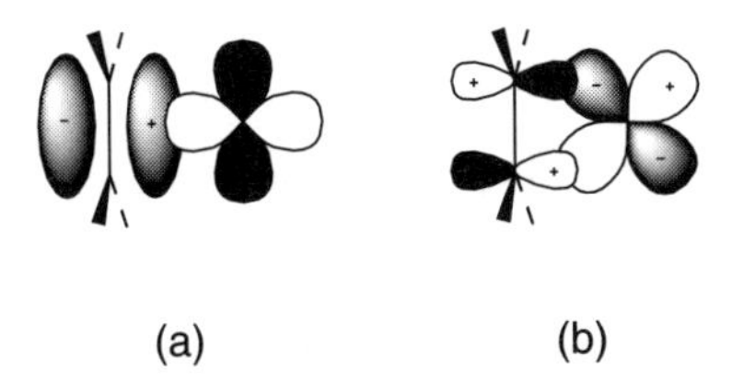

Fig. 1. Dewar–Chatt–Duncanson model for bonding in π complexes of C_2H_4.

With strongly electron-withdrawing alkenes such as C_2F_4 or $C_2(CN)_4$ there is a large amount of back donation, which weakens the C–C bond so that its length is similar to that of a single bond. The geometry of the ligand then also changes from the planar configuration associated with sp^2 hybridization, to a nonplanar form more characteristic of single-bonded sp^3. The result (**4**) can be viewed as a metallocyclic compound with two M–C σ bonds.

Pt C C

4

Bonding in sandwich compounds such as ferrocene arises through interaction of the delocalized π MOs of the ring with orbitals of the metal, and cannot be treated in a localized fashion (see Topic C6). As in alkenes, both donor and acceptor interactions are involved. Other ligands such as CO can be present, as in the 'piano-stool' structure **5** or the metal–metal bonded dimer **6**.

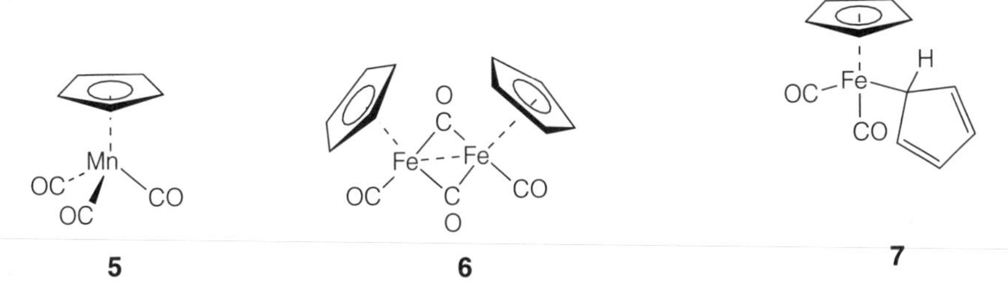

The **18-electron rule** can be a useful guide to stable organometallic compounds, especially when π-acceptor ligands are present, although it has the limitations referred to in Topic H9. Compounds **3**, **5** and **6** obey this rule, but **1** without π bonding ligands has an electron count of only 12. **Metallocenes** $[M(\eta^5\text{-}C_5H_5)_2]$ are known for the 3*d* series elements V–Ni, with 15–20 valence electrons, respectively. Ferrocene (M = Fe with 18 electrons) is by far the most stable of these, cobaltocene (M = Co with 19 electrons) being a very strong reducing agent that easily forms the 18-electron ion $[Co(\eta^5\text{-}C_5H_5)_2]^+$. Compounds with more than 18 valence electrons are uncommon, and thus one can understand the unusual structure of $[Fe(\eta^5\text{-}C_5H_5)(\eta^1\text{-}C_5H_5)(CO)_2]$ (**7**), as two pentahapto ligands would give an electron count of 22. Reactions of organometallic compounds often involve 16-electron intermediates formed by the loss of one ligand (e.g. CO) from an 18-electron parent compound.

Preparative methods

Preparative methods for organometallic compounds are exceedingly diverse but the following are generally useful.

- **Reduction of metal salt in the presence of the ligand**:

  $CrCl_3 + Al + 2C_6H_6 \rightarrow AlCl_3 + [Cr(\eta^6\text{–}C_6H_6)_2]$

- **Reaction of a transition metal salt with a main-group organometallic compound**. C_5H_5 is often delivered as the sodium salt $Na^+(C_5H_5)^-$:

  $FeCl_2 + 2Na^+(C_5H_5)^- \rightarrow 2NaCl + [Fe(\eta^5\text{–}C_5H_5)_2]$

  In other cases a Grignard reagent or aluminum alkyl may often be used:

  $WCl_6 + 3Al_2(CH_3)_6 \rightarrow W(CH_3)_6 + 6AlCl(CH_3)_2$

- **Metal vapor synthesis**. Vaporizing the metal (e.g. by electron-beam heating) helps by providing the sublimation energy required; metal atoms are then condensed in the presence of the ligand on the sides of the vessel, cooled in liquid nitrogen. This method is good for compounds that cannot be made by other routes, or ones stable only at low temperatures. For example,

  $Ti(g) + 2C_6H_6 \rightarrow [Ti(\eta^6\text{–}C_6H_6)_2]$

Insertion and elimination

Among the many reactions of organometallic compounds, ones involving insertion and elimination of ligands are important in applications to synthesis and catalysis. An example of a **carbonyl insertion** is:

$$Mn(CO)_5CH_3 + CO \rightarrow Mn(CO)_5C(O)CH_3$$

in which a Mn–CH_3 bond is replaced by Mn–C(O)–CH_3. The terminology is misleading as it is established by isotopic labeling that the incoming CO is not the one inserted. The first step is a reversible **alkyl migration** leading to a 16-electron intermediate

$$Mn(CO)_5CH_3 \rightleftharpoons Mn(CO)_4\text{–}C(O)CH_3$$

which then picks up another CO molecule.

Many other unsaturated ligands can 'insert' into M–C or M–H bonds; for example, alkanes as in:

$$L_nM\text{–}H + CH_2{=}CH_2 \rightarrow L_nM\text{–}CH_2\text{–}CH_3$$

Such reactions are often reversible, the backwards process leading to **elimination** of a ligand. The reverse of alkene insertion is the β-hydride elimination reaction referred to above.

Organometallic compounds are used widely as homogeneous catalysts in the chemical industry (see Topic J5). For example, if the alkene insertion reaction continues with further alkene inserting into the M–C bond, it can form the basis for **catalytic alkene polymerization**. Other catalytic cycles may include **oxidative addition** and **reductive elimination** steps as described in Topic H9. *Figure 2* shows the steps involved in the **Monsanto acetic acid process**, which performs the conversion

$$CH_3OH + CO \rightarrow CH_3CO_2H$$

In the catalytic cycle on the right-hand side, the 16-electron species **A** undergoes oxidative addition of CH_3I to form **B**. Carbonyl insertion then proceeds via **C** to

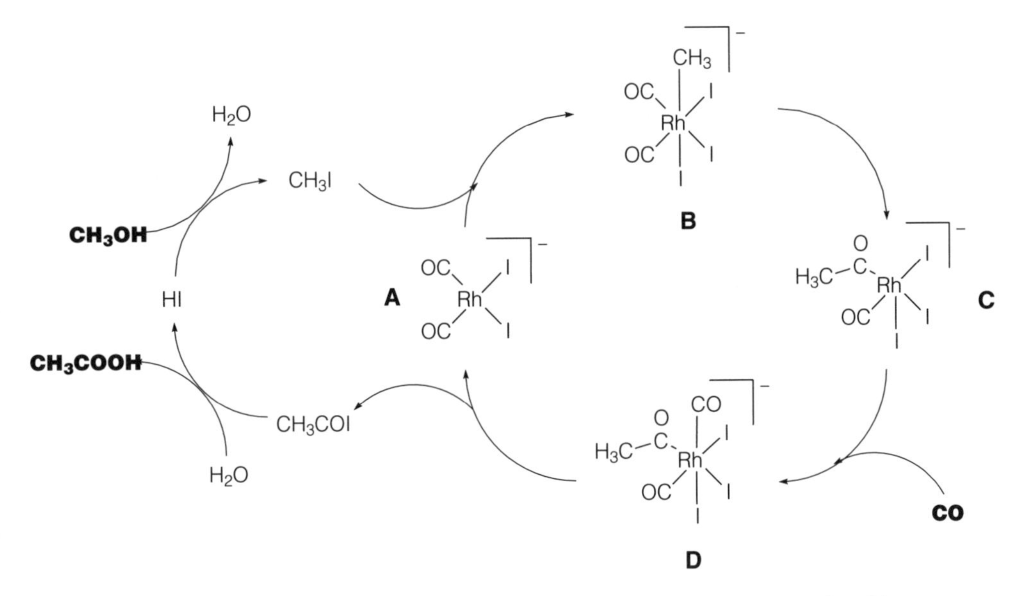

Fig. 2. Reaction steps involved in the catalytic Monsanto acetic acid process.

give **D**, which regenerates **A** by reductive elimination of CH_3COI. The organic steps on the left-hand side of *Fig. 2* can be varied to give different overall reactions, for example, converting $CH_3CO_2CH_3$ into $(CH_3CO)_2O$.

I1 LANTHANUM AND THE LANTHANIDES

Key Notes

The elements

The elements (sometimes called rare earths) are found together in nature and are electropositive metals. Chemistry is dominated by +3 state with ions in $(4f)^n$ configurations, and is similar for all elements.

Oxidation state +3

A wide range of +3 compounds is formed as well as aqua ions. The ionic radius decreases gradually across the series, leading to changes in solid structures, and an increase in stability of complexes in solution. Organometallic compounds are more ionic than in the *d* block.

Other oxidation states

Sm, Eu and Yb form many compounds in the +2 oxidation state. With the other elements, compounds in this state are formed only with large anions and are often metallic. Ce, and to a lesser extent Pr and Tb, show the +4 state.

Related topics The periodic table (A4) Actinium and the actinides (I2)

The elements

The **lanthanides** are 14 elements following **lanthanum** in the periodic table, and associated with the filling of the seven orbitals of the 4*f* shell. The symbol Ln is often used to denote these elements collectively. Atomic configurations are complex with electrons in 4*f*, 5*d* and 6*s* orbitals outside the Xe core. The first three ionization energies are relatively low, leading to electropositive metals with chemistry dominated by the Ln^{3+} state in solution and in ionic solids. All Ln^{3+} ions have electron configurations $(4f)^n$ (see list in *Figure. 1*), but the 4*f* orbitals are highly contracted in size and do not overlap significantly with neighboring atoms. Unlike the case with the *d* orbitals in the transition elements, spectra and magnetism associated with 4*f* orbitals in Ln^{3+} compounds are very similar to those found in free gas-phase ions. Ligand field and chemical bonding effects associated with incomplete 4*f* orbitals are very small and hardly detectable in chemical trends. The chemistry of all Ln^{3+} ions is therefore very similar and differentiated only by the gradual contraction in radius associated with increasing nuclear charge. The **lanthanide contraction** is also important for the transition elements of the 5*d* series (see Topics H1 and H5).

The oxidation states +2 and +4 are found for some elements, following the trend in ionization energies across the series, which show patterns analogous to those found in configurations of *p* and *d* electrons (see Topics A5 and H1). The third ionization energy rises from La to Eu (see *Fig. 1*) and then a drop occurs after the half-filled shell (Eu^{2+}, $4f^7$). The rise then continues to Yb, and drops at Lu because the 4*f* shell is filled and the electron ionized is in 5*d*. Fourth ionization energies (which are substantially larger) show a similar pattern displaced by one element, thus rising from Ce to Gd and falling to Tb.

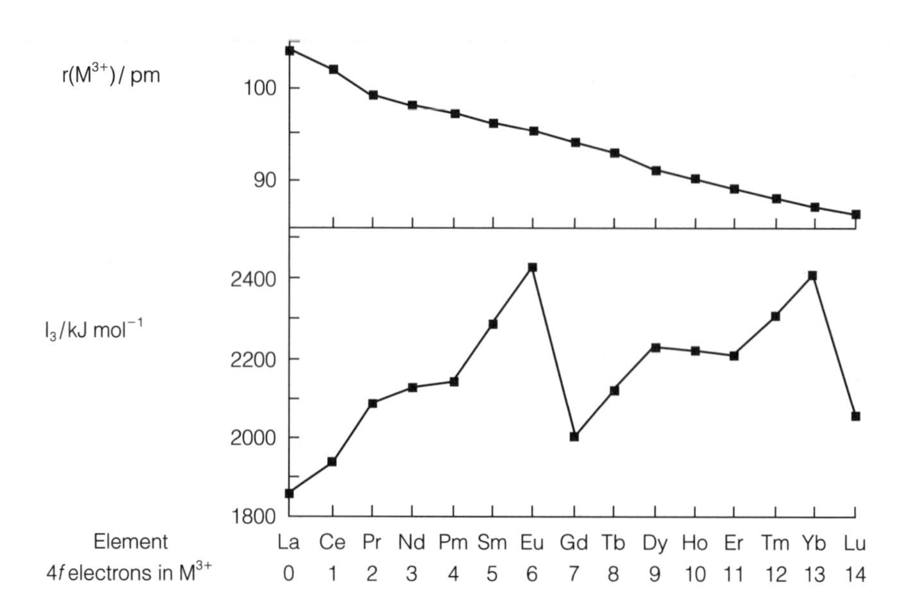

Fig. 1. Ionic radius of M^{3+}, third ionization energy I_3, and number of 4f electrons in M^{3+} for the elements La–Lu.

Promethium is a radioactive element with a half-life of 2.6 years and does not occur naturally. The other elements, known sometimes as the **rare earth elements**, are always found in association, principally in the minerals monazite ($LnPO_4$) and bastneasite ($LnCO_3F$). The electropositive and reactive elements can be obtained by reduction of $LnCl_3$ with Ca, and are sometimes used together as 'mischmetal'. Specialist applications of individual lanthanides depend on the spectroscopic properties of Ln^{3+} ions (e.g. Nd in lasers) and on the magnetic properties of some of the elements (e.g. Sm). The ions can be separated by ion-exchange chromatography from aqueous solution, using the variation of complexing properties across the series (see below).

Oxidation state +3

The Ln^{3+} state is the most stable under normal conditions for all elements in the series. Halides $\mathbf{LnX_3}$ and oxides $\mathbf{Ln_2O_3}$ are known for all elements, as well as an extensive range of oxo salts including mixed and hydrated compounds such as $Ln_2(SO_4)_3.3Na_2(SO_4).12H_2O$. Ionic radii vary from 104 pm (La^{3+}) to 86 pm (Lu^{3+}) and this relatively large size for 3+ ions (cf. Al^{3+} 53 pm) is associated with correspondingly high coordination numbers in solid compounds. LnF_3 compounds for earlier elements have nine-coordination, Ln_2O_3 are seven-coordinate. For later Ln elements the decrease in radius leads to changes in structure with reduction in coordination.

The **aqua Ln^{3+} ions** show slight acidity, which increases from La to Lu as the radius decreases but is still much less than for Al^{3+}. Strong complexes are formed with hard oxygen donor ligands, and especially chelating ones such as EDTA (see Topic E3) or β-diketonates (L–L = $[RC(O)CHC(O)R]^-$ **1**), which give eight-coordinate complexes $[Ln(L\text{–}L)_4]^-$. Complex strengths generally increase across the series as the radius decreases, and this may be used to separate a mixture of Ln^{3+} ions. For example, in an ion-exchange chromatography column with a complexing

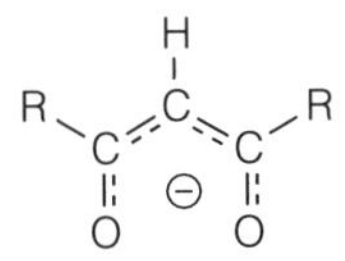

$$\mathrm{R{-}C({=}O){\cdots}C(H){\cdots}C({=}O){-}R}^{\ominus}$$

1

agent present in aqueous solution, the earlier lanthanides, which are less strongly complexed, are retained preferentially on the column and elute more slowly.

The **organometallic chemistry** of lanthanides is much more limited than in the *d* block (see Topic H10). Compounds such as $(C_5H_5)_3Ln$ and $(C_5H_5)_2LnX$ (X = Cl, H, etc.) have more ionic character than for transition elements, and compounds with neutral ligands such as CO are not stable. Some interesting chemistry has, however, been found with compounds such as $(C_5(CH_3)_5)_2LuH$ where a bulky ligand is combined with a small lanthanide. For example, the **methane activation** reaction

$$(C_5(CH_3)_5)_2LuH + CH_4 \rightarrow (C_5(CH_3)_5)_2LuCH_3 + H_2$$

occurs under mild conditions in solution.

Other oxidation states

According to the ionic model the relative stability of Ln^{2+} and Ln^{3+} compounds is determined by a balance between the third ionization energy (I_3) of the lanthanide, and the difference of lattice (or solvation) energies associated with the two ions (see Topics D6 and G1). The I_3 value for lanthanides is small enough that most Ln^{2+} compounds are unstable with respect to disproportionation to Ln and Ln^{3+}. The exceptions are of two kinds. For **Sm, Eu and Yb**, I_3 is large enough to stabilize a number of compounds such as SmO, EuF_2 and $YbCl_2$. The aqueous Ln^{2+} ions are strongly reducing, especially so for Sm and Yb. On the other hand, compounds with large anions have small lattice energies and so disproportionation is less favorable. Thus LnS and LnI_2 are known for all Ln. Many of these compounds are metallic in appearance and highly conducting, which suggests an unusual electron configuration as 4*f* orbitals on one atom cannot overlap sufficiently with orbitals on other atoms to form bands (see Topic D7). A formulation such as $(Ln^{3+})(S^{2-})(e^-)$ is sometimes given, implying a $(4f)^n$ configuration appropriate to Ln^{3+} with one electron delocalized in a band (formed probably from overlapping 5*d* orbitals). For compounds of Sm, Eu and Yb this peculiarity disappears, and, for example, EuS and YbI_2 are not metallic but have 'normal' Ln^{2+} ions.

Ln^{4+} compounds are known only for elements with the lowest I_4 values. Ce^{4+} is known in aqueous solution and forms many compounds such as CeO_2. Pr^{4+} and Tb^{4+} are more strongly oxidizing, giving fluorides LnF_4, and being present together with Ln^{3+} in mixed-valency oxides such as Pr_6O_{11} (which is actually non-stoichiometric).

I2 ACTINIUM AND THE ACTINIDES

Key Notes

Nuclear properties

All actinide elements of the 5*f* series are radioactive. Th and U are long lived and occur in minerals that also contain their radioactive decay products. Elements beyond uranium are made artificially, by bombardment with neutrons or with nuclei. Uranium and plutonium are used as nuclear fuels.

Chemical properties

Early actinides show a variety of oxidation states. The +6 state is common for U but becomes progressively more strongly oxidizing. Later actinides are more similar to lanthanides, with the +3 state being common.

Related topics

The nuclear atom (A1)
The periodic table (A4)
Lanthanum and the lanthanides (I1)

Nuclear properties

Following actinium (group 3) are the 14 elements of the **actinide series** (represented by the symbol An) associated with progressive filling of the 5*f* shell and so analogous to the lanthanides. All are radioactive, their longest-lived isotopes being shown in *Table 1*. The progressively shorter half-lives reflect the decreasing stability of heavy nuclei, resulting from the changing balance between the attractive strong interaction and the repulsive Coulomb forces (see Topic A1). Most actinide nuclei undergo α **decay** by emitting ^{4}He, but for heavier elements **spontaneous fission** into two fragments is an increasingly important alternative decay route.

Table 1. Longest-lived isotopes of actinides

Atomic no.	Element	Isotope	Half-life
89	Actinium	^{227}Ac	21.77 years
90	Thorium	^{232}Th	1.40×10^{10} years
91	Protactinium	^{231}Pa	3.28×10^{4} years
92	Uranium	^{235}U	7.04×10^{8} years
		^{238}U	4.47×10^{9} years
93	Neptunium	^{237}Np	2.14×10^{6} years
94	Plutonium	^{244}Pu	8.26×10^{7} years
95	Americium	^{243}Am	7370 years
96	Curium	^{247}Cm	1.65×10^{7} years
97	Berkelium	^{247}Bk	1380 years
98	Californium	^{251}Cf	898 years
99	Einsteinium	^{253}Es	20.5 days
100	Fermium	^{257}Fm	100.5 days
101	Mendelevium	^{257}Md	5.2 h
102	Nobelium	^{259}No	1.0 h
103	Lawrencium	^{256}Lr	28 s

Only thorium and uranium have half-lives long enough to survive since the formation of the Earth (see Topic J1). Thorium is found together with lanthanides in the phosphate mineral monazite ($LnPO_4$), and uranium occurs as **pitchblende** U_3O_8 and **carnotite** $K_2(UO_2)_2(VO_4)_2.3H_2O$. Uranium is principally used as a nuclear fuel, as the isotope ^{235}U undergoes **neutron-induced fission**, the nucleus splitting into two smaller fragments together with more neutrons, which can thus initiate a **chain reaction**. The energy liberated (about 2×10^{10} kJ mol^{-1}) is vastly greater than that obtainable from chemical reactions.

^{232}Th, ^{235}U and ^{238}U are the first members of **radioactive decay series**, forming other radioactive elements with atomic numbers 84–91, which are therefore present in small amounts in thorium and uranium ores. Each series ends with a different stable isotope of lead (^{208}Pb, ^{207}Pb and ^{206}Pb, respectively) and the proportions of these present in natural lead samples varies detectably. This variation can be used to give geological information, including an estimate of the age of the Earth.

Transuranium elements beyond U do not occur naturally on Earth but can be made artificially. The neutron irradiation of ^{238}U in nuclear reactors produces ^{239}U, which rapidly undergoes β decay to ^{239}Np and thence to ^{239}Pu. Further neutron irradiation produces heavier actinides in progressively smaller amounts, up to Fm. The remaining elements Md, No and Lr cannot be obtained in this way but have been produced in exceedingly small quantities by bombardment of lighter actinides with nuclei such as ^{4}He and ^{12}C using particle accelerators. (Note that the longest-lived isotopes listed in *Table 1* are not necessarily the ones most easily made.) Similar methods have been used to make **transactinide elements** with atomic number up to 110, presumably forming part of a 6*d* transition series. However, the very small quantities made (often a few atoms only) and their very short half-lives make chemical studies almost impossible.

Chemical properties

Unlike the 4*f* orbitals in the lanthanides, the 5*f* orbitals in the earlier actinide elements are more expanded and so can be engaged in chemical bonding. This leads to a pattern of chemistry more analogous to that found in the *d* block, with the possibility of **variable oxidation states** up to the maximum possible determined by the number of valence electrons (see Topic H1). Most thorium compounds contain Th^{IV} (e.g. ThO_2) and with uranium the states from +3 to +6 can be formed. UO_2 is frequently nonstoichiometric, and the natural mineral U_3O_8 probably contains U^{IV} and U^{VI}. **Uranium hexafluoride** is made industrially using ClF_3 as a fluorinating agent (see Topic F9):

$$U(s) + 3ClF_3(g) \rightarrow UF_6(l) + 3ClF(g)$$

Being volatile, it is used to separate the isotopes ^{235}U and ^{238}U for nuclear fuel applications. Many other U^{VI} compounds contain the **uranyl ion** UO_2^{2+}, a linear unit with bonding involving both 5*f* and 6*d* orbitals: examples include the mineral carnotite (see above) and $Cs_2[UO_2Cl_4]$ where uranyl is complexed to four chloride ions.

The maximum attainable oxidation state in the series is +7, in the mixed oxides Li_5AnO_6 (An = Np, Pu). With increasing atomic number high oxidation states become more strongly oxidizing, as in the *d* block. This trend is illustrated in *Figure. 1*, which shows a Frost diagram with the oxidation states of some actinides found in aqueous solution (see Topic E5). The oxocations AnO_2^{+} and AnO_2^{2+} are characteristic for An^{V} and An^{VI} with An = U, Np, Pu and Am, but the slopes of the

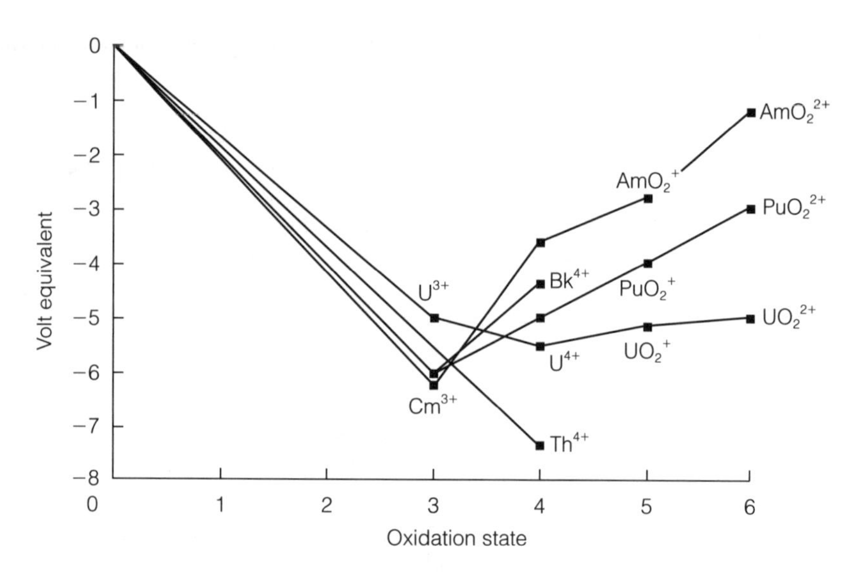

Fig. 1. Frost diagram showing the oxidation states of some actinides in aqueous solution at pH = 0.

lines in the diagram show their increasingly strong oxidizing character. Complex solution equilibria are possible: with Pu, for example, all states from +3 to +6 can be present simultaneously. The different redox stability of U and Pu is important in **nuclear fuel reprocessing**, one function of which is to separate unused uranium from ^{239}Pu, which is itself used as a nuclear fuel. Dissolving the spent fuel elements in aqueous HNO_3 gives Pu^{IV} and U^{VI}. Subsequent separation steps then depend on differences in complexing power and solubility of these ions.

The **organometallic chemistry** is much less extensive than that of the *d* block (see Topic H10), and differs from that of the lanthanides by virtue of the large sizes of the early actinides, and their wider range of accessible oxidation states. Uranium has been much more investigated than other elements. Typical compounds include the cyclopentadienyl (Cp = η^5–C_5H_5) compounds [$AnCp_3$], [$AnCp_4$] and mixed Cp-halides such as [$AnCp_3Cl$]. Particularly interesting is the sandwich compound [$U(\eta^8–C_8H_8)_2$] with two planar cyclooctatetraene rings known as **uranocene**; analogs are formed with neighboring actinides. Although formally they can be regarded as compounds of An^{4+} with the aromatic 10 π electron ring $[C_8H_8]^{2-}$ (see Topic C6) there is some covalent bonding involving actinide 5*f* and 6*d* orbitals.

Later actinides show a much more restricted range of oxidation states, and are more similar to the lanthanides. The +4 state is found in AnO_2 and AnF_4 as far as Cf. It becomes progressively more oxidizing for later elements, but with a break at Bk^{4+} (which is more easily formed than Cm^{4+} or Cf^{4+}) following the half-filled 5*f* shell and so analogous to the occurrence of Tb^{4+} in the lanthanides. From Am to Md the +3 state is most stable in solid compounds and aqueous solution. Near the end of the series, however, the +2 state appears more stable than in the lanthanides and is the normal one for No. This difference must reflect a different balance of ionization energies and lattice or solvation energies, but the data required to understand it in detail are not available. With only a few atoms available, and with very short half-lives, chemical investigations of later actinides depend on

tracer techniques using a stable element of presumed similar chemical behavior to act as a carrier. For example, the presence of No^{2+} can be inferred from its precipitation (and subsequent detection by its radioactivity) along with Ba^{2+} as $BaSO_4$ under conditions where other oxidation states form soluble compounds.

J1 ORIGIN AND ABUNDANCE OF THE ELEMENTS

Key Notes

Patterns of abundance

O and Si are the commonest elements in the Earth's crust, Fe and O in the Earth as a whole, and H and He in the Solar System and the Universe. Nuclear reactions (controlling the amounts of elements made) and subsequent chemical reactions are both important in determining these abundances.

The origin of the elements

The lightest elements H and He were formed at the origin of the Universe. Nearly all others have been made by nuclear fusion reactions inside stars. Fusion of He makes C and O, and then heavier elements up to Fe. Elements heavier than Fe are formed by neutron bombardment of lighter nuclei. Nuclei with even atomic and/or mass numbers tend to be commoner than those with odd ones.

Formation of the Earth

The Earth was formed from solid dust particles containing metallic elements such as iron, together with silicates and other solids. The abundant elements on Earth are ones that are both made in large quantities in nuclear reactions, and also condensed efficiently to form solids.

Related topics

The nuclear atom (A1)

Geochemistry (J2)

Patterns of abundance

Information on the abundance of elements comes from diverse sources. Most elements are obtained from minerals in the **Earth's crust**. The availability of elements therefore depends on the **crustal abundance**, which can be estimated by analyzing representative samples of minerals. The abundances of elements vary enormously, from common ones such as oxygen and silicon (respectively 46% and 27% by mass) down to ones such as Os, Ir and Xe (one part in 10^{10} or less). The commonest elements are listed in *Table 1*.

The crust is thin, and rests on the Earth's **mantle**, which in turn surrounds the metallic **core**. As these inner regions are not directly accessible, information on their composition comes from less direct sources, including **meteorites**, which fall from space, and which are derived from one or more planets that broke up in the early stages of formation of the Solar System. Estimates of the overall abundance

Table 1. The most abundant elements in the crust, the whole Earth and the Solar System (mass fraction, with elements listed in order of decreasing abundance within each range)

Abundance range	Earth's crust	Whole Earth	Solar System
>10%	O, Si	Fe, O, Mg, Si	H, He
1–10%	Al, Fe, Ca, Mg, Na	S, Ca, Ni, Al	–
0.1–1%	K, C, Ti, H, Mn, P	Cr, Na, P, Co, Ti	O, C, Fe, Si, Mg, N, Ne

of elements in the whole Earth show some differences from the crust (see *Table 1*). Iron is the dominant element in the core and has a similar abundance to oxygen in the Earth as a whole.

The **Solar System** is dominated in mass by the Sun. Estimates of elemental composition can be obtained from the spectrum of sunlight, which shows atomic absorption lines. Hydrogen and helium are by far the most abundant elements, followed at a level of less than 1% by oxygen and carbon. This pattern of abundances is typical of the Universe as a whole, which is dominated by H and He in an atomic ratio of about 10:1, all other elements together making up only 1%.

Two very different factors are important in determining the abundance patterns shown in *Table 1*. The overall abundance in the Universe and in the Solar System depends on how elements were made by **nuclear reactions**. The very different distribution in the Earth and its constituent parts is a consequence of subsequent **chemical differentiation** of elements during the formation of the planets.

The origin of the elements

The synthesis of elements requires **nuclear reactions**, of which the most important type is the **fusion** of two light nuclei to make one of higher charge and mass. The attractive **strong interaction**, which holds protons and neutrons together, operates only over very short distances (around 10^{-15} m) and is opposed at longer range by the electrostatic repulsion between positively charged protons. To get two nuclei close enough together for fusion requires enormously high energies, which are normally found only at extreme temperatures (above 10^7 K) in the interior of stars. Under such conditions the chemical properties of elements are irrelevant, as no compounds can exist, atoms being in highly ionized states stripped of their electrons.

It is thought that the Universe began about 15 billion years ago in a state of extraordinarily high temperature and pressure known as the **big bang**. It rapidly cooled, and exotic elementary particles originally present formed protons, neutrons and electrons. Some protons and neutrons combined to form nuclei of deuterium (^{2}H, the heavy isotope of hydrogen; see Topic F2), which then fused to form ^{4}He nuclei. Because of the rapidly falling temperature nuclear reactions ceased after about 3 min, and only very tiny amounts of elements heavier than helium were formed. Calculations based on the assumed conditions agree very well with the observed abundance of hydrogen and helium in the Universe. The dominance of these elements forms one of the strongest pieces of evidence for the big bang model.

As hydrogen and helium cooled, local gas concentrations formed and contracted under gravitational forces. Release of gravitational potential energy heated the center of each concentration to the temperature (around 10^7 K) where nuclear fusion reactions restarted. The energy output of all stars, including our Sun, comes from such reactions. Fusion of hydrogen nuclei produces helium, and forms the energy source for stars throughout most of their lifetime. When hydrogen is used up in the center of a star, further gravitational contraction raises the temperature to about 10^8 K and ^{4}He nuclei themselves start to fuse. The main products of this stage are ^{12}C and ^{16}O, the most abundant nuclei in the Universe after H and He. Exhaustion of He gives higher temperatures and further fusion reactions, producing elements up to around iron. ^{56}Fe has the highest binding energy of all nuclei, fusion reactions producing heavier nuclei being endothermic. Elements such as Co and Ni just beyond Fe are produced in equilibrium at the enormously high temperatures (above 10^9 K) at the center of a star in the final stages of its life, but beyond this point successive elements are formed by a process of **neutron capture**.

Neutrons are produced as side products of some of the fusion reactions. They may be captured by nuclei, followed by a radioactive β decay process, which leads to an element of higher atomic number (see Topic A1). Successive capture and decay processes are thought to have produced all the heavy elements, probably including some transuranium elements (see Topic J2) that have subsequently decayed.

When no further exothermic nuclear reactions are possible in the center of a star, it collapses under gravitational attraction, which releases enough energy to cause a gigantic explosion known as a **supernova**, which throws most of the material 'cooked' by nuclear reactions into space. Studies of supernovae in nearby galaxies show atomic spectral lines confirming the presence of these elements.

Calculations based on these ideas can account for the abundance of elements, and of their different isotopes, observed in the Universe. The nuclei made in greatest numbers are the most stable ones, generally having even numbers of protons and neutrons. Beyond ^{12}C and ^{16}O the most abundant are ^{20}Ne, ^{24}Mg, ^{28}Si, ^{32}S and ^{56}Fe. For this reason (which has nothing to do with chemistry) elements in odd-numbered groups in the periodic table tend to be less common than in even-numbered ones, a pattern that is apparent in the composition of the whole Earth shown in *Table 1*.

Formation of the Earth

Gases thrown out by a supernova cool, and may subsequently be incorporated into new stars. The formation of planetary systems may be common in the Universe. Studies suggest that the Earth and other planets formed about the same time as the Sun (4.5 billion years ago). While the Sun formed at the center, chemical reactions in the cooler outer regions of the gas concentration produced solid particles, which gathered under gravitational forces, first into small bodies known as **planetesimals**, and subsequently into the planets. In the outer regions of the Solar System temperatures were low enough to form 'ices' of water, solid methane, carbon dioxide and ammonia, which are constituents of the giant planets Jupiter and Saturn. The inner planets such as Venus, Earth and Mars formed at higher temperatures, and their composition is dominated by elements that form metallic solids, such as iron and nickel, and ones with stable involatile oxides, such as SiO_2. Many other electropositive elements were incorporated as silicates, and some also formed sulfides and halides. The molecular compounds of H, C and N were still gaseous at the temperature at which the Earth was formed, so that these elements largely escaped, except for relatively small amounts of H_2O, CH_4, CO_2 and NH_3, which were trapped in solid silicates. Noble gas elements (group 18) are rare on Earth.

Abundant elements on Earth are therefore ones which were both made efficiently in nuclear reactions in stars, and also formed involatile metals or compounds when the Solar System was formed. Subsequent heating by radioactive decay allowed the denser metals (Fe, Co and Ni combined with some S) to melt and sink towards the center, forming the core. Silicates and other complex oxides remained as the dominant constituents of the outer layers.

J2 GEOCHEMISTRY

Key Notes

Element classification

Lithophilic elements are those present on Earth in oxide (mostly complex, e.g. silicate) and halide minerals, chalcophiles in sulfide minerals, and siderophiles in metallic form.

Crust formation

The crust is formed by melting and recrystallization of minerals in the mantle. Compatible lithophilic elements (Mg, Fe, Cr) are commoner in the mantle, incompatible ones (e.g. Na, K, Al) in the crust. Chemical reactions in molten rocks and in water at high temperature lead to the concentration of many elements in particular minerals.

Weathering and sedimentation

The breakdown of rocks by water and CO_2 gives insoluble resistates (e.g. Al, Ti, Sn oxides) and soluble ions. Some ions oxidize to form solids (e.g. Fe_2O_3); others pass into the ocean and eventually form evaporite minerals (e.g. NaCl).

Atmosphere and oceans

The atmosphere was formed by outgassing of minerals. O_2 comes from photosynthesis. Ions common in sea water are ones that do not form insoluble salts.

Related topics

Chemical periodicity (B2)

Origin and abundance of the elements (J1)

Element classification

Geochemistry is the study of chemistry in the Earth's natural environment. Most elements available to us come from the solid rocks of the Earth's **crust**. Underlying the crust is the **mantle** of rather similar composition to the crust, inside which is a **metallic core**. Overlying the crust are the **atmosphere** and the aqueous environment or **hydrosphere** of the oceans, lakes and rivers.

The chemical processes taking place in the crust are especially important as they have formed **ores**, concentrated mineral deposits that are exploited industrially as sources of specific elements and their compounds. *Figure 1* summarizes the principal chemical forms in which each element occurs. At least half the elements occur in the crust as oxides (mostly complex ones such as silicates) or less commonly as halides, and are called **lithophiles**. All the highly electropositive metals are in this class. **Chalcophiles** are elements present in sulfide minerals; these include some elements chemically similar to sulfur (Se, As) together with less electropositive metals of the later transition and post-transition metal groups (see Topics G1 and H1). A few metallic elements of low reactivity are found in **native** (uncombined) form on Earth. They are known as **siderophiles** and are commoner in the Earth's metallic core. A few nonmetallic elements (N, noble gases) occur in uncombined form. As *Fig. 1* shows, some elements have intermediate behavior and fall in more than one class. For example, iron is found in both lithophilic (Fe silicates, Fe_2O_3, etc.) and chalcophilic (FeS_2) states.

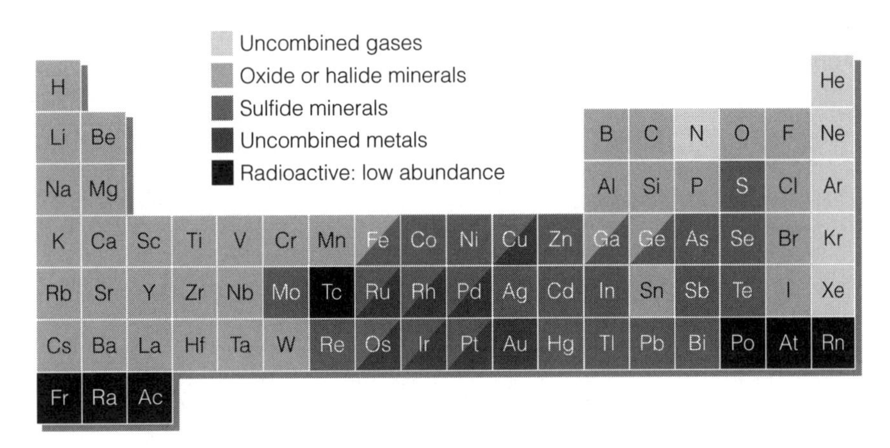

Fig. 1. The periodic table showing the principal types of chemical compound occurring for elements at the Earth's surface. Oxides include many complex forms, especially silicates.

Crust formation

New crust is formed by **tectonic processes** caused by upwelling convection currents in the mantle, driven ultimately by heat from radioactive decay of elements in the Earth (see Topic A1). The melting of rocks and subsequent recrystallization leads to fractionation of some lithophilic elements. Magnesium tends to remain in the mantle, and with it some other **compatible elements**, which form ions of fairly similar charge and size to Mg^{2+} (e.g. Fe^{2+} and Cr^{3+}). **Incompatible elements** (e.g. Na, K, Ti) do not remain with the magnesium silicate but pass easily into the melt and hence are more concentrated in crustal rocks.

Whereas the rocks of the mantle contain mostly **orthosilicates** with nonpolymerized SiO_4^{4-} ions, and **chain silicates** such as $MgSiO_3$ (see Topics D5 and F4), the minerals of the crust mostly contain more highly polymerized silicate units. Some of the commonest crustal rocks are **feldspars**, three-dimensional **framework silicates** consisting of corner-sharing [SiO_4] groups, like SiO_2 but with some Si is replaced by Al. Some idealized formulae are $KAlSi_3O_8$ and $CaAl_2Si_2O_8$, but in reality these minerals are much more complex, with many other elements present in small concentrations.

Many less common elements (e.g. Ga and Ge) are incorporated to some extent into the crystal structures of major minerals, and thus may be rather thinly spread over the crust. Others are concentrated by forming individual minerals. Native gold and cinnabar (HgS) were known in antiquity although Au and Hg are very rare elements. On the other hand, the less rare Ga and Ge were not discovered until the late 19th century.

The chemical processes leading to different minerals are diverse. Highly incompatible lithophilic elements (e.g. Li, Be, Zr and lanthanides) are concentrated in the final stages of solidification of molten rocks, known as **pegmatites**. Many sulfide minerals (e.g. of Cu, Zn, Mo and Pb) are formed by **hydrothermal processes**, in which water circulates deep in the crust and at high temperatures and pressures, and forms soluble complexes of these elements with anions such as Cl^- and HS^-, which may subsequently precipitate solids when they cool.

Weathering and sedimentation

Sedimentary processes begin with **weathering** of rocks, a chemical breakdown produced by the action of water and atmospheric CO_2. A typical reaction is the weathering of potassium feldspar ($KAlSi_3O_8$) to form the **clay mineral** kaolinite:

$$2KAlSi_3O_8 + 2CO_2 + 3H_2O \rightarrow 2K^+ + 2HCO_3^- + Al_2Si_2O_5(OH)_4 + SiO_2$$

CO_2 acts to provide acid in this reaction, and weathering is accelerated by living organisms that provide CO_2 through respiration and decay. A further step in this process leads to very insoluble $Al(OH)_3$:

$$Al_2Si_2O_5(OH)_4 + H_2O \rightarrow 2Al(OH)_3 + 2SiO_2$$

Rocks are therefore transformed by weathering, with soluble ions such as K^+ being washed out and insoluble **resistates** remaining. Some important sources of elements are of this form, including bauxite $Al(OH)_3$, rutile TiO_2 and cassiterite SnO_2.

The action of atmospheric oxygen on soluble ions may produce insoluble **oxidates** such as $Fe(OH)_3$ and MnO_2 from Fe^{2+} and Mn^{2+}, respectively. Other elements pass into the ocean and become deposited in various ways: as **biogenic deposits** such as $CaCO_3$ and SiO_2, which originate as the shells of marine organisms (see Topic J3), or as **evaporites** such as NaCl produced by evaporation of salt lakes.

Atmosphere and oceans

The atmosphere was originally formed by outgassing of crustal minerals that decomposed under heating. N_2 and CO_2 were probably the main original constituents. Water vapor condensed to form the liquid oceans. O_2 is a very unusual constituent of our own atmosphere by comparison with other planets. Nearly all of it comes from photosynthesis, the process by which green plants obtain their organic carbon from CO_2 with the help of energy from sunlight (see Topics J3 and J6).

The major dissolved constituents of the oceans are ions that do not form very insoluble compounds. Large amounts of many common elements such as Ca and Si are carried into the sea in soluble form by rivers, but many are precipitated either by inorganic or by biological processes (see above). The remaining ions of high abundance (Na^+, Cl^-, Mg^{2+}) form soluble compounds and are removed only by evaporation.

J3 BIOINORGANIC CHEMISTRY

Key Notes

The elements in biology	Around 25 elements are known to be essential for life. There are 11 major elements with a concentration greater than one part in 10^4, the others being known as trace elements.
Major elements	Major elements form constituents of biological molecules (C, N, O, P, S), ions either in solution or complexed to biomolecules (Na, K, Mg, Ca, Cl), and solids such as bones (e.g. calcium phosphate).
Trace metals	Essential *d*-block elements (e.g. Fe, Zn, Cu) are mostly constituents of metallo-enzymes, which act in the transport and chemistry of O_2, and perform many catalytic functions including redox and acid–base reactions.
Toxic and medicinal elements	Some strongly complexing nontransition metals (e.g. Hg, Pb) are very toxic. Elements used in medicine include Li, Pt, Au and radioactive Tc.
Related topics	Chemical periodicity (B2); Environmental cycling and pollution (J6)

The elements in biology

Life is sometimes thought of as 'carbon chemistry', but around 25 elements are essential for life. It is normal to divide these into **major elements** and **trace elements** according to their concentration (greater or less than one part in 10^4 by mass). *Table 1* shows elements classified in this way and according to their diverse roles. Nearly all known elements can be detected in the human body by modern analytical methods, but most are presumed to be there adventitiously without playing an essential role. To establish whether an element is essential is therefore difficult, especially as some essential elements (e.g. Co, for which the biochemistry is well studied) are present in much lower concentrations than many adventitious ones such as Rb or Sr. The active research field of **bioinorganic chemistry** aims to understand the role of elements, especially metallic ones fulfilling specialized functions.

Major elements

Most major elements of life (except P) are readily available in sea water, which may resemble the environment in which life began. They fulfill three main functions. **Nonmetallic elements** (except Cl) are components of covalently bound molecules and ions. H, C, N, O and often S are constituents of proteins, and nucleic acids (DNA and RNA) contain P as well. The chemical forms of these elements

Table 1. Essential and toxic elements in biology

Chemical form and function	Major elements	Trace elements
Covalently bound atoms in biomolecules and polymers	H, C, N, O, P, S	$B^{a,b}$, $As^{a,b}$, $Si^{a,b}$, Se, I
Aqueous ions, sometimes complexed to biomolecules	H^+, Na^+, Mg^{2+}, Cl^-, K^+, Ca^{2+}	–
Constituents of inorganic solids	C, O, P, Ca	F, Si,[a] Fe
Constituents of metalloproteins or other specialized molecules	–	V, Cr[b], Mn, Fe, Co, Ni, Cu, Zn, Mo, W[a]
Inessential toxic metals	–	Be, Cd, Hg, Tl, Pb

[a]Essential for some species, not necessarily humans.
[b]Chemical form unknown.

differ. S is normally present in reduced (R–S–H) form (R representing organic groups), but the facile oxidation to R–S–S–R gives **disulfide bridges**, which play a structural role in proteins. P, on the other hand, is always fully oxidized as **phosphate**. DNA is essentially made of phosphate diesters $(RO)_2PO_2^-$ complexed with Mg^{2+}. Adenosine di- and triphosphate, ADP (**1**: the organic part denoted Ad) and ATP, respectively, are used in energy storage in all types of cell. The forward reaction

$$AdO(PO_2^-)O(PO_3^{2-}) + H_2PO_4^- \rightleftharpoons AdO(PO_2^-)_2O(PO_3^{2-}) + H_2O$$

has ΔG around $+35$ kJ mol^{-1} under physiological conditions of concentration and pH. Metabolic energy input (e.g. from oxidation of glucose) is used to drive it. The reverse reaction then provides energy for essential functions such as muscle contraction or the action of the ion pumps mentioned below. An active human may turn over an amount of ATP equivalent to his or her body weight every day.

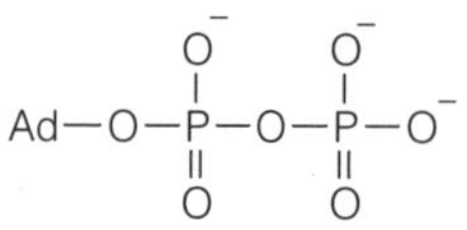

```
            O⁻    O⁻
            |     |
Ad—O—P—O—P—O⁻
            ||    ||
            O     O
```

1

Bicarbonate and phosphate ions are also present in aqueous solution and act as buffers to maintain pH.

Metal cations are present in aqueous solution, and are often complexed to anionic biomolecules. For example, Mg^{2+} is essential for the function of DNA and for the synthesis and utilization of ATP. Special 'ion pumps' maintain much lower concentrations of Na^+ and Ca^{2+} within cells than in extracellular fluids, and local changes in the concentrations of these two cations are used for signaling. The passage of nerve signals is associated with an influx of Na^+. Ca^{2+} forms complexes with carboxylate groups and acts to alter the conformation of many macromolecules; in particular this ion plays a role in muscle contraction.

Another role for some elements is in **inorganic solids**. Internal skeletons (bones, teeth) are composed mostly of apatite $Ca_5(PO_4)_3(OH)$ whereas external shells of molluscs are mostly calcium carbonate. Silica (SiO_2) is used as a protective solid by many single-celled marine plants, and in the brittle hairs of grasses and stinging nettles. Fe_3O_4 is used to store iron, and, as it is magnetic, by 'magnetotactic' bacteria to sense the direction of the Earth's magnetic field.

Trace metals

Essential elements of the *d* block are mostly components of specialized **metalloproteins**, which provide coordination sites with appropriate ligand atoms (O, N or S) in favorable geometrical arrangements for complexing a particular metal ion (see Topic E3). The important element iron is present in **heme proteins** containing the group shown in *Fig. 1a* and in **iron–sulfur proteins** containing groups such as the 4Fe–4S cluster shown in *Fig. 1c*. Many metalloproteins are enzymes with a catalytic role.

Dioxygen chemistry

Absorption of sunlight in green plants by **chlorophyll** (which contains magnesium; see Topic G3) provides energy for photosynthesis, which converts CO_2 and H_2O into organic compounds and dioxygen. Respiration by both plants and animals provides metabolic energy by oxidation of organic matter using atmospheric O_2. Both photosynthesis and respiration involve complex electron-transfer chains, including redox reactions of organic (e.g. quinones) and inorganic compounds (e.g. Fe proteins). The terminal step in photosynthesis is

$$2H_2O \rightarrow O_2 + 4H^+ + 4e^-$$

and is performed by a unit containing four Mn ions. The O_2-consuming unit in respiration (known as **cytochrome *c* oxidase** as it obtains electrons from the

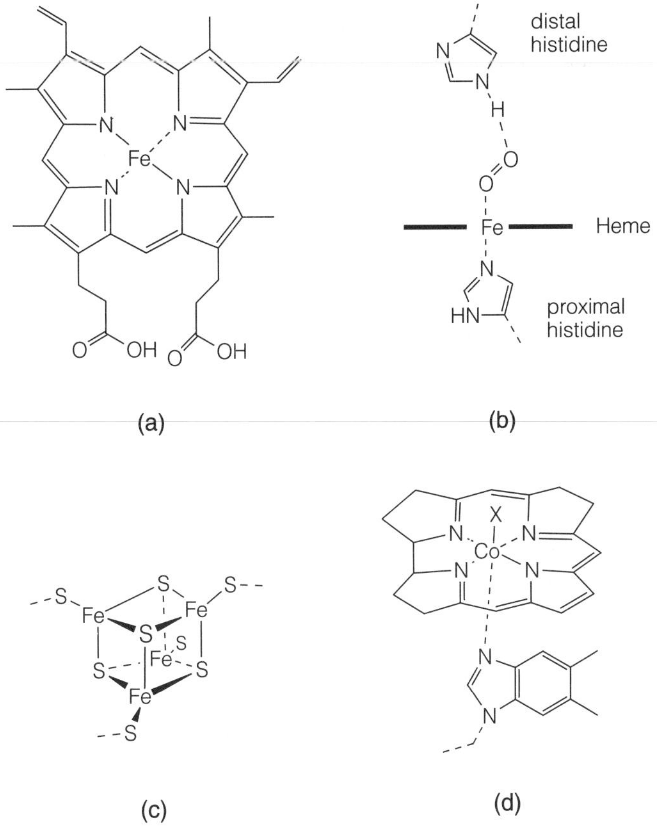

Fig. 1. Fe and Co in biomolecules: (a) heme; (b) O_2 binding in hemoglobin (see text); (c) [4Fe–4S] center in iron–sulfur proteins; (d) cobalamin (B_{12}) structure.

heme-iron protein cytochrome *c*) contains both heme Fe and Cu at the active site. The many-electron redox step is facilitated by the presence of two or more transition metals with a number of accessible redox states, $Mn^{II/III/IV}$, $Cu^{I/II}$ and $Fe^{II/III/IV}$ with the **ferryl** ($Fe^{IV}=O$) state which is unusual for Fe (see Topic H3).

Intermediate oxidation states of O, peroxide (O_2^{2-}) and superoxide (O_2^-) are generally toxic and are undesirable side products of the above reactions. **Superoxide dismutase** is a Cu–Zn-containing enzyme catalyzing the disproportionation of O_2^- to O_2^{2-} and H_2O; various **catalases** and **peroxidases** act to decompose peroxide.

Oxygenase enzymes catalyze specific oxidation reactions by O_2. Cytochrome P-450 enzymes perform reactions such as R–H → R–OH and involve a ferryl intermediate. Copper-containing oxygenases are generally found outside cells (rather than inside as with Fe; the difference is thought to reflect the later adoption of Cu in evolution) and are especially important in reactions that form connective tissue such as collagen.

Dioxygen transport

Around 65% of Fe in the human body is a component of **hemoglobin**. The protein molecule contains four heme sub-units. The 'resting state' has high-spin Fe^{2+} but coordination of the strong π acceptor ligand O_2 changes it to the low-spin form (see Topic H2). This is important for the action of hemoglobin, as the uptake of O_2 by one heme sub-unit increases the O_2 affinity of the others, the cooperative effect making the uptake and release more efficient. High-spin Fe^{2+} is slightly too large to fit comfortably in the heme ring, but the low-spin ion is smaller and so O_2 coordination causes the Fe to shift into the ring center. The 'proximal' histidine in the Fe coordination sphere (see *Fig. 1b*) also moves and acts as a means of communicating between sub-units. Another feature of hemoglobin is designed to lessen its affinity for other ligands. The position of the 'distal' histidine shown in *Fig. 1b* forces coordination in a nonlinear geometry. This is favorable for O_2 but not for strongly competing species such as CO and CN^-; although these are still very toxic they would be even more so otherwise.

Other metals

In terms of abundance in the human body, **zinc** is the most important trace element after iron. As it occurs only as Zn^{2+} (see Topic G4) it is not redox active, but coordination of molecules to the 'soft' cation is involved in many enzymes for **acid catalysis**. These include **carbonic anhydrase**, which converts HCO_3^- to CO_2, and enzymes for hydrolysis and condensation reactions of biopolymers. Zinc-containing proteins also perform regulatory functions; for example, **zinc finger proteins** recognize specific DNA sequences and are involved in gene function.

Nitrogen fixation is performed by specialized microorganisms, which reduce atmospheric N_2 to biological nitrogen compounds. Normal **nitrogenase** enzymes contain molybdenum and iron, but less common forms with vanadium are known.

Part of the cobalt-containing **coenzyme B_{12}** is shown in *Fig. 1d*. The group **X** is variable; X = CH_3 gives **methyl cobalamin**, which appears to be the only biological compound containing an organometallic metal–carbon bond (see Topic H10). It performs alkylations and radical-induced rearrangements of biological compounds.

Toxic and medicinal elements

Also listed in *Table 1* are some notably toxic elements. Toxicity is a relative term, and many essential elements are toxic either in excess (e.g. Fe) or if present in the wrong chemical form (e.g. elemental P or Cl). Many of the very toxic elements listed in *Table 1* are **heavy metals**, that is, post-transition elements of periods 5 and 6 (see Topics G4 and G6). These elements have strong complexing ability and an especially strong affinity for sulfur. They may displace essential elements such as Ca and Fe, and may also disrupt protein structure by breaking S–S bridges. Once attached to suitable ligands they are hard to displace. **Chelation therapy** is a treatment for heavy metal poisoning using chelating ligands that bind very strongly and can remove the elements in complexed form (see Topic E3).

Metallic elements used in medicine include lithium for treating manic depressive illness, platinum complexes such as *cis*-platin (**2**), which act as antitumor agents by combining with DNA and inhibiting cell division, and gold compounds used to treat arthritis.

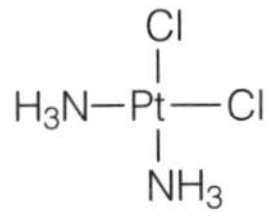

Cl
|
H_3N—Pt—Cl
|
NH_3

2

Radioactive isotopes are generally harmful to life because of the damaging effects of ionizing radiation. Elements that are retained by the body and/or concentrated in specific organs (e.g. Pu in the liver and in bones, I in the thyroid gland) are especially dangerous. On the other hand, many radioactive isotopes are used in medicine for diagnostic (tracer) and occasionally therapeutic (cancer treatment) purposes. One of the most useful is **technetium** (Tc), an artificially made element with no stable isotopes. The synthesis of Tc complexes designed to 'target' particular organs in the body is an active research area.

J4 Industrial chemistry: Bulk inorganic chemicals

Key Notes

Production — Major inorganic chemicals include gases (NH_3, N_2, O_2, Cl_2, H_2), acids (e.g. HCl, HNO_3, H_2SO_4), alkalis (NaOH, Na_2CO_3) and phosphates. One important sector is the chlor-alkali industry, which produces Cl_2, NaOH and Na_2CO_3 from NaCl.

Uses — The petrochemical and metallurgical industries and agriculture (fertilizers) are major users of inorganic chemicals, as are glass and paper making and the production of domestic products such as detergents and bleaches.

Related topic — Industrial chemistry: catalysts (J5)

Production

Table 1 shows a selection of the major inorganic chemicals that are produced in annual quantities of many millions of tonnes. Substances made in comparable amounts that are not listed include fuels and organic chemicals produced from petroleum, and construction metals such as iron. In many cases the basic chemical reactions used to produce the compounds in *Table 1* are simple, although catalysts are often required (see Topic J5). The design of processes to make the most economical use of energy and raw materials, and to minimize polluting wastes, is, however, not straightforward. The raw materials needed include air (for N_2 and O_2), sulfur (mined as native S or obtained from processing sulfide minerals), natural gas and oil (a source of energy and H_2), NaCl, and calcium carbonate and phosphate. It is interesting to consider some details of the **chlor-alkali industry**, one of the oldest sectors of the chemical industry, which links the production of Cl_2, NaOH and Na_2CO_3. The source material, NaCl, is used in greater amounts than any other raw material in the entire inorganic chemical industry.

Cl_2 is produced by electrolysis of NaCl (see Topic B4). A molten NaCl–$CaCl_2$ mixture gives metallic Na and Ca at the other electrode; these metals are used industrially; for example, as reducing agents for production of other electropositive metals such as Ti. Much greater quantities of NaCl are electrolyzed in aqueous solution to give NaOH, with H_2 as a byproduct. Cells with a mercury electrode give Na–Hg amalgam as the initial product, which is then reacted with water; modern diaphragm cells giving aqueous NaOH directly are cheaper to run and avoid the use of toxic mercury.

For economic reasons the amount of NaOH produced depends on the demand for Cl_2. However, NaOH and Na_2CO_3 are interchangeable in many uses (e.g. glass and paper manufacture), and so any short-fall in NaOH production can be made up by the carbonate. Some Na_2CO_3 is obtained from natural deposits but it is also made synthetically by the **ammonia–soda** or **Solvay process**. The overall reaction,

Table 1. Production and uses of some major inorganic chemicals

Substance	Production	Uses
H_2SO_4	Oxidation of S	Production of fertilizers, industrial and household chemicals
NH_3	$H_2 + N_2$	Production of HNO_3, fertilizers and organic N compounds
HNO_3	Oxidation of NH_3	Production of fertilizers, organic chemicals and explosives
O_2	Liquid air distillation	Steel making, metallurgy, welding, production of industrial chemicals
N_2	Liquid air distillation	Iron and steel making, industrial processing, refrigeration
CaO, $Ca(OH)_2$	Heat on $CaCO_3$	Steel, glass and paper production; water treatment
NaOH	NaCl electrolysis	Chemical manufacture, other Na compounds
Cl_2	NaCl electrolysis	Chlorinated organic compounds and polymers, bleaching and water treatment
H_3PO_4 and phosphates	$Ca_5(PO_4)_3F + H_2SO_4$	Fertilizers, metallurgical treatment, industrial chemicals, detergents and food products
Na_2CO_3	Natural; also ammonia–soda process	Glass and paper making, industrial and household chemicals
HCl	Byproduct of Cl_2 + organics	Metallurgical treatment, manufacture of organic and inorganic chlorides
H_2	Byproduct, also hydrocarbons + H_2O	Production of NH_3, organic chemicals, margarine
TiO_2	Natural; purification via $TiOSO_4$ or $TiCl_4$	Paints, Ti production
NH_4NO_3	$NH_3 + HNO_3$	Fertilizers
$(NH_2)_2CO$ (urea)	Liquid $CO_2 + NH_3$	Fertilizers

$$2NaCl + CaCO_3 \rightarrow CaCl_2 + Na_2CO_3$$

is thermodynamically unfavorable but can be achieved in several steps:

(i) $NaCl + NH_3 + H_2O + CO_2 \rightarrow NH_4^+(aq) + Cl^-(aq) + NaHCO_3(s)$
(ii) $2NaHCO_3(s) \rightarrow Na_2CO_3(s) + CO_2(g) + H_2O(g)$
(iii) $CaCO_3(s) \rightarrow CaO(s) + CO_2(g)$
(iv) $CaO(s) + 2NH_4^+(aq) + 2Cl^-(aq) \rightarrow Ca^{2+}(aq) + 2Cl^-(aq) + 2NH_3(g) + H_2O$

NH_3 and CO_2 are recycled. The equilibrium is shifted in favor of the products by providing heat in the endothermic stages (ii) and (iii). The overall energy required comes from burning coke, mixed with $CaCO_3$ in stage (iii), and some heat is also recycled from the exothermic steps (i) and (iv).

Uses

Many of the substances listed in *Table 1* are used primarily for further processes in the inorganic or organic chemical industry. Some important areas of application are summarized below.

Petrochemicals

Petrochemicals processing to give polymers and other organic compounds uses large amounts of many inorganic chemicals, including acids and alkalis (mostly

H_2SO_4 and NaOH), Cl_2 and HCl for production of chlorinated compounds, NH_3 and HNO_3 for organic N compounds, and H_2 for reduction and cracking. Smaller but significant quantities of inorganic compounds are also required as catalysts (see Topic J5) and as additives for polymers.

Metallurgical production and processing

Metallurgical industry uses the gases O_2 (e.g. for welding, and for oxidation of impurities such as P and S in steel production) and N_2 and Ar (as an inert blanket to prevent oxidation), refractory materials such as MgO and CaO for furnace linings, and acids such as HCl and H_3PO_4 for 'pickling' or removing oxide films.

Agriculture

Intensive agriculture requires a supply of plant nutrients containing K, P and N. K is derived from natural sources (mostly KCl). P is provided in the form of phosphates, a major use of H_2SO_4 being in the treatment of natural calcium phosphate minerals to obtain more soluble forms. The major sources of N in fertilizers are ammonium nitrate, urea (a nonexplosive alternative) and ammonium phosphate, which provides P simultaneously. Agricultural products consume over 80% of all industrially produced NH_3, itself the highest volume synthetic chemical in terms of molar quantity.

Many pesticides used in agriculture are based on organophosphorus compounds, PCl_3 and $POCl_3$ being important intermediates in their production.

Glass making

Silicate glass is based on a random network of corner-sharing SiO_4 units interspersed with ions such as Na^+ and Ca^{2+} (see Topic D5). Its manufacture, by heating together SiO_2, $CaCO_3$ and Na_2CO_3, is a major user of Na_2CO_3. Borates are added when increased thermal resistance is required, for example, for 'Pyrex' cooking utensils.

Paper making

The digestion of plant cellulose fibers to make paper is facilitated by a variety of inorganic compounds including NaOH, Na_2SO_4 and SO_2. Bleaching agents (which act by oxidizing colored organic material) are required for white paper and include Cl_2 and chlorine compounds such as $Ca(OCl)_2$ and ClO_2, as well as hydrogen peroxide and O_2.

Soaps, detergents and bleaches

Traditional soap is the sodium salt of long-chain carboxylic acids, made from NaOH and animal fats. Replacing Na by K reduces the melting point and is used for liquid soaps. Synthetic detergents are based on sulfonate (RSO_3^-) rather than carboxylate (RCO_2^-) salts and often contain many additives. 'Builders', used to complex or remove Ca^{2+} from hard water, include condensed phosphates such as sodium tripolyphosphate $Na_5P_3O_{10}$, although these are now regarded as environmental pollutants and are being replaced by zeolites, which act as ion exchangers (see Topic D5). For similar environmental reasons, traditional bleaches based on chlorine compounds such as NaOCl are being increasingly replaced by peroxides, often in the form of the peroxoacid salts such as perborates or percarbonates, which liberate H_2O_2 on heating (see Topic F7).

J5 INDUSTRIAL CHEMISTRY: CATALYSTS

Key Notes

Introduction

Catalysts speed up reactions, and provide selectivity for the desired product. Homogeneous catalysts (often organometallic compounds) act in the same phase as the reactions, heterogeneous catalysts (often transition metals on an oxide support) in another phase.

Dihydrogen chemistry

Heterogeneous catalysts are used in the production of synthesis gas (H_2–CO) and from it methanol, for ammonia synthesis and for large-scale hydrogenation. Organometallic catalysts provide more selectivity for specialized hydrogenation reactions.

Oxidation reactions

Heterogeneous oxidation catalysts are used in the production of H_2SO_4 and HNO_3 and for the selective oxidation of hydrocarbons.

Alkene polymerization

Alkene polymerization is performed by Ti compounds (Ziegler–Natta catalysts) and by organometallic compounds of early transition metals.

Gasoline and automobile catalysts

Petroleum products undergo catalytic hydrodesulfurization and re-forming processes. Catalysts in automobile exhaust systems convert pollutants (hydrocarbons, CO, NO) to CO_2 and N_2.

Related topics

Stability and reactivity (B3)
Organometallic compounds (H10)
Industrial chemistry: bulk inorganic chemicals (J4)

Introduction

A **catalyst** is a substance that speeds up a chemical reaction but is not itself consumed. Catalysts do not alter the thermodynamics of a reaction or the position of equilibrium, but act by providing an alternative pathway of lower activation energy. A high proportion of industrial chemical processes, inorganic and organic, use catalysts. They allow many reactions to be performed at lower temperatures than without a catalyst, and also provide **selectivity** in producing a specific product in reactions where several products are feasible thermodynamically. Enzymes (which often contain metallic elements; see Topic J3) are uniquely selective biological catalysts.

A catalyst present in the same phase as the reactants (generally liquid) is called **homogeneous**, one in a different phase is **heterogeneous**. Most heterogeneous catalysts are solids, and act by **adsorption** of gaseous or liquid reactants on a surface. Homogeneous catalysts are specific molecules, often organometallic compounds, that can be tailored in a more specific way to give a required product than is possible with heterogeneous catalysts. On the other hand, it is generally harder to separate the products from homogeneous catalysts. Another difference is that heterogeneous catalysis is usually performed at higher temperatures than homogeneous catalysis. Metallic elements used as heterogeneous catalysts are often in the form of small particles on a **support** such as Al_2O_3 or SiO_2. This provides a high surface area for the active catalyst, and reduces the tendency for **sintering** (i.e. coalescence into larger particles) at the operating temperature.

A catalytically active substance must be able to bind the reactant molecules and release the products. For example, hydrogenation with a metallic Ni catalyst proceeds via adsorbed hydrogen atoms formed by dissociation of H_2 on the metal surface (**1**). Transition metals and their compounds are effective catalysts because of their ability to coordinate molecules and to change oxidation state (see Topic H1). An appropriate degree of reactivity is often provided by elements from groups 9 and 10, either in metallic form or as organometallic compounds. More reactive elements (e.g. early transition metals) are more likely to bind reactants irreversibly, less reactive ones (e.g. Au) not to bind them at all.

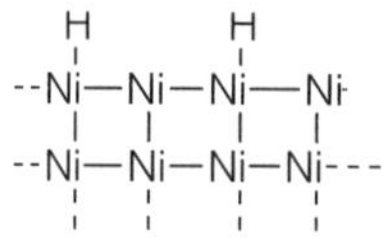

```
   H         H
   |         |
--Ni—Ni—Ni—Ni
   |    |    |    |
--Ni—Ni—Ni—Ni---
   |    |    |    |
```

1

Dihydrogen chemistry

H_2 is mostly produced by **steam re-forming** of hydrocarbons. The simplest reaction is

$$CH_4 + H_2O \rightleftharpoons CO + 3H_2$$

In this endothermic reaction, the equilibrium is driven to the right by using elevated temperatures (900°C). The usual catalyst is Ni supported on Al_2O_3. The H_2–CO mixture is known as **synthesis gas** or **syngas**, and may be used directly for further reactions, the most important being methanol synthesis:

$$CO + 2H_2 \rightarrow CH_3OH$$

A Cu–ZnO–Al_2O_3 catalyst is used, which is less active than ones involving metals in earlier groups, so that the C–O bond is not broken. Methanol is an important stage in further synthesis; see, for example, the Monsanto acetic acid process catalyzed by Rh compounds (Topic H10).

Another important reaction for synthesizing useful organic compounds is **hydroformylation**:

$$RCH{=}CH_2 + CO + H_2 \rightarrow RCH_2CH_2CHO$$

Homogeneous catalysts based on cobalt carbonyl compounds were previously used, but have been largely replaced by the **Union-Carbide process** with $[(Ph_3P)_2Rh(CO)Cl]$ as the catalyst.

Two major uses of H_2 are for ammonia synthesis and the hydrogenation of unsaturated vegetable oils to make margarine. The former reaction (see Topic F5) is exothermic and the equilibrium constant therefore decreases with temperature. In the **Haber process** a potassium-promoted iron catalyst is used, this relatively reactive metal being necessary to adsorb and dissociate the very stable N_2 molecule at moderate temperatures (400°C). The yield of NH_3 is improved by working at high pressure. Large-scale **catalytic hydrogenation** of alkenes is normally carried out with a Ni–SiO_2 catalyst. More selective hydrogenation for specialized purposes (e.g. pharmaceuticals where specific isomers are required) is possible with homogeneous catalysts; for example, **Wilkinson's catalyst** (**2**), which undergoes oxidative addition of H_2 under mild conditions (see Topic H9).

PPh_3
|
Ph_3P–Rh—Cl
|
PPh_3

2

Oxidation reactions

Catalytic oxidation steps are involved in the manufacture of the major industrial chemicals sulfuric and nitric acid. HNO_3 manufacture starts with ammonia, which is oxidized to NO (see Topic F5). A very active Pt–Rh catalyst is used, the contact time being minimized to avoid forming the thermodynamically more favorable products N_2 and N_2O. To make H_2SO_4 it is necessary to oxidize SO_2 to SO_3 (see Topic F8). The catalyst is vanadium pentoxide, V_2O_5. Oxides of transition and post-transition metals are also used for **selective oxidation of hydrocarbons**; for example, from alkenes to carbonyl compounds and acid anhydrides, which are used for polymer manufacture. The action of these catalysts appears to depend on the ability of the metallic element to change oxidation state and coordination number. Oxygen is transferred to the adsorbed reactant molecules from the catalyst surface, which is then reoxidized in another step.

Alkene polymerization

Domestic and industrial plastics are mostly formed by polymerizing alkenes:

$$nRCH{=}CH_2 \rightarrow (\text{–}C(R)H\text{–}CH_2\text{–})_n$$

The reaction is exothermic and can be initiated by free radicals (often from peroxo compounds), but organometallic catalysts give more controllable results. Most widely used are **Ziegler–Natta catalysts** made by mixing Al_2Et_6 (where Et is the ethyl group) with $TiCl_4$. Solid $TiCl_3$ is formed and catalysis occurs at surface Ti–Et groups, to which alkene molecules coordinate and undergo insertion into the Ti–R bond (see Topic H10). An advantage of these catalysts is that they may form **stereoregular polymers** where all the R groups in $\text{–}C(R)H\text{–}CH_2\text{–}$ have the same stereochemical configuration. This gives stronger materials with higher melting points than the random stereochemistry resulting from radical polymerization.

A new generation of catalysts is based on cyclopentadienyl compounds of early transition metals such as **3**, which in the presence of aluminoxane $(MeAlO)_n$ forms the active species $[(\eta^5\text{-}C_5H_5)_2ZrCH_3]^+$.

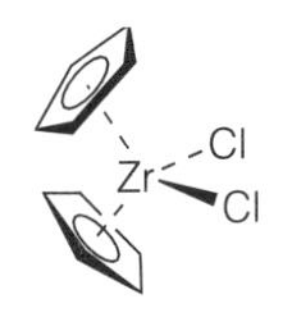

3

Gasoline and automobile catalysts

Natural petroleum contains organic sulfur compounds, which must be removed before further processing, as they block active sites in some catalysts and so act as **poisons**. When burnt they also give the environmental pollutant SO_2 (see Topic J6). **Hydrodesulfurization** is the reaction in which organic sulfur is converted to H_2S, which is easily removed. Catalysts based on mixed Co–Mo sulfides are used. Subsequent processing of petroleum involves **catalytic cracking** and **re-forming** in which long-chain hydrocarbons are reduced to shorter ones, together with isomerization processes giving a more desirable mixture of compounds. **Bifunctional catalysts** for these reactions contain metals such as Pt that are active for hydrogenation, and zeolites (see Topic D5) as **acid catalysts** providing H^+ to give carbocations that readily isomerize.

Catalysts for automobile exhaust systems are designed to remove environmental pollutants such as unburned hydrocarbons, CO formed from incomplete combustion and oxides of nitrogen. **Three-way catalysts** are based on Pt and Rh together with various additives that together perform a complex series of reactions, including removal of hydrocarbons by oxidation and steam re-forming (see above), and

$$2CO + 2NO \rightarrow 2CO_2 + N_2$$

Their operation depends on the absence of poisons such as lead compounds, and on a fuel injection system that provides an almost perfect stoichiometric ratio of fuel and oxygen to the engine: this is achieved by a feedback system using a sensor that monitors the O_2 content of the exhaust gases, based on an electrochemical cell using the ionic conductor ZrO_2 as a solid electrolyte (see Topic D7).

J6 ENVIRONMENTAL CYCLING AND POLLUTION

Key Notes

Introduction

The cycling of elements is driven by energy fluxes that produce circulation of the crust, oceans and atmosphere, and that allow photosynthetic and photochemical transformations. The presence of liquid water and of life contribute to the complexity of these processes.

The carbon cycle

Carbon is cycled by both inorganic processes (involving CO_2, HCO_3^- and carbonates) and by photosynthesis and respiration. The slow burial of fossil fuels has been accompanied by the production of O_2, but the current burning of fossil fuels is increasing CO_2 in the atmosphere and leading to global warming.

Other nonmetallic elements

S and N are cycled by life and by atmospheric photochemistry through many oxidation states. Natural Si and P compounds are involatile and less mobile in the environment. Environmental problems include acid rain, and pollution by soluble phosphates and organochlorine compounds.

Heavy metals

Compounds of Cd, Hg and Pb are potentially serious pollutants. Their use (especially that of Pb, which has been widespread) is declining.

Related topics Geochemistry (J2) Bioinorganic chemistry (J3)

Introduction

The cycling of substances through the environment is driven by energy fluxes within the Earth and at its surface. The radioactive decay of elements in the mantle and core drives tectonic processes that lead to crust formation, volcanic activity, and hydrothermal processes in aqueous solutions deep within the crust (see Topic J2). Absorption of solar energy drives the physical circulation of winds and ocean currents. It also fuels the physicochemical **hydrological cycle,** which entails the evaporation of water from oceans and lakes, and subsequent rainfall giving rivers that flow into the sea. Solar energy has in addition some direct chemical consequences, through **photosynthesis** by green plants, and **atmospheric photochemistry**, which depends on reactive species produced by absorption of UV radiation. Human activity contributes to these cycles through the **burning of fossil fuels** and the extraction and use of elements in technology.

The existence of liquid water and the presence of life are two features that make the chemistry of the Earth's surface uniquely complex among the known planets. Biological processes cycle some elements (especially C, N, O and S) through different oxidation states, and photosynthesis has given us both a strongly oxidizing atmosphere and buried fossil fuels. Hydrological cycling entrains many other substances, through the chemical breakdown of rocks and by evaporation from the oceans.

Elements respond to these driving forces in ways that depend on their chemical characteristics. Volatile molecules formed by nonmetallic elements enter the atmosphere from volcanic emissions, as 'waste products' of life, and from human energy use and industry. Some volatile compounds are rapidly oxidized by photochemical processes, and some are quickly washed out by dissolving in rainfall. Elements (especially metallic ones) that do not form volatile compounds under normal conditions are confined to the solid and liquid parts of the environment. Soluble ions (e.g. Na^+, Cl^-) are removed from rocks in weathering processes and end up in sea water. Other elements (e.g. Al, Ti) that form very insoluble oxides or silicates are by comparison highly immobile.

Some pollutants from human activity are natural substances (e.g. CO_2) produced in excessive amounts that unbalance the natural cycles. Others are synthetic (e.g. organochlorine compounds) and are harmful either because they are toxic to life, or because they interfere with natural chemical processes (e.g. in the ozone layer).

The carbon cycle

The environmental cycling of carbon compounds involves a flux of over 2×10^{14} kg C per year, much larger than for any other substance except water (about 5×10^{17} kg per year in the hydrological cycle). Understanding the carbon cycle has become especially urgent as the atmospheric CO_2 content is currently increasing, producing **global warming** through the trapping of IR radiation in the atmosphere. *Figure 1* shows a summary of the main processes, with estimates of the reservoirs (square boxes) and annual fluxes (round-cornered boxes) in units of 10^{12} kg C.

Atmospheric CO_2 is cycled in about equal amounts by two different processes: (i) the conversion into soluble bicarbonate HCO_3^- and the subsequent regeneration of CO_2 when water evaporates; (ii) the conversion into biological carbon compounds by photosynthesis, and reoxidation to CO_2 by respiration. Anaerobic

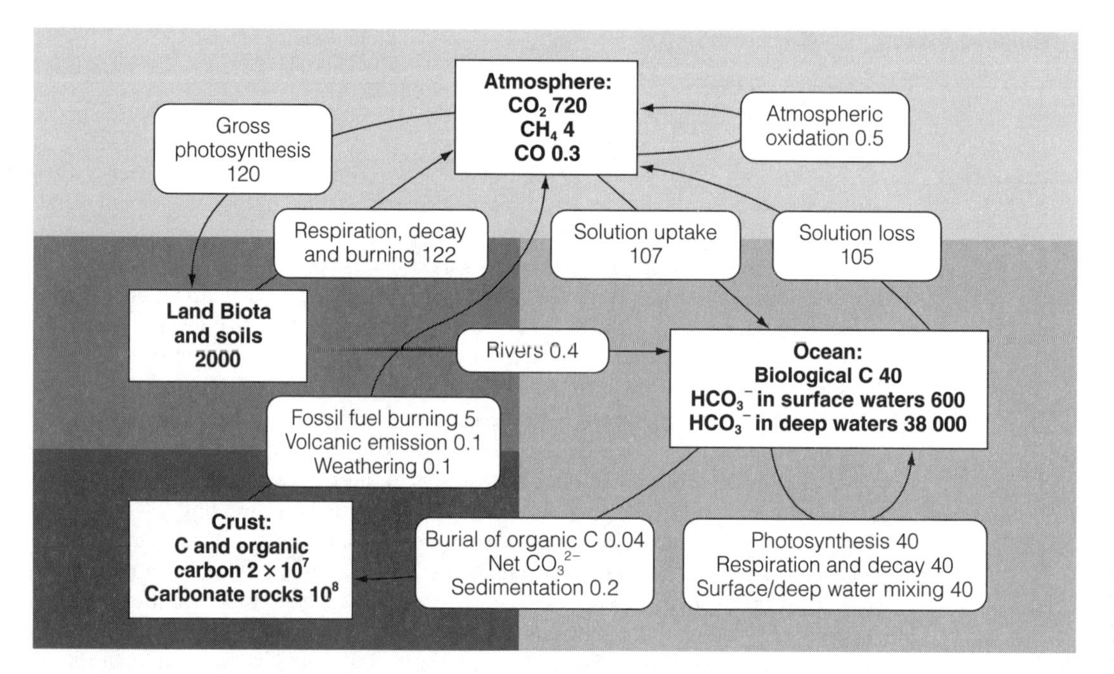

Fig. 1. The carbon cycle, showing reservoirs (square-cornered boxes) and annual fluxes (round-cornered boxes) in units of 10^{12} kg C.

decay of vegetation, ruminant animals such as cows, and other natural processes produce small amounts of CH_4 and CO, which are oxidized in the atmosphere to CO_2. Some parts of the cycle operate with much larger reservoirs of carbon, but also much more slowly: they include the mixing of surface bicarbonate with deep ocean waters, the production of sedimentary carbonate rocks (mostly from $CaCO_3$ shells and skeletons of marine organisms) and the eventual decomposition of carbonates by heating deep in the crust to regenerate CO_2.

Different parts of the natural cycle must be very nearly in balance, although over a period of millions of years some organic carbon has been buried before reoxidation, giving **fossil fuels** containing reduced carbon in the crust. Dioxygen from photosynthesis has passed into the atmosphere, but over geological time most of it has been used up in oxidizing surface rocks (principally Fe^{II} to Fe^{III} compounds, and sulfides to sulfates), only a small fraction remaining as free O_2. The burning of fossil fuels has reversed this natural trend and currently transfers around 5×10^{12} kg C per year into the atmosphere as CO_2. Parts of the cycle outside human control may be responding to take up some of this extra input, but the capacity of either surface ocean waters or life to accommodate it in the short term is very limited, and the burning of land vegetation contributes to the problem by reducing photosynthesis. Although excess CO_2 must ultimately return to the crust as carbonate minerals, that can happen only over time scales measured in thousands or even millions of years.

Other nonmetallic elements

Nitrogen and sulfur

N and S have a diverse and important environmental chemistry, associated in both cases with the wide range of oxidation states possible. Biological **nitrogen fixation** converts atmospheric N_2 into organic compounds needed by life. A high proportion is recycled within the biosphere, but some microorganisms convert it into nitrate (**nitrification**) and others reduce nitrate to N_2 (**dinitrification**), both processes being used to obtain metabolic energy. Denitrification thus recycles N back into the atmosphere. The major human perturbations to the cycle come from the use of nitrate fertilizers (which can lead to undesirable concentrations of NO_3^- in drinking water) and high-temperature burning of fossil fuels, which produce NO and NO_2. These gases are air pollutants, locally because they are toxic and take part in photochemical processes that generate other noxious compounds, and on a wider scale because they oxidize to nitric acid, which contributes to acid rain.

The biological and atmospheric redox chemistry of sulfur is also complex. The main natural inputs to the atmosphere come from biological decay (mostly H_2S) and emissions of dimethyl sulfide $(CH_3)_2S$ by marine organisms, together with volcanic emissions (mostly SO_2). These natural sources are now exceeded by the emission of SO_2 from burning sulfur-containing fossil fuels. Most atmospheric sulfur compounds oxidize rapidly to sulfuric acid, which is the major component of acid rain. The comparison of N and S is interesting, as the total atmospheric inputs of the two elements are similar in magnitude ($1–2 \times 10^{11}$ kg per year). The oxidation and removal of sulfur compounds is much more rapid than for the very stable N_2 molecule, and so the atmospheric concentrations are enormously different (about 1 p.p.b. for sulfur compounds, 78% for N_2).

Silicon and phosphorus

Si and P occur naturally only in fully oxidized forms (SiO_2 and silicates, phosphates), which are involatile and have low solubility in natural waters. Phosphorus is one of the most important elements of life (see Topic J3) and in aquatic environments the one that is often in shortest supply. Pollution by soluble polyphosphates (e.g. from detergents; see Topic J4) can seriously upset the ecological balance of lakes, leading to uncontrolled growth of algae and depletion in dissolved oxygen.

Halogens

These elements occur naturally in halide minerals. CaF_2 is very insoluble in water, but other halide ions are easily washed out of rocks and are abundant in sea water. Volcanic emissions contain small amounts of HF and HCl but these gases are very soluble and washed out of the atmosphere quickly. Marine organisms produce small quantities of methyl compounds such as CH_3Cl, which are oxidized and also washed out. Some synthetic organohalogen compounds pose environmental problems because natural chemical processes break them down very slowly. Organochlorine compounds of concern include dioxins and persistent insecticides such as DDT, and volatile chlorofluorocarbons (CFCs) used as aerosol propellants and in refrigerators. CFCs resist photochemical breakdown in the lower atmosphere and can enter the stratosphere where short-wavelength UV radiation splits them to produce Cl atoms, which then act as catalysts for the decomposition of UV-absorbing ozone.

Heavy metals

The heavy post-transition metals such as Cd, Hg and Pb are toxic because of the very strong complexing ability of 'soft' cations such as Hg^{2+} (see Topics G4, G5 and J3). They have low concentrations in natural waters because they form insoluble sulfides. Compounds that are either more soluble in water or volatile pose an environmental hazard. Of these elements, **lead** has been the most widely used, in pipes for drinking water, in paints and (in the form of tetraethyl lead $Pb(C_2H5)_4$) as a gasoline additive to improve combustion. As the toxic hazards have been more clearly recognized, these uses have been phased out.

Mercury also had many applications, including in hat-making (where the symptoms of mercury poisoning gave rise to the saying 'mad as a hatter') but its industrial usage (e.g. for NaCl electrolysis; see Topic J4) has also declined. Cases of acute mercury poisoning have resulted from eating fish from water polluted by industrial Hg compounds. Some organisms convert inorganic compounds into ones containing $[CH_3Hg]^+$, which are especially toxic as they pass more easily through the nonpolar constituents of cell membranes. It is likely that methylcobalamin (see Topic J3) is involved in this transformation.

FURTHER READING

Text-books on inorganic chemistry differ greatly in their balance of conceptual and descriptive material. All the books listed under the General heading below (except that by Emsley, which is a useful compilation of data) include some discussion of general concepts.

General

Cotton, F.A., Wilkinson, G. and Gaus, P. L. (1995) *Basic Inorganic Chemistry*, 3rd edn., Wiley, New York, USA.

Douglas, B., McDaniel, D.H. and Alexander, J.J. (1983) *Concepts and Models of Inorganic Chemistry*, 2nd edn., Wiley, New York, USA.

Emsley, J. (1991) *The Elements*, 2nd edn., Clarendon Press, Oxford, UK.

Huheey, J.E. (1993) *Inorganic Chemistry: Principles of Structure and Reactivity*, 4th edn., Harper Collins, New York, USA.

Mackay, K.M. and Mackay, R.A. (1989) *Introduction to Modern Inorganic Chemistry*, 4th edn., Blackie, Glasgow, UK.

Owen, S.M. and Brooker, A.T. (1994) *A Guide to Modern Inorganic Chemistry*, Longman, Harlow, UK.

Porterfield, W.W. (1984) *Inorganic Chemistry: A Unified Approach*, 2nd edn., Academic Press, San Diego, USA.

Raynor-Canham, G. (1996) *Descriptive Inorganic Chemistry*, W.H. Freeman, New York, USA.

Sharpe, A.G. (1992) *Inorganic Chemistry*, 3rd edn., Longman, Harlow, UK.

Shriver, D.F. and Atkins, P.W. (1999) *Inorganic Chemistry*, 3rd edn., Oxford University Press, Oxford, UK.

Section A

Atkins, P. W. (1998) *Physical Chemistry*, 5th edn., Oxford University Press, Oxford, UK (Ch. 11, 12).

Cox, P. A. (1996) *Introduction to Quantum Theory and Atomic Structure*, Oxford University Press, Oxford, UK.

Cox, P. A. (1989) *The Elements: Their Origin, Abundance and Distribution*, Oxford University Press, Oxford, UK (Ch. 2).

Section B

Alcock, N.W. (1990) *Bonding and Structure: Structural Principles in Inorganic and Organic Chemistry*, Ellis Horwood, Chichester, UK.

Johnson, D.A. (1982) *Some Thermodynamic Aspects of Inorganic Chemistry*, 2nd edn., Cambridge University Press, Cambridge, UK.

Leigh, G.J. (1990) *Nomenclature of Inorganic Chemistry: Recommendations 1990*, Blackwell Scientific, Oxford, UK.

Mingos, D.M.P. (1998) *Essential Trends in Inorganic Chemistry*, Oxford University Press, Oxford, UK.

Smith, D.W. (1990) *Inorganic Substances: A Prelude to the Study of Descriptive Inorganic Chemistry*, Cambridge University Press, Cambridge, UK.

Section C

DeKock, R.L. and Gray, H.B. (1980) *Chemical Structure and Bonding*, Benjamin-Cummings, Menlo Park, USA.

Kettle, S.F.A. (1995) *Symmetry and Structure: Readable Group Theory for Chemists*, 2nd edn., Wiley, Chichester, UK.

Murrel, J.N., Kettle, S.F.A. and Tedder, J.M. (1978) *The Chemical Bond*, Wiley, Chichester, UK.

Section D

Cox, P. A. (1987) *The Electronic Structure and Chemistry of Solids*, Oxford University Press, Oxford, UK.

Müller, U. (1993) *Inorganic Structural Chemistry*, Wiley, Chichester, UK.

Smart, L. and Moore, E. (1996) *Solid State Chemistry*, 2nd edn., Chapman and Hall, London, UK.

Wells, A.F. (1985) *Structural Inorganic Chemistry*, 5th edn., Clarendon Press, Oxford, UK.

West, A.R. (1984) *Solid State Chemistry and its Applications*, Wiley, Chichester, UK.

Section E

Burgess, J. (1978) *Metal Ions in Solution*, Ellis Horwood, Chichester, UK.

Jensen, W.B. (1980) *The Lewis Acid–Base Concepts: An Overview*, Wiley, New York, USA.

Gutmann, V. (1968) *Coordination Chemistry in Nonaqueous Solution*, Springer, Berlin, Germany.

Section F, G, H, I

Cotton, F.A. and Wilkinson, G. (1988) *Advanced Inorganic Chemistry*, 5th edn., Wiley, New York, USA.

Elsenbroich, Ch. and Salzer, A. (1992) *Organometallics: A Concise Introduction*, 2nd edn., VCH, Weinheim, Germany.

Greenwood, N.N. and Earnshaw, A. (1997) *Chemistry of the Elements*, 2nd edn., Butterworth-Heinemann, Oxford, UK.

Kettle, S.F.A. (1998) *Physical Inorganic Chemistry: A Coordination Chemistry Approach*, Oxford University Press, Oxford, UK.

Nicholls, D. (1974) *Complexes and First-Row Transition Elements*, Macmillan, London, UK.

Seabourg, G.T. and Loveland, W.D. (1990) *The Elements Beyond Uranium*, Wiley-Interscience, New York, USA.

Section J

Cox, P. A. (1989) *The Elements: Their Origin, Abundance and Distribution*, Oxford University Press, Oxford, UK.

Cox P. A. (1995) *The Elements on Earth: Inorganic Chemistry in the Environment*, Oxford University Press, Oxford, UK.

Kaim, W. and Schwederski, B. (1994) *Bioinorganic Chemistry: Inorganic Elements in the Chemistry of Life*, Wiley, Chichester, UK.

Thompson, D. (1995) *Insights into Speciality Inorganic Chemicals*, Royal Society of Chemistry, London, UK.

Thompson, R. (1995) *Industrial Inorganic Chemicals: Production and Uses*, Royal Society of Chemistry, London, UK.

Williams, R.J.P. and Frausto de Silva, J.J.R. (1991) *The Biological Chemistry of the Elements: The Inorganic Chemistry of Life*, Clarendon Press, Oxford, UK.

INDEX